T0269033

Brain Computation as Hierarchical Abstraction

Computational Neuroscience

Terence J. Sejnowski and Tomaso A. Poggio, editors

For a complete list of books in this series, see the back of the book and http://mit-press.mit.edu/Computational_Neuroscience

Brain Computation as Hierarchical Abstraction

Dana H. Ballard

The MIT Press
Cambridge, Massachusetts
London, England

This book was set in ITC Stone Serif Std by Toppan Best-set Premedia Limited, Hong Kong.

Library of Congress Cataloging-in-Publication Data

Ballard, Dana H. (Dana Harry), 1946– author.
Brain computation as hierarchical abstraction / Dana H. Ballard.
    p. ; cm. — (Computational neuroscience)
Includes bibliographical references and index.
ISBN 978-0-262-02861-5 (hardcover : alk. paper)—978-0-262-53412-3 (pb.)
I. Title. II. Series: Computational neuroscience.
[DNLM: 1. Brain—physiology. 2. Mental Processes—physiology. 3. Models, Neurological. 4. Nerve Net—physiology. 5. Neural Networks (Computer) 6. Neurons—physiology. WL 337]
QP357.5
612.8′23343—dc23
                              2014021822

Contents

Series Foreword

Computational neuroscience is an approach to understanding the development and function of nervous systems at many different structural scales, including the biophysical, the circuit, and the systems levels. Methods include theoretical analysis and modeling of neurons, networks, and brain systems and are complementary to empirical techniques in neuroscience. Areas and topics of particular interest to this book series include computational mechanisms in neurons, analysis of signal processing in neural circuits, representation of sensory information, systems models of sensorimotor integration, computational approaches to biological motor control, and models of learning and memory. Further topics of interest include the intersection of computational neuroscience with engineering, from representation and dynamics, to observation and control.

Terrence J. Sejnowski
Tomaso Poggio

Preface

The 1950s saw a huge step in the development of computers with the introduction of the IBM 701, the FORTRAN programming language, and the integrated circuit, but arguably the major landmark in their promotion as a model of human information processing came with Lindsay and Norman's seminal book in 1972.[1] Nonetheless, the idea of a computer being a model for human thought was not greeted with enthusiasm by the biological community, and indeed the general reaction was very negative. The main impediment was that conceptualizations of computation were grounded in the ways that silicon computing approached them, and the animal brain is nothing like a silicon computer in implementation.

Change in this perspective was given a major impetus in 1982 with the publication of David Marr's *Vision*,[2] which promoted the distinction between the computational problems that the brain was faced with and its neural implementation of solutions. The computational problem now could be studied in the abstract without having knowledge of the detailed workings of the very complex cellular and molecular underpinnings.

Nonetheless, an ultimate account of brain functioning has to address the primary signaling method of the voltage spike in nerve cells. A major step forward in this direction was and Churchland and Sejnowski's *Computational Brain*,[3] which addressed computation with respect to the brain's overall complexity, particularly its organization at multiple spatial scales. But there is still the issue of computational abstraction. Even though the signaling in the brain is all about spikes, their myriad of different functions is unlikely to be interpretable without the concept of computation at different spatial and temporal scales. To appreciate this point, we can cross over to silicon computing. Silicon computer architectures depend crucially on the abstraction of the low-level realization of computation in hardware through an elaborate succession of more abstract descriptions of the same in levels of software. Without these levels, there would be no computers in their present form. In this context, it is surprising that thinking about brain computation tends to eschew computational hierarchical descriptions. Thus, while it is true that the voltage spike is the basic way neurons communicate with each other, the contexts of spikes can be so different that it is extremely unlikely that they can ever be interpreted without understanding their hierarchical venues.

For the brain, increasing abstraction levels buy a crucial survival advantage: the ability to predict the future at ever larger spatial and temporal scales. In addition, the complexities surrounding brain computation spectacularly can be reduced if the entirety of the computation can be factored into different levels of abstraction. This book focuses on these issues. After the two introductory chapters of part I, the book is organized into three main parts: neural, embodiment, and social. Each of these is similarly factored into composite abstraction levels. The aim is to show that there is a natural correspondence between the computational issues and the anatomic levels of organization in the brain.

It is a wonderful time to be thinking about a comprehensive picture of brain computation, as there is a constant flood of new insights both from the experimental and theoretical sides. Nonetheless, the dynamic nature of this boon introduces challenges for a book of this kind. As a consequence, the book is an admixture of models that are generally believed to be settled and other models that are very speculative. Without the more speculative parts, it would be impossible to paint a coherent picture, as they provide helpful scaffolds for discussing problems that the brain has to solve. I have tried to indicate wherever things are on the edge.

In the modern scientific arena, research progress in developing both computational and biological understanding is racing ahead at breakneck speed, with the result that in any subject, one quickly enters territory accessible only to the specialist. Nonetheless, to paint a picture in a single book, many important details have to be abstracted away, from both the computational and biological sides. The net result is that the descriptions of either side may disappoint the specialist, but the focal intent is to point out connections that may promote new understanding.

Acknowledgments

This book developed over the course of many years, and as a result there are many people and institutions to thank. Sometime in 1982, at lunch with my mentor Jerry Feldman and colleague Chris Brown at the University of Rochester's faculty club, the topic of really big unsolved problems came up, and when the brain made a short list of three, I decided that was a lot more interesting than the biomedical image processing I was doing. I had a leave to study anatomy in cortical hierarchies with Paul Coleman and was hooked. A few years afterward, the field of machine learning was born, or reborn to some, and a plethora of new statistically driven algorithms such as reinforcement learning, backpropagation, and support vector machines suddenly appeared on the scene to provide new ways of thinking about brain computation. It has taken a while, but these algorithms have since matured to the point where the connections between them and the brain's underlying complexity are being made rapidly.

The idea and impetus for the book started at the University of Rochester. The scale of the university makes it ideal for researchers from different fields to interact, and umbrella structures such as the Center for Visual Science and umbrella funding from the National Center for Research Resources of the National Institutes of Health made it normal for researchers from different disciplines to interact on a daily basis. I will always be grateful for the conversations with Jerry, Chris Brown, Robbie Jacobs, Dave Williams, Dick Aslin, Peter Lennie, Tania Pasternak, Ed Freidman, Charley Duffy, Daeyeol Lee, Marc Schieber, and Gary Paige.

In the course of a move to the University of Texas at Austin in 2006, I spent 6 months on leave at the University of Sydney where most of the book took shape. It assumed its current form with another semester leave in 2011 at the Queensland Brain Institute in Brisbane, and I am grateful for help I received there from my sponsor Mandayam Srinivasan and host

Peter Bartlett, as well as Jason Mattingly, Ada Kritikos, Judith Reinhard, Charles Claudianos, Janet Wiles, and Geoff Goodhill.

The University of Texas at Austin's Center for Perceptual Systems, directed by Bill Giesler, has provided a wonderfully stimulating intellectual setting for testing the book's ideas. I am grateful for feedback from Bill as well as colleagues Alex Huk, Jonathan Pillow, Larry Cornack, and Eyal Seidemann as well as Nicholas Priebe and Ila Fiete from Neuroscience.

I have been extremely fortunate to have worked with many extraordinarily talented PhD students and Postdocs whose creativity appears throughout the book. Special thanks go to Joseph Cooper, Rahul Iyer, Dmitry Kit, Rajesh Rao, Polly Pook, Justinian Rosca, Garbis Salgian, Virginia DeSa, Xue Gu, Constantin Rothkopf, Nathan Sprague, Weilie Yi, Chen Yu, Andrew McCallum, Michael Swain, Steven Whitehead, Shenghuo Zhu, Janneke Jehee, Greg Zelinsky, and Jochen Treisch.

In fall 2012, the book was polished off during participation in the ZiF program on Attention at the University of Bielefeld. I very much appreciate the helpful discussions I had with Helge Ritter, Wolfgang Einhuser-Treyer, Gernot Horstman, and Werner Scheider.

Most of all I am grateful for my longtime collaboration with Mary Hayhoe.

The book would not have been possible without the leaves, and I am most grateful to the Department of Computer Science of the University of Texas for, particularly to the Chair Bruce Porter, who has been an unfailing supporter of this work.

I am very grateful to my editor at MIT Press, Bob Prior, who was hugely encouraging toward the idea of a book with a hierarchical perspective, and also to Chris Eyer for his essential help with all aspects of its production. Katherine Almeida and her copyediting staff did a fantastic editing job, with the result that the book is enormously more polished.

Ultimately, what makes research possible is funding, and in the process I have been generously supported by the National Science Foundation and by the National Institutes of Health through National Eye Institute grants and, particularly at Rochester, through a National Center for Research Resources grant, which, together with startup funds from the University of Texas at Austin, funded the technological developments described in the book.

I Setting the Stage

The vast differences between silicon circuitry and the brain's neural circuitry can easily lead to the conclusion that they have nothing in common; however, not only is this not the case, but also computational tools turn out to be essential for understanding brain function. The hierarchical organization of the brain finds many parallels in the hierarchical organization of silicon. What chapter 1 stresses most is that to be comprehensible, the computation done by the brain must be organized into functional hierarchies, as is done in silicon.

The staggering complexity of the brain itself can be daunting, but the enormous research focus on the brain in recent times has crystallized an overview of its function. This overview has many lacunae, where pieces are missing or still not completely understood, but nonetheless, a broad picture is emerging, which is characterized in chapter 2. At the level where the brain is organized into its major subsystems, the interactions between them are increasingly well-defined.

1 Brain Computation

To say the brain is a computer is correct but misleading. It's really a highly specialized information-processing device—or rather, a whole lot of them. Viewing our brains as information-processing devices is not demeaning and does not negate human values. If anything, it tends to support them and may in the end help us to understand what from an information-processing view human values actually are, why they have selective value, and how they are knitted into the capacity for social mores and organization with which our genes have endowed us.

—David Marr, *Vision* (W. H. Freeman, 1982, p. 361)

The human brain is a candidate for the most complex structure of any kind in the universe. It is a truly remarkable information-processing device that can learn the structure of the world, including intricate social interactions with other intelligent agents necessary to build and execute successful plans for survival and procreation. It is also the most complicated part of the body. Although the brain represents only about 2% of total body weight, it is estimated that 40% of the human genome is used in putting the brain together. How are we to understand the brain? And what would it mean to understand the brain? The emergent thinking is that this enormous capability and complexity can be made comprehensible through computational science.

The acceptance of the brain as a computational device is recent. With the explosion of computer technology in the 1950s came suggestions that the brain is some kind of computer, and these suggestions were not greeted with enthusiasm. At that time, the National Institutes of Health had no explicit study section to address computation in the brain. However in the modern day, the situation has changed dramatically. The field of computational neuroscience has been born and endorsed. Dozens of scientific meetings worldwide are devoted to computational brain models, and the number is growing rapidly.

If the brain is a computer,[a] it is almost certainly unlike any one we've seen before, and so even the computer expert has to be open to very unusual and radical ways of getting things done. We are only just beginning to understand how these kinds of differences are handled, but enough has been learned to offer a coherent framework. Thus, the thrust of this book is to show how the kinds of things that we associate with thinking all have computational and neural underpinnings, and the thrust of this chapter is to get started by outlining the main issues, of which there are three.

1. *Hierarchies* The hierarchical structure of the brain can be related to the hierarchical organizing principle of computation. We now know that the brain evolved in layered structures and that later layers usefully exploit the structure of the earlier layers to great advantage. This observation leads directly to the principle of *computational abstraction hierarchies*, which composes the essential organizing backbone of computers. So important is this principle for understanding brain function that it dictates the central organizational structure of this book. In part I, chapter 1 introduces the computational issues, and chapter 2 provides an overview of brain function, focusing on the mammalian forebrain. Later chapters are organized in a sequence of increasing computational abstraction. In part II, chapter 3 describes basic neuron function, with a focus on timing issues. Chapter 4 describes the cortex, the forebrain's essential memory system that makes everything else possible.[b] Chapter 5 describes the basal ganglia, a collection of brain subsystems that can be very loosely thought of as analogous to a parallel processor. With part III we jump abstraction levels. Chapter 6 shows how behavioral programs can use the basic brain architecture to interrogate the world for crucial information. Chapter 7 shows how motor programs can use this information to act in the world to achieve goals. Chapter 8 shows how the crucial ability of multiplexing can be handled, wherein the brain can manage different programs simultaneously, each trying to achieve different objectives. With part IV we jump abstraction levels again. Chapter 9 focuses on properties that can be experienced by the user, starting with decision making. Chapter 10 focuses on emotions. Finally with all this structure in place, we can consider properties of consciousness in chapter 11, which is the final chapter.

2. *Slow circuitry* The brain must have very ingenious ways of coping with its tardy neural circuit responses, which are more than a million times slower than switching times in silicon circuits. Consequently, it is a huge mystery how the brain can accomplish all of its cognitive tasks fast enough to act successfully in the world. In this chapter, we introduce in outline the

main tricks the brain uses, and the details will come later. Each chapter will have to deal with the timing issue in some way. But even understanding these will not settle the issue completely. Our understanding of how the brain compensates for its slow circuitry remains a work in progress.

3. *The enterprise* This issue, which will be touched upon lightly in this chapter, concerns the enterprise of a computational brain theory itself. Assertions of great progress do not move skeptics. They allow that although computation can be a model of the brain, and a poor one at that, it cannot truly be the ultimate description of brain function because human brains operate outside of the computational realm. To counter this negativity, we will have to touch on the very nub of the issue, and that is: What is computation? There are ways of deciding what is computation and what is not, even though, because they appeal to abstract issues involving infinities, they aren't very interesting to the computational convert. Nonetheless, because anticomputational arguments have been widely circulated,[4, 5] they need to be addressed.

Given the surge in focus on brain computing, what kinds of explanatory power can we expect? Current computation is far from predicting, and probably will never be able to predict, individual acts such as Michelangelo carving the *Pieta*—although it would have more to say about Romeo wooing Juliet—but for a broad theoretical understanding of why we think the way we do and why we feel the way we do, computation is rapidly becoming the best alternative. It's not that we will ever be able to predict exactly what we'll do in any situation, but we will be able to predict the kinds of things we are likely to do under different circumstances with an increasing fidelity.

Computational descriptions of humans make us uneasy in a way that descriptions by other disciplines are spared. We don't rail against physicists or chemists for their models of us. Perhaps the main reason that computation is singled out is that it is associated with mindless robotic behavior, the very antithesis of the rich tapestries of affect and contemplation associated with being human. However, this connection is an overinterpretation that takes the instances of current computers to be coextensive with the umbrella discipline of computational science. It is true that conventional robots can be driven by silicon-based computers, but computation has much deeper things to say, in particular some stunning things, about how brains work. It may well be that ultimately the best way to understand our collection of human traits, such as language, altruism, emotions, and consciousness, is via an understanding of their computational leverage. We shouldn't worry that this understanding will preclude us celebrating all our humanness in the usual ways. We will still fall in love and read Shakespeare with this knowledge, just as knowing a chair is made up of atoms doesn't preclude the joy of a comfortable sit down.

The job of understanding the brain has been characterized as one of reverse engineering.[6] There are brains galore sitting in front of us; we just have to deduce what makes them work. This in turn involves figuring out what the parts are and how they interact with each other. What makes the task daunting is that the brain exhibits complexity that is nothing short of staggering. Somewhere from 10 billion to 100 billion brain cells act in ways unlike any other cells in the body, forming tangled networks of interconnections, and each of these cells is itself a small factory of thousands of proteins that orchestrate a myriad of internal functions. Faced with all these webs of interconnected detail, reverse engineers have the enormous challenge of breaking the overall system into manageable pieces. To meet this challenge, computation proves an extremely useful tool. Evolution is a great tinkerer, always ready to exploit a solution to a problem that appears anywhere in the dynamics of life, but with the result that, in retrospect, the solutions can be hard to anticipate or analyze from a first-principles perspective. Computation, being very abstract, has the great versatility of being able to fit many different settings once the appropriate identifications are established. Thus to use a crude analogy, an abstract level might be described in terms of a high-level programming language, whereas a lower level of abstraction might look more like an assembly-level language. The job of the computationally minded reverse engineer is to posit these kinds of distinctions and splice them together.

Most of the time, the researchers doing computational sleuthing would take their models for granted and not spend much time wondering what other possibilities there could be. The fact is, to date we do not have any good alternatives to thinking about the brain other than as doing computation, particularly when we focus on accounting for behavior. The central construct of computer science is the *algorithm*,[7] expressed in very rich symbolic languages that have constructs that are analogous to those of a recipe in a cookbook. An algorithm's steps can be repeated a certain number of times (e.g., "stir until thickened") until a test is satisfied, partial results can be saved and combined (e.g., "add the marinade to the mixture"), and steps can be conditional (e.g., "if you are allergic to cow's cheese add goat's cheese at this point and continue"). Mathematics and especially physics have at times competed with computation with rival explanations of brain function, so far without result because they do not have the construct of the programming language. The parts of mathematics and physics that are accessible and useful turn out to be the parts that readily can be specified by algorithms. In the same way, chemistry and especially biology and molecular biology have emerged as essential descriptive constructs, but when we try to characterize the aspects of these disciplines that are important for cognition, more and more often we turn to algorithmic constructs. The main factor is that the primitives we need to describe brain function that involve the direction of physical and mental behaviors naturally seem to fit onto the algorithmic notions of iteration, decision making, and memory that compose the essence of computation.

The central thrust of this book is about computation, but at this crucial introductory juncture we must emphasize an extremely important point. While we argue that computation will ultimately prove necessary in understanding the brain, it will never displace the extraordinary experimental and analytical work that leads to the discovery of the fundmental descriptions and workings of the brain's basic biology. The mere thought of the magnitude of this enterprise, which involves chemistry, biology, and physics and includes many other ancillary disciplines, can be daunting and overwhelming. The deepest hope is that the essential inclusion of computational principles will play a complementary and important role in the drive to understand the most complex device nature ever created.

1.1   Introducing the Brain

The *tropical pitcher* carnivorous plant has an astonishing design. Insects drawn to it by its fragrant nectar slip on its sides, fall into a slippery pit, and cannot get out. The plant can then digest them at leisure. Think about this for a moment. Is the plant doing computation or not? The sequence of operations that lead to an insect's demise are well suited to a symbolic description, but perhaps the line has to be drawn here. The realization that we come to is that it is animals that do the computation characterized by brains. Animals move, and to avoid pitfalls they necessarily have to sense aspects of the environment and change direction to improve their survival chances. Tiny bacilli have flagella to follow sensed gradients left by nutrients but can spin randomly in the absence of these to try and stumble into another nutrient cache. So it's animals that have the brains, and they use them to navigate dynamic environments.

Our focus is the human brain, and it has a very specialized architecture compared to other brains of a much earlier evolutionary heritage. Its basic brain architectural plan starts with its vertebrate heritage, but humans are mammals and mammals represent a radical point of departure in brain design. While the brains of other animals have precursor elements, mammals have a very integrated *forebrain* that contains specialized subparts for creating and using new programs. The mammalian forebrain itself is an exquisitely complex structure that has evolved over millennia to perform an enormous number of very specific new functions related to animal survival and procreation. The mammalian forebrain is a breakthrough system that is very complex, and it is likely that its workings will only be properly understood by studying its features from many different vantage points.

Here let us sketch the largest anatomic viewpoint, and that is how the forebrain is situated with respect to antecedent structures and what properties they confer on its overall organization. Figure 1.1 will serve to orient the reader. What the figure makes immediately obvious is that the forebrain has the largest percentage of the total brain volume. The functions of the various parts of the forebrain will be dissected in the next chapter, but collectively they perform in large part the basic sophisticated planning and

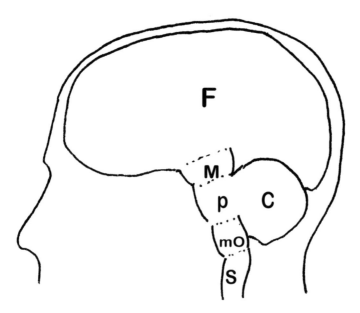

Figure 1.1
A cutaway view of the brain showing a major division between the forebrain (F), which handles simulation, planning, and acting, and the lower brain structures that handle basic life support functions. Midbrain (M): basic drives. Pons and cerebellum (P and C): sensorimotor calibration. Medulla oblongata (mO): management of body organs and life support operations. Spinal cord (S): motor functions and organ regulation circuitry.

acting that we associate with being human. The crucial point to note here is that they depend on the parts of the brain that evolved before them, and those subsystems of the brain evolved to keeping us alive and kicking. The brain's evolutionarily earlier functionality comprises a huge network of primordial structures that are heavily optimized for a myriad of life-support functions that relate both to its internal structures and the external world. If certain parts of these are damaged, the consequences can be dire. But to return to the main point, in order to think about what the forebrain does computationally, it helps to understand what its predecessor structures do. Let's start from the periphery and work our way inward.

Reflexes in the Spinal Cord

One of the most important of the lower brain functions is the *reflex* that connects sensory information to motor actions. Figure 1.2 shows the basic features of a reflex. Sensory information is almost directly connected to *efferent* (meaning "conducting away" as opposed to *afferent*, "conducting toward") neurons for a fast response. This means that the neural response is not dependent on adjudication by the forebrain, a long way away in space and time. Although we might be tempted to think (recalling experiences of withdrawing a hand from a hot plate or being whacked on the knee's patella by a well-meaning physician) that reflexes are simple, they are not, and they represent coordinated muscle patterns.[8] Concatenations of such circuitry can produce oscillatory patterns that are the basis of an animal's library of complicated posture changes. In experiments with cats without a forebrain, the basic spinal cord circuits are enough for the cat to walk and run, with only slight assistance in support. Moreover, the cat's feet will retract from encounters with small obstacles and step over them unaided. The sensory motor systems of the human spinal cord are even more sophisticated and have many additional capabilities that allow the ready programming of complex movements.

Reflexes can get an animal a long way. A praying mantis can get dinner by snapping at another unsuspecting insect as can a frog. In each case, some visual feature of the stimulus is enough to set off the reflexive snap. Primitive robots have demonstrated that a variety of behaviors can be packaged to generate useful behaviors. IRobot's Roomba vacuum cleaner makes extensive use of a reflexive level of behavior.

Life Support in the Medulla Oblongata

At the next level, proceeding from the spinal cord toward the forebrain, is the medulla oblongata. The primary function of this cell complex is to regulate vital internal processes such as heart rate, breathing, and the many

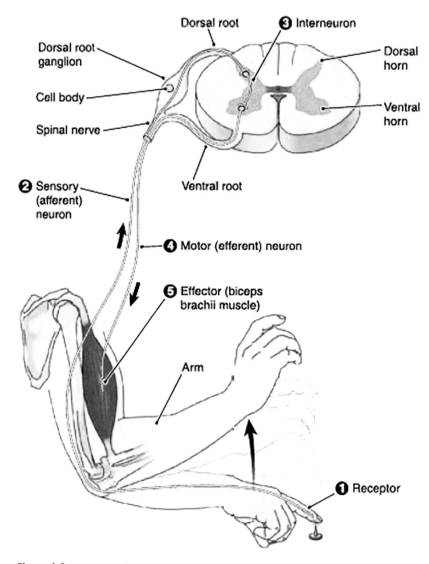

Figure 1.2
A basic sensorimotor reflex. Transmission through banks of neurons is sufficiently slow that the brain uses "hard-wired" circuits to achieve fast computation that cannot be done with the forebrain in the control path. Withdrawing a finger from a heat source is a very familiar reflex, but there are many others that are essential that control the body's complicated dynamic systems in a timely way. Reprinted with permission from http://encyclopedia.lubopitko-bg.com/.

steps in digestion. If this area is damaged, then some functions can be taken over by machinery in a hospital's intensive care unit, but, if the patient does not recover, these functions cannot be duplicated in the long term.

Sensorimotor Calibration in the Pons and Cerebellum

Next to the medulla oblongata are the pons and cerebellum, which are in charge of sensory motor calibration. An easy way to understand the importance of this function is to focus for a moment on infant development. As an infant grows, the body size changes drastically, but the visual system's optics is relatively invariant. This means that the number of steps needed to reach a distant object, as measured by the visual system, is continually changing. So the connections between the visual measurements and the motor output system have to be continually adjusted. Another example is when, as an adult, you carry a heavy backpack. The effective mass of your upper body has changed, and so the forces needed to balance have to be recalculated. It is not inaccurate to think of the cerebellum as a huge sensorimotor input-output table that, for every sensory stimulus, registers the appropriate parts of the motor system to innervate. Thus, the cerebellum is the first stage that realizes a very important evolutionary step: a map of the body's sensorimotor parts that is an abstraction of the body's more concrete neuromusculature. In effect it is a kind of regulatory model of the more concrete structures. In sensorimotor calibration, the adjustments are made to the model and then signaled to the spinal cord. It's as if evolution discovered a programmable "patch panel." Rather than signal an enormous number of changes throughout the body, they can be made far more succinctly within a more local programmable circuit that uses abstractions of the body's parts.

Unlike the medulla oblongata, damaging your cerebellum is not fatal, but there can be very substantial costs. An adult without one loses the exquisite ability to readjust to different loads, and even simple sensorimotor coordination such as touching one's nose with a forefinger becomes very laborious.

Chemical Regulation in the Midbrain

The medulla oblongata is a command center for regulating body function. Given the notion of an abstraction for sensorimotor control in the cerebellum, why not have one for the body functions as well? The midbrain does this via an armamentarium of *neurotransmitters*, chemicals that can modify circuit function, of which five are crucially important and have elaborate private circuits covering large parts of the forebrain. Imagine for a moment the issue of regulating all the different body systems, each with its own preferred setpoint and mechanism. Would it not be better to summarize these with a small set of neural mechanisms that code the optimal setpoints for each of the group of systems? Evolution discovered a general way of regulating the state of the body with chemical signals.

Programs in the Forebrain

All the previous structures are an enormous benefit for the forebrain, as they provide a set of sophisticated primitives as well as contexts that can be used as part of a neural "programming language" to realize very complex behaviors. The forebrain is all about making and storing programs, and compared to other animal brains, the mammalian brain has by far the most sophisticated neural machinery for doing this.

When thinking about the virtues of human intelligence, there is often a tendency to jump to its more exotic capabilities such as understanding calculus, but even the simplest ordinary behaviors are a huge advance over reflexive behavior. Suppose you are ready to make a peanut butter and jelly sandwich. The plate is empty, and it is time to get two slices of bread. So the next thing to do is to get them. You can effortlessly distinguish this state of affairs from the state you'd have been in if the slices were already on the plate. For either of these states, you know what to do next. Furthermore, you know how to sequence through a set of actions that will produce the sandwich. You can handle frequent errors (such as spilling jelly on the table) as well as the unexpected (the jelly jar lid is especially difficult to remove). And you would not repeat a step that you had already carried out. After a series of such steps, the sandwich is ready. You have just exhibited a dexterity beyond any current robot.

You probably were not aware that in making the sandwich, you have executed a complex program with many features that can be described as computation. Specifically:

• There has to be a way of defining the state of affairs, or *state*. Even though the sandwich construction that has jelly on the bread might seem obviously different than without, defining the notion of state succinctly in general turns out to have many subtleties.
• There has to be a way of transiting between states. To make a sandwich, coordinated actions consisting of coordinated movements have to be carried out.
• There has to be a way of assembling these states and action transitions into programs. The sandwich recipe has to be remembered as a unit.

The core computations of all these tasks are done by the forebrain, the focus of the next chapter. Furthermore, sandwich making is just one of millions of things you know how to do. And the way most of these are encoded are in the forebrain's "memory" system that in many ways is much closer to the computer concept of memory than what we mean by human memory colloquially. When we describe how this neural memory functions later on, we will see that one way of compensating for the very slow neural circuitry is to remember how to do enormous numbers of things—basically all of the things you do—more or less exactly.

1.2 Computational Abstraction

A good way to start thinking about computational abstraction is to introduce the enormous contribution of David Marr, who made seminal contributions in defining the enterprise of computational neuroscience. He and Poggio showed how the perception of depth in random-dot stereograms could be explained in purely geometric and computational terms.[9] But perhaps more importantly, Marr pointed out that the study of computation could be factored into three parts (see ref. 2):

- the formal statement of the problem that needed to be solved;
- an algorithm for solving it; and
- the implementation of that algorithm in the brain's neural circuits.

Thus, Marr's triage is a series of constraints. The formal problem statement is defined at a logical level and ignores the methods for solving it. The algorithm is defined at the computational level and ignores the biological details of implementing it. The implementation level takes up the problem of making correspondences between abstractions in the algorithm and corresponding biological mechanisms.

This tri-part factorization of a computational problem continues to prove enormously helpful in thinking about the brain's algorithms, particularly because at this point in time, as we move toward more abstract problems, there are still huge gaps in our knowledge of how the nervous system carries them out. At the time, this triage was a breakthrough as it opened up thinking about the brain's computation in abstract algorithmic terms while postponing the reconciliation with biological structures, and this mode of thinking remains extremely useful for at least two reasons. One is that there are many areas of brain function where we still do not have enough information to make satisfying detailed connections. A second reason is that despite such a gulf, the abstract properties of an algorithm can suggest new ways of thinking about biological data.

It is important to keep in mind that the Marr triage is a prescription to help reverse engineers organize their thoughts. In effect, it is a sequence of specifications on the way to a detailed theory. But there is another way to think hierarchically, and it is to that way we now turn. Basically, there is only one known way to design large complex systems, as pointed out by Alan Newell[10] and his long-time colleague Herb Simon, and that is to introduce hierarchical organization, where each higher level in the hierarchy is successively more abstract. Let's make sure we have properly drawn the distinction between the Marr hierarchy and the intrinsic hierarchy inherent in complex systems. A simple example from programing languages will get us started. Let's multiply two numbers together in the language Python (where * stands for multiplication):

$z = x * y$.

Because Python is a high-level language, we do not have to worry about the details, but some part of the computer does. Before the computation can be carried out, it has to be translated into a lower-level assembly language that is closer to the machine's architecture. So we might have

LOAD x, A
MULT A, y
STORE A, z

At the higher level, we did not have to worry about where the multiplication happens, but the lower level knows that the multiplication only works for data in special registers, such as the one here denoted by A. The crucial point here is that the two descriptions are equivalent in a strong sense in that the higher-level description can be translated into the lower-level description. You can immediately see that the Newell and Simon hierarchies are a very different way of characterizing hierarchies than that of Marr. Note that the Marr strategy can be used at any level.

To gain further purchase in the brain's management of program complexity, it helps to set the stage by visiting other examples of hierarchies, so let's take another look at hierarchies in the brain's anatomy, hierarchies in silicon computer organization, and then introduce a hierarchy for human behavior. Given the state of knowledge, this last hierarchy should be regarded as very provisional. Nonetheless, we hope to give it some credence by enfleshing its parts with experimental data as well as algorithms in later chapters. Let's elaborate on these points.

Anatomic Levels of Abstraction

You have already been introduced to the brain's top level of abstraction, whereby the brain is divided into major anatomic and functional subdivisions. Now let's move inside the forebrain's cerebral hemispheres. At the scale of a few centimeters, there is a predominantly two-dimensional organization of "maps" consisting of repetitions of very characteristic circuits,[11] wherein subdivisions of cells have identifiable roles. For example, there will be several areas responsible for computing different aspects of visual motion across the visual field of view that have their own map. At an abstraction level below that, subdivisions within these maps represent circuits of cells responsible for motion computations in a small area of the visual field. Such circuits, in turn, are made up of different varieties of neurons, each of which has an elaborate set of connections to other nearby neurons. Going to an even smaller scale, the connections to an individual neuron are defined by synapses that regulate the charge transfer between cells. Figure 1.3, from reference 3, summarizes this information. The different anatomic levels are easy to appreciate because they have a characteristic physical appearance that can readily be captured by various imaging techniques that operate at different scales.

Levels of Abstraction for Silicon

Now let's cross over to the idea of *brain computation* needing and using levels of abstraction. This is first easiest to appreciate by seeing how computational abstraction is essential in silicon computers. The basic levels are summarized in table 1.1. At the top, any standard computer has an operating system. This is the program that does the overall management of what computation happens when. It determines when jobs are sent to the printer, when input is read in, and when to clean up the memory after a bunch of running programs have left it in disarray (unsurprisingly this is called "garbage collection" in the jargon). But among all these jobs, the operating system runs the program you may have written—the user program. Your program contains instructions of things to do, but to the operating system these are just data, and when it determines your program's turn, the instructions are carried out systematically by the computer's processor. Of course, when you wrote such a program, you would not have used a language that the lowest level of the machine understands, but instead you would have chosen a high-level language such as Java or C. The reason is that the lower-level instructions are too detailed, being appropriate for the minutiae of the machine. The instructions you program with are translated into two lower levels: first of all, as we just saw, assembly language, which

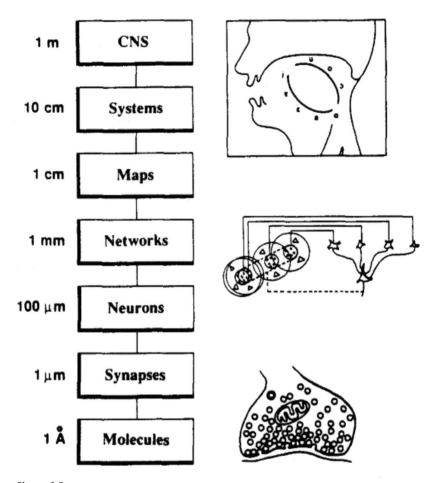

Figure 1.3
The organization of the brain into units at different spatial scales. The scale for the neuron characterizes its central body, or *soma*. Its other parts project over much larger distances. Subsystems have well-defined roles in behavior. For example, the hippocampus is responsible for new memories. Each such subsystem tends to have very characteristic large-scale organization, usually organized with respect to body topology. These subsystems also have characteristic circuits. Neurons also exhibit specialized types such as basket cells for short-range inhibition and pyramidal cells for long-range excitation. Messages are sent to individual cells through thousands of contacts termed *synapses*. Reprinted from Churchland and Sejnowski (1992).

Table 1.1
Basic levels of computational abstraction of a standard computer

Description	Function
Operating system	Control the running of other programs; manage input and output
User program	A particular program with a specialized objective; written in a high-level language
Assembly language	The translation into basic machine instructions that are for the most part hardware independent
Microcode	Machine instructions that can be interpreted by a particular machine's hardware
Circuits	Adding, shifting, etc.
Gates	Basic logical operations; e.g., AND and OR

Note: The standard computer uses many levels of abstraction in order to manage the complexity of its computations. Its hardware levels also can be divided into abstraction levels consisting of circuits that are composed of basic switches, or *gates*.

addresses elemental operations but postpones the details of how these are done, and then finally microcode, a language that can be understood by the machine's hardware, which in turn uses hierarchical levels that start with circuits composed of logic gates. The bottom line is that just to run your program, many different levels are needed, and indeed it is almost impossible to see how the resultant complexity could be handled if they were dispensed with.

From a historical perspective, computer software was designed from the bottom up. Original computer instructions were in the form of an assembly language, and the more abstract user program languages were added later, as were the constructs used by modern operating systems. An old saw declares that it takes the same time to debug 10 lines of code no matter what language it is written in. It is easy to understand why almost everyone writes in the highest-level language possible: you get the most leverage. Nonetheless, everything the computer does ultimately is carried out by the logic gates at a low hardware level of abstraction.

Although at first encounter, the idea that the computations of cognition have to be organized this way might seem to be eccentric, reading pioneers Simon and Newell one concludes that there is unlikely to be a non-hierarchical alternative. Like almost any strong statement, however, this one needs to be tempered, and perhaps the most important caveat is the following.

The language-translation view could easily be misconstrued as implying that the higher-level abstraction, when translated to the level below, can account for all the lower-level representation. The opposite is true, as typically the lower level has details (such as particular registers in our earlier example) that are suppressed when abstracting. In biology, this feature is compounded many times, as there may be many housekeeping details needed to make the circuitry viable that are unnecessary at the more abstract levels. This points to the virtue of the abstractions, as trying to understand a composite of high-level functions and low-level functions can easily be too demanding a task if confined to the lowest level.

Neural Computation Levels of Abstraction
If people had not designed the computer in the first place and kept track of the addition of its hierarchies, it would be an enormous job to figure out what it did just by looking at its outputs at the gate level, yet that daunting vista magnified is just what is facing us in understanding the brain. Despite the fact that so much has been learned, most of the important issues as to how computation is accomplished remain unsettled. Even the basic output of neurons has still to be satisfactorily decoded. Nonetheless, the concept of processing hierarchies provides the reverse engineer with enormous leverage. So much so that you probably are guessing correctly what comes next. We will use these insights about hierarchies from anatomy and computation as constraints with which to formulate neural computation levels of abstraction.

Just by looking at table 1.1, you can intuit why we are unlikely to understand the brain without a similar triage of functions into computational levels of abstraction. We can understand the machinery that generates a voltage spike (the main signaling mechanism) in a neuron, as well as how networks of neurons might use those spikes. However, we cannot really understand them together. For clarity, we have to keep them separate. And when we come to model something several levels more abstract, such as altruistic behavior, we have to ignore all this low-level machinery entirely and model an entire brain as a small number of key parameters. We should not think of this technique of using abstraction levels as a disadvantage at all. Instead, it is a tremendous benefit. By telescoping through different levels, we can parcellate the brain's enormous complexity into manageable levels.

To make these points more concrete, let's reconsider the job of making a peanut butter and jelly sandwich. In the typical kitchen, peanut butter and jelly would be found in their respective jars, and bread would come presliced. You might be tempted to think that making the sandwich is a trivial task, but its relative straightforwardness belies the complex of neural computations that are needed to put things together. Thus, this task can be used to illustrate the different kinds of computational levels that the brain must address. Working from the bottom to the top, one needs to appreciate that the raw visual and motor information extracted is essentially unusable by the brain's programs as it is too unstructured. The visual information extracted at the retina is summarized into about a million spatial samples of lightness and color that have no explicit indication of what object the samples come from. This kind of information must be extracted in the forebrain, which creates elaborate visual data indexes that allow the fast identification of image samples. Similarly, the peripheral motor codes contain a huge number of signals for the contraction of muscle fibers, with no explicit indication of the total coordination patterns necessary for purposive movement. These are also created in the forebrain's elaborate indexing structure, which has special partitions devoted to motor representations. Neither of these representations, along with companion representations of other sense information, comes for free, but instead must be created through elaborate computation at what table 1.2 calls the *data abstraction* level.

Table 1.2
Levels of computational abstraction

Description	Abstract Function	Example Function
Evaluation	Strategic decisions	Evaluate current task suite. Hungry? What are the nourishment options?
Scheduler	Multitask management	Regulate different sandwich-making programs. Jelly jar lid off now?
Programs	Solve a single task	Spread peanut butter on bread. Peanut butter is viscous and spreads easily.
Routines	Individual fixations used to guide posture changes	Find location of bread slice. Vision locates the bread loaf.
Data abstraction	World sensory data coded to emphasize intrinsic organization	Compact codes for sensorimotor signals: Activate codes for color and texture of bread.

Note: To manage complexity, the brain also has to resort to different levels of computational abstraction. While the ultimate abstraction has not been precisely determined, we can describe tentative organization based on the tasks that the brain has to direct.

Having elaborate indexes for sensory input and motor output is a big step forward, but a crucial but subtle step is to process this data structure to obtain vital information. Consider filling a cup with coffee. The sensory system can represent the coffee cup and the coffee going into the cup, but how do we know when the cup is full? There are several available cues such as the increased weight of the cup and the closeness of the fluid level to the cup brim, but the point is that these must be tested to ascertain the cup's fullness. So what is needed here is another computational abstraction level that takes the sensed data measurement system for granted and interrogates its results. Let us term this the *routines* level.

At this point in the discussion, we have defined an elemental data abstraction level and another level to test that data, so the next step is to compose these tests into sequences. Consider putting the peanut butter on the bread. The knife goes in the jar for a glob of peanut butter, removes it, and then it is spread on the bread; if there is still not enough peanut butter on the bread, then the brain sends the knife back for more, and so on. We have just defined another abstraction level, the *programs* level. At this level, we can take the details of the tests for granted and worry about how they are melded together into an action sequence that accomplishes a larger goal.

When you start to think along these lines, you quickly realize that there are many more possible abstraction levels above the program level, but in this illustration let us stop at two more. The particular program for spreading peanut butter can be defined at the program level, but what of multitasking? Perhaps you were boiling water for tea to go with the sandwich, so that the kettle is on. At the same time, the phone rings. Should you stop what you are doing and answer it or let it go to the message recorder? You only have one body, so somehow you must juggle its position in space and time to make the best use of its resources. The need to manage motivates another computational abstraction, the *scheduler*, which takes programs as primitives and manages their execution. The various steps in each program can be scheduled in an efficient way to get everything done.

There are other levels to think about, but for the moment let's introduce one more, the *evaluation* level. Is the current task suite of making a meal the most important or is there a more pressing demand? To catch a bus, perhaps the sandwich should go in the fridge for later. To make these kinds of judgments, the brain needs some kind of scoring function so that the adjudication can be managed systematically. This is not the only computational taxonomy one can think of, and one can debate the necessity of any aspect of the hierarchy of table 1.2. But the point is that if we take the lessons of biology and silicon computers to heart, we are unlikely to get away with a "flat" neural computation description. The far more likely arrangement is that the brain is composed of many more abstract neural networks that leverage the results of less abstract networks in the process of getting things done. And if we do not acknowledge and address the need for such hierarchies, then the overall neural organization is likely to appear very confusing.

1.3 Different than Silicon

The brain is nothing like a conventional computer and is staggeringly more complex, even though at an abstract level the brain has to solve some of the same kinds of problems. Nonetheless, the huge number of differences between silicon circuits and neurobiological structures means that the biological solutions must be of a hugely different character. Let's introduce some fundamental characteristics that show just how shockingly different brain computation must be.

The major factor separating silicon and cells is time. The switching speed of silicon transistors, which limits the speed at which a processor can function, is in the nanosecond regime. In contrast, neurons send messages to each other using voltage pulses or, in the jargon, "spikes." Neurons can send these spikes at a top speed of several hundred spikes per second, but in the main processing areas the average rate is 10 spikes per second, with 100 spikes per second regarded as a very high rate. Figure 1.4 shows some typical representative spike sequences. For a long time, it was thought that a spike was a binary pulse, but recent experiments suggest ways in which it

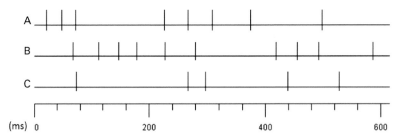

Figure 1.4
Three spike trains from an integrate and fire neuron model show the characteristic low firing rate behavior with random inter spike intervals. Such features can pose additional challenges in explaining the control real time behaviors. Courtesy of Liz Stuart, University of Plymouth Visualization Laboratory.

could signal an analog value, so let's assume that its message is of the order a byte per pulse. Even with this assumption, nerve cells communicate 10 million times slower that silicon transistors. Given 10^{11} nerve cells, only about 10^{10} are sending spikes at 10 Hz. It would be easy to compress these data by a factor of 10, so that roughly 10^3 seconds of your brain's neural firing (more than enough for a thought or two) could be saved on 10 terabytes of storage. The task for brain scientists is to break this code.

Code breaking will ultimately require a collection of many different insights, but to introduce just one as an example, let's bring to mind the metaphor of a old-fashioned player piano. Such a piano uses special sheet music in the form of thick paper with perforations. As the drum rotates, the perforations depress pistons pneumatically, causing piano keys to be struck. Think of the piston depressions as "spikes." In an analogous way, the neural spike code can be a sparse discrete code; the body makes the music.

The slow communication rate is sandwiched from above by the time for the fastest behavior. Evidence from behavioral experiments suggests that essential computations take about 200 to 300 milliseconds. This means that the average neuron has to get its computation done with two to three spikes. From these analyses, the consensus is that the way the brain must do it is to have most of the answers precomputed in some tabular format so that they just have to be looked up. Evidence in favor of this view comes from the rich connectivity between nerve cells. Each neuron connects to about 10,000 other neurons, compared to a gate's connectivity to just a handful of gates on a silicon chip. As shown in figure 1.5, the size of the gates in silicon are comparable to the processes of a living cell. It is the neuron's huge connectivity that gives it one of its biggest advantages.

Another factor that the brain uses to overcome the speed handicap is the power of nerve cells themselves. The exact computing power of a neuron is unknown, but a good guess is that it is at least much more powerful than a transistor. The synapses that connect it to other cells are closer to transistors, but again more powerful as their action can be modified by neurotransmitters. Thus, the neuron itself has been likened to a microprocessor, albeit a specialized one. Because the brain has approximately 100 billion nerve cells—much less than the U.S. fiscal debt in 2004 dollars, but still a lot—that can work simultaneously, the parallel computing power is obviously one source of compensation for the brain's slow circuitry.

With the incredibly slow circuitry, there is no hope of implementing the strategies used by the billion times faster silicon circuitry. In fact, even the

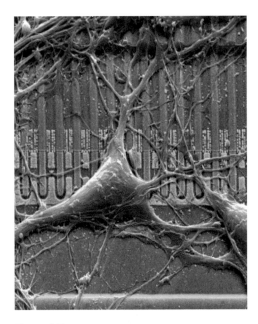

Figure 1.5
An exotic electron micrograph of a neuron artificially grown on a silicon wafer reveals the comparable scales of the two technologies. The raised clump is the neuron's body, or *soma*. One of the spidery processes coming out of the soma is its axon, which connects it to an average of 10^4 other cells. In contrast, silicon transistor connections between gates are limited to a handful. Figure courtesy of Peter Fromherz, Max Planck Institute.

basic ways of measuring silicon performance have to be thrown out the window. To understand this provocative claim, we have to take a look at how the standard algorithm's accounting is done.

Even if you have no training in computer science, it's easy to understand what an algorithm can do because, as noted earlier, it is very much like a recipe in cooking. There are standard steps, and it's important to remember where you are in the process. You may have to repeat operations, as in stirring, and you may have to test something to see if you are done and keep going if you are not. Recipes can be thought of as algorithms for cooking.

Let's introduce the silicon computer's traditional bookkeeping methodology for counting the steps in a recipe. On serial silicon computers, most algorithms are dominated by the size of the input. For example, consider sorting a list of n numbers. Here is a recipe: Go through the list and move the biggest number to the top. Then go through the $n - 1$ remaining numbers and pick the second largest and move it to the penultimate position. After you get down to fixing the last two elements, you are done. This is called "bubble sort" because the larger numbers bubble up to the top. This basic algorithm would take

$$\frac{n(n+1)}{2}$$

steps because the total number of steps is the sum of the numbers from 1 to n. In practice, we don't sweat the factor of 1/2 or the 1, and we say "of the order" or $O(n^2)$ operations.[c]

Of course, computer science majors all know that there is a faster algorithm that takes advantage of the fact that two *sorted* lists of length n can be merged in $O(n)$ steps. Here is the algorithm that uses the merging property:

Sort(List)
if List has one element
   return the resultant list

else if a List has two elements
   Sort them
   return the resultant list

else Merge(Sort(Front-of-List),Sort(Back-of-List))

You start out with the original list and contract to merge its two sorted sublists. Of course, they too need to be handled in the same way, but the key is that they are only half as long as the original list. For each list of more than two elements, an IOU is created. These IOUs are resolved when we get down to two- or one-element lists. Once this happens, the many outstanding merging processes that need sorted sublists can be completed, resulting in a sorted list. To see the idea, try it on a piece of paper with small lists of say four to six numerical elements.

A careful accounting shows that this only requires, in the notation, $O(n \log n)$ operations, which of course is better than the easier-to-understand $O(n^2)$ algorithm that was considered first. Why is all this analysis important? Because basically if the number of operations in an algorithm is known, along with the time each operation takes, the total time an algorithm takes can be calculated. Carrying such calculations through, most of the best algorithms that are fine for silicon computers working at clock speeds of 1 GHz, or a billion operations per second, are not possible for the brain's much slower neural circuits. The clever sorting algorithm just described is the best that can be done but still won't do for a model of brain computation. The main reason is that the neurons that are the candidates for the principal computing elements are very slow, more than a million times slower than silicon. A reasonable size for n is 1 million for human vision, and if we assume neurons are computing at 10 binary "bits" per second, you can see why an $O(n \log n)$ algorithm could not be a candidate. An algorithm that had to poll each of these cells serially would take an impractical 100,000 seconds.

From the perspective of naively counting steps as we did with the silicon computing examples, it would seem that the situation is hopeless. Given the number of steps in simple algorithms and the slowness of executing steps with neurons, there seems to be no way that computations can finish in time. But the brain has very powerful tricks up its sleeve.

1.4   The Brain's Tricks for Fast Computation

If the brain is doing computation, then at some point we have to be able to explain how that computation gets done in time to direct our daily activities. In computational science, this question has to be resolved by coming up with the brain's specific algorithms. Of course, our main point is that an overarching hierarchical structure of brain computation makes the computation at each level easier to specify. But there is still the issue of describing what computation goes on.

Although we still do not quite know how to specify the brain's processors in detail, we suspect that nerve cells will be at the center of the answer. So the method is to define models of neurons that represent best guesses of what is important about them and then define algorithms that use these abstractions. When one starts to do this, one quickly finds out that the brain's algorithms must be very unlike those for conventional silicon computers. This difference is most apparent when considering the standard way of evaluating algorithms, and that is to see how long they take to complete. But even this measure must be changed for brain algorithms. For silicon processing, the traditional way of counting operations is called "worst case." We want to guarantee that the algorithm will take no longer than some temporal bound. But the brain doesn't care about the worst case because it is always under time pressure. If something is taking too long, there is always the option of giving up and moving on. And in this spirit, the brain also uses lots of dramatic economies. Let's introduce the main ones.

1. *Parallel computation* The nerve cell can be thought of as the brain's basic computing unit, with the result that the brain has at least an astonishing 10^{10} processors. If there was some way of exploiting this huge capability for simultaneous processing, then the brain could compete with silicon speeds. Fortunately for much of the sensory and motor circuitry, this parallelism is possible. For example, in visuomotor processing, raw image data flows in parallel through banks of neurons, where each bank is able to compute a successively more abstract representation. Thus, a moving black and yellow stimulus becomes, after less than 10 of these banks, a neural code for "tiger," which can immediately be passed on to banks of motor circuits elaborating a "flight" response.

2. *Using probability* The $O(n \log n)$ algorithm for sorting is provably the best there is. It can sort *any* sequence in its allotted time. But this is not the case for all problems. For some algorithms, such as finding the best round-trip route through a set of cities, getting the shortest possible path is very expensive. You have to examine all of the paths, and that is exponentially many. Naturally, this is prohibitively expensive for humans.

But although humans can get themselves end-played in fatal situations, in the vast majority of cases that does not happen. The normal environments we inhabit are very rich in alternatives, and as a consequence, if we just want a reasonably good solution, this can be had at a reasonable cost. Thus, one of the main ways of speeding things up is to use probably, approximately correct (PAC) algorithms. The PAC way of accounting was pioneered by Valiant[12] and is standard issue for thinking about brain computation.

It turns out that probability is enormously helpful in coming up with fast estimates even when it uses not very reliable data. Suppose that you are wondering if a coin you have is fair. You flip it five times and observe HTHHT. Based on these data, you can't be very sure. But if you flip the coin 200 times and observe 70 heads and 130 tails, then you can be extremely sure the coin is biased. As we will see, the brain has vast networks that allow approximate estimates of individual nerve cells to be pooled quickly. This kind of probabilistic reasoning allows questions to be answered incredibly quickly.

3. *Oscillations at different frequencies* Although one can be more confident of the bias upon seeing a head-to-tail total coin flips ratio of 70:130, the computation has used up a lot of time. Take another gander at figure 1.4

and count the spikes to determine the spike rate. You can come up with an estimate, but you have used almost half a second on the figure. A much faster, and at this point controversial, way to do this would be to code the estimate as the delay from the zero phase point of a reference frequency. To unpack the last sentence, let's go back to figure 1.4 and, focusing on one of the traces, superimpose a grid on the timescale with ticks 20 milliseconds apart. No draw a short line rightward from each tick to the nearest spike. The idea is that this short interval could be a number. So using this convention, the ratio can be sent in one spike! Of course, the accuracy of this coding strategy depends on the ability of the brain to time spikes with great precision, but this is something the brain can do.

Not only can the brain achieve the requisite timing, but also accumulating evidence suggests that it can do this for a number of distinct frequency ranges, and these ranges have specific computational functions, as reported in table 1.3.

There is a lot to be said about the particulars of how these different frequencies are used, and the full appreciation for their functional properties is still a work in progress, nonetheless we will paint a precis of the situation here.[13, 14] One difficult problem the brain has is to slice and dice the continuous nature of sensory motor commerce into a form that can be interpreted by internal codes. The frequency that demarcates the beginning and end of such an episode is the θ (theta) frequency. For humans, the length of an episode can be arbitrary, say as in planning a long trip. For more near-term behaviors that involve real-time control of the body, such as reaching for a coffee cup on a nearby table, evidence suggests that the β (beta) frequency is used.

Table 1.3
Oscillation frequencies associated with computational functions in the brain

Frequency	Range (Hz)
θ	4–7
α	8–12
β	13–39
γ	40–90

Note: Recent evidence is revealing that different temporal frequencies of oscillation created by neurons have computational roles. Oscillations in the brain have long been known, but their involvement in computation has only much more recently been appreciated.

The jury is still out on the use of the α (alpha) frequency. Some evidence suggests it has a role in the timing of behavior. Another idea, perhaps a minority view, suggests that it has a role in maintaining calibration in forebrain circuitry. The circuits are never "off," and when they are not actively involved in a computation, α might be used to calibrate their dynamic range.

When thinking about brain circuitry, should we think of the neurons in one huge holistic computation or is there a way that the computation is broken down into more or less independent parts? Parsimony favors the compositional view because, if it were true, the brain could achieve enormous diversity in composing different collections of component parts. To appreciate a problem, at least for the discussion, adopt the cloak of a compositionalist and imagine one of the circuits that is essential among the huge network of the rest of the brain circuitry. How does it keep its function separate? There have been various ways suggested to do this, but they all require some technical artifice. One is to assume that the circuit can somehow be tuned to a distinct frequency in the γ (gamma) range. This idea is relatively new, but evidence is accruing for the importance of γ in this role.

4. *Bounded input and output sizes* In the analysis of sorting algorithms on silicon computers, the assumption is that the dominant factor is the size of the input. Where the size of the input can grow arbitrarily, this is the correct analysis. For example, suppose we pick a give cost for bubble sort so that now there is no "Big O," but instead we know that the cost is exactly $1,000n^2$ on a given computer. Now your colleague gets a computer that is 1,000 times faster so that the cost is exactly n^2. You have to use the old computer but can use the merge-sort algorithm. So now even though you have the better algorithm, your colleague wins when

$$1,000 \, n \log n > n^2.$$

Thus, the standard "Big O" analysis breaks down for biological systems as the number of inputs and outputs are for all practical purposes fixed. When this happens, it pays to optimize the hardware. To pursue the example of vision, suppose now that each image measurement could be sampled in parallel. Now you do not have to pay the 1,000,000 factor. Furthermore suppose that you wanted to use this parallel "bus" of image measurements to look something up. Now you only need $O(\log n)$ measurements. Furthermore, the brain ameliorates this cost as well by using a pipeline architecture so that the log factor is amortized quickly over a series of stages. Each stage can be thought of as answering one of 20 questions so that by the time the process exits, the answer has been determined.

5. *Special-purpose sensors and effectors* The design for the light-sensing photoreceptors used by vision is believed to have started with the properties of sea water. It turns out that the visible spectrum is especially good at penetrating water and so could be exploited by fish. Once the hardware was discovered, it worked on land as well.

In the same way, the human musculoskeletal system is especially designed for the human ecological niche. We cannot outrun a cheetah, but we can climb a tree slightly better, and for manual coordination it's no contest. The niche is even more sharply illustrated by current robotics. Although silicon computers can easily out-calculate humans, the design of robotic bodies is still very much inferior to that of human bodies. Furthermore, the general problems that these human bodies solve seemingly effortlessly are still very much superior to their robotic counterparts except in a few special cases. To appreciate this further, try wearing gloves and going about your normal everyday activities. You will quickly find that you are frustrated in nearly every situation that requires detailed hand coordination. If this does not convince you, put on boxing gloves! You will still be better off than a robot with a parallel-jaw gripper robot arm.

Of course, there are difficult situations that can overwhelm a human body, such as staring into the Sun or trying to jump a canyon on a motorcycle. But these situations are for *Darwin Award* contestants. For almost all the problems of everyday existence, the human body is an exquisite design that works just fine.

6. *Amortized computation* One contributing factor to fine motor coordination that we have just discussed is the design of the physical system. To date, no robot can even come close to the strength-to-weight capabilities of the human musculoskeletal system. But there is another factor, too, which is that the process of designing the control algorithms that work so well happens over many years. Babies learn to control their arms sequentially. They'll lock all the outboard joints and try movements. When the nervous system has a model of this reduced system, they'll unlock the next more distal joint and try again. The new system has fewer variables than it would if starting from scratch. Naturally, it is essential to have parents that patiently care-take while this process is happening, but the essential feature is that the computation is amortized over time. A wonderful analogy is Google. To make fast Web searches, overheated warehouses

of server computers crawl the Web around the clock to find and code its interesting structure. The coded results are what makes the response to your query lightning fast. In the same way, efficient motor behavior reflects a multiyear process of coding the way the body interacts with its physical surroundings. The result is that reaching for a cup is fast and effortless, and carrying it upstairs without spilling its liquid contents is a snap.

Combinations of all these tricks, plus others that await discovery, are what allows the brain to do its job fast enough to keep up with real-time behavioral demands. The chapters ahead use the computational abstraction formalism to index the different collections of tricks used at different levels. For the moment, we will turn to take a look at some of the pessimistic views.

1.5   More Powerful than a Computer?

Is the brain just a computer or is it somehow much more powerful? Many readers would be agnostic to the answer to this question, but it is fundamental. Is computation a *superb theory*, to use Penrose's term (see ref. 4), or is it just a useful engineering model that produces helpful answers some of the time? To answer this question as to whether or not the brain could be a computer, we must first understand computation in the abstract. This is because the popular notion of computing is irretrievably tied to silicon machines. Furthermore, these machines have evolved to augment human needs rather than exist on their own and as such have not been made to exhibit the kinds of values inherent in biological choices. Thus, an immediate reaction to the idea that brains are kinds of computers is to reject the idea as baseless, with the rejection based on the limitations and peculiarities of modern silicon computers. To counter such intuitions will take a bit of work starting with a formal characterization of computation. We need to describe what a computer is *abstractly* so that if a brain model can be shown to be incompatible or compatible with this description, then the issue is settled. It could turn out that brains are unreachably better than formal computation. I don't think so for a moment, but the crucial point is to frame the question correctly.

In framing the question, one has to put one's reverse engineer hat aside and ask a fundamental question: Can computation be the root of a *theory* of brain function? As Penrose points out, there are many grades of theory, Newtonian mechanics and quantum mechanics being graded by him as superb theories for their enormous predictive scopes. Could computation wind up being a superb theory also? The jury is still out. One way to settle the question would be to show that humans have abilities that are more powerful than those of computers. If this could be done, then of course computation would lose its potential for a superb rating. So let's take a brief look at formal computation to sketch the prospects.

Turing Machines

What is computation? Nowadays, most of us have an elementary idea of what a computer does, so much so that it can be difficult to imagine what the conceptual landscape looked like before its advent. In fact, the invention or discovery of formal computation was a remarkable feat, astonishing in retrospect. Enter Alan Turing, a brilliant mathematician who led the team that broke the German Enigma code during World War II. Turing's approach, which resulted in the invention of formal computation, was constructive: He tried to systematize the mechanisms that people went through when they did mathematical calculations. The result was the Turing machine, a very simple description of such calculations that defines computation (box 1.1). Although there have been other attempts to define computation, they have all been shown to be equivalent to Turing's definition. Thus, it is the standard: If a Turing machine cannot do it, it's not computation.

The steps a Turing machine (TM) goes through in the course of accomplishing even simple calculations are so tedious that they challenge our intuitions when we are confronted with its formal power: Any computation done by any computer anywhere can be translated into an equivalent computation for a TM. Of course, it could easily be the case that the TM would not finish that computation in your lifetime, no matter how young you are, but that is not the point. The computation can be simulated.

Box 1.1
TURING MACHINE

> All computation can be modeled on a universal machine called a Turing machine (TM). Such a machine has a very simple specification, as shown in the figure below. The machine works by being in a "state" and reading a symbol from linear tape. For each combination of state and tape symbol, the machine has an associated instruction that specifies a triple consisting of the new state, a symbol to write on the tape, and a direction to move. Possible motions are one tape symbol to the left or right. Although the TM operation appears simple, it is sufficiently powerful that if a problem can be solved by any computer, it can be solved by a TM.

Box 1.1
(continued)

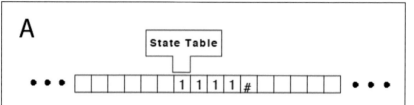

A

State Table

● ● ● | | | | | | 1 | 1 | 1 | 1 | # | | | | | ● ● ●

Situation		Action		
Read	State	Move	New State	Write
0	0	R	0	0
1	0	R	1	0
1	1	R	1	0
0	1	R	Halt	0

B

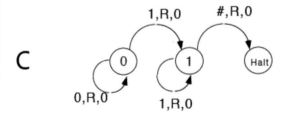

C

1,R,0   #,R,0

0   1   Halt

0,R,0   1,R,0

(A) A Turing machine program for the very simple function of erasing a series of 1's on the tape. (B) The program can be represented by a table that shows what to do for each state and input. (C) Equivalently, a TM program can be described by a state transition diagram in which the nodes of a graph are states, and arcs are labeled by the symbol read, the direction of motion, and the symbol written. Despite the extreme modesty of its structure, a TM is sufficiently powerful to be able to emulate all the operations of any other computer, albeit much less efficiently.

Box 1.1
(continued)

> This example shows a very simple program for erasing a block of contiguous 1's. The head is moved along the tape serially, replacing each 1 with a 0. When the end symbol is encountered, the program terminates. You can define any number of auxiliary symbols to help write the program or alternately find ways around using them. Here, for example, you could avoid the # symbol just by terminating after you see the second 0. For more of a challenge, try to add two numbers together, and then for even more of a challenge, try to multiply two numbers together. Remember that they are in a unary representation, just like that used by convicts marking jail time. To get you started, think of the initial tape as containing, for example,
>
> 0000#1111#000#111#0000#000000000
>
> Your program will probably find a 1, replace it with a 0, and then go and put it in the answer region, repeating the process until all the 1's were used up. For multiplication, the answer would be
>
> 0000#1111#000#000#0000#11111111111100
>
> This program will require even more sawing back and forth on the tape.
>
> As you try to write more complex programs, you will quickly be overwhelmed by the tedium associated with the low-level description used by the TM. But the point is that in principle, any program in any computer language has a TM equivalent.

An important point that cannot be overstressed is the breakthrough of the TM architecture that makes explicit the utility of thinking about programs in terms of a collection of states and actions that can be taken when "in" a state. In terms of everyday behavior, if you are making a cup of tea and you are in a state where {*the kettle is nearby and empty*}, then presumably the next action is to {*put water in the kettle*}. The power of thinking in this way cannot be overestimated. In chapter 5, when we describe ways of formalizing the brain's programs, it will be in terms of the *state, action* terminology.

To return to the formal computational theme, TMs are not without controversy, as they do have limitations. In general, a TM cannot tell whether the program of another arbitrary TM will halt or keep going forever. Of course, it can for some TMs but not in general. Furthermore, TM calculations cannot use random numbers because the very definition of a random number is that a TM cannot decide whether it is random or not. And it cannot use real numbers either because there are infinitely many more real numbers than TMs. Experienced programmers know that when they have randomness in their programs, the program is using pseudo-random numbers; that is, numbers that behave enough like random numbers to get useful answers to programs that need them. Similarly, programs use integers to sample the space of real numbers, again getting numbers close enough to the real thing for the calculations to have meaning. Finally, as we will elaborate in a moment, a TM cannot use everyday formal logic either without some form of limitation, as if it tries to prove a theorem that is not true, there is a possibility that the program will run forever and not halt.

The question is: Are these limitations important or do we need a more powerful model such as physics? What are the prospects for a physics-based computing? Some scientists think big, and one is Seth Lloyd.[15] He calculated the operations that would be needed to simulate the universe since its inception. The estimate is that you need no more than 10^{120} operations on 10^{90} bits for the memory. These numbers are more than the accessible universe has, but that is because any simulation will have some overhead. Also, you might not get the same universe if you started with a different random number seed; indeed, if the process depended on truly random numbers, you might not get our universe at all. One important take-home message from this vantage point is that to the extent that the universe can be described in terms of computation, then presumably our small brains can, too! But a larger point is that perhaps there is a prospect of harnessing quantum computing to solve difficult problems. Potentially many solutions could be coded as probabilistic quantum states in a machine that could then pick the best one. While this is intriguing, the technical problems in making this work at any useful scale are enormous. For an introduction, see Lloyd's book (see ref. 15).

1.6 Do Humans Have Non-Turing Abilities?

Given that real numbers and random numbers are the death knell for TMs, one quick way to distance humans and machines would be to show that in fact humans have the ability to use either or both. This is a tricky task because one has to show that the infinite precision of real numbers is realized by the humans. An intriguing way station toward this task shows that a neural-like model that can use real numbers is in fact more powerful than Turing computation, as is done by Selgelmann[16] in a model where model neurons have the ability to realize real numbers as input. However, a crucial next step would be to show that real neurons can in fact do this. The real world is riddled with noise sources that limit the precision of analog signals.

A more pessimistic view of computational prospects is represented by Roger Penrose, who has written three books with the theme that TMs are too weak a model to encompass the brain (see ref. 4; Penrose also holds out hope for the quantum computation and proposes a way in which it might be utilized by cells that is wildly speculative). But what of his arguments that TMs do not have sufficient power? One of Penrose's main arguments settles around Gödel's theorem.

Penrose argues that as humans understand this theorem that points to a fundamental weakness of logic (in proving statements about arithmetic), but computers are forced to *use* logic, ergo humans think out of the logical box and are more powerful than TMs. However, if we understand a proof, it has to be logical. The trick is that the referents of the proof are highly symbolic sentences. Gödel's brilliant insight, wonderfully described by Nagel and Newman,[17] was that when these were about mathematics, they could be reduced to arithmetic. Hence, the referents of the logical statements are regularized, and no special machinery is necessary. We are not saying it is easy; after all there has only been one Kurt Gödel! However, any graduate student in computer science or mathematics can easily understand the logic of the proof.

One central, potentially confusing issue that Gödel's theorem addresses successfully and that we touched upon when discussing hierarchies earlier is that of managing concepts at different levels of abstraction. When working a given level, one has to be careful to stay within that level to have everything make conceptual sense. You can switch levels but you have to be careful to do the bookkeeping required to go back and forth (see ref. 10). The failure to do this can lead to "strange loops" and the delight in experiencing the ambiguity that comes with it.[18] However, keeping this straight diffuses the alleged mystery in understanding Gödel's theorem. If we take the vantage point of the coded formula, the proof is such that anyone with advanced training in logic can understand it. It is only when we simultaneously try to entertain consequences of the uncoded logical formulas together with statements at a different level of abstraction that use their coded form that things get confusing.

Turing Machines and Logic

At this point, there might be one last puzzling question in your mind. For one thing, the standard computer is constructed with network logic gates, each of which implements an elementary logic function such as the logical

AND of two inputs: If they are both TRUE, then the output is TRUE else the output is false. In contrast, we have the exotic result of Gödel that there exist true statements in arithmetic that cannot be proved. So what is the status of logic vis-à-vis Turing machines? The resolution of this ambiguity is that two kinds of logic are involved. The fixed architecture of the Turing machine can be modeled with *propositional logic*, whereas the logic we implicitly use in everyday reasoning is modeled with *predicate logic*. The crucial difference is that the latter has variables that can arbitrarily correspond to tokens that represent real-world objects. Thus, to express the familiar "All men are mortal," we need predicate logic to have a variable that can take on the value of any particular man. We write: for all x the proposition $P(x)$ (being a man) being true implies that the proposition $Q(x)$ (being mortal) is also true. Formally,

$\forall x\, P(x) \supset Q(x)$.

To prove the theorem "Socrates is a man therefore Socrates is mortal," we can use the fact that the formula must work for any x, so we can assign it *Socrates* and use the standard method in elementary logic of constructing a truth table that reveals that whenever the preconditions of the theorem are true, the consequent is true.

A whole scientific field of endeavor termed *automatic theorem proving* uses a host of other fancier techniques. However, the main difficulty is hinted at by the simple example. To use the process, a particular binding— here x = *Socrates*—must be chosen. This is ridiculously easy in the example, but in a general case there will be many variables and many, many bindings for each variable. Nonetheless, they can be organized so that they can be tried in a systematic way. The upshot is that if the theorem is true, an automated theorem prover using resolution will eventually produce the *nil* clause and prove the theorem in a finite number of steps, but if it is not true, it will grind through possibilities forever. This kind of situation will be familiar to any programmer: Is the program almost done and about to stop at any moment or is it going to grind on forever owing to some bug?

Now for the punch line on theorem proving and TMs. Silicon computers use gates that implement *propositional logic* formulas such as AND, OR, and NOT. This is different than the *first-order predicate logic* used by mathematicians to prove theorems. To do the latter, a silicon computer has to simulate predicate logic using its propositional logic hardware. If humans are TMs, they have to do the same, with the same formal consequences.

As an epilogue, the same point can be illustrated with human deduction. Consider figure 1.6. In this deck, letters are on one side, and numbers

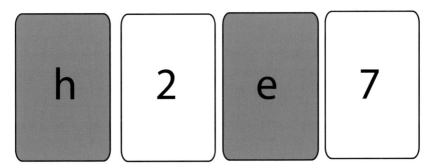

Figure 1.6
Cards have letters on one side and numbers on the other. Which two should you turn over to check whether every even-numbered card has a consonant on the other side?

are on the other. Your task is to test the claim that in this deck, "every even-numbered card has a consonant on the other side." You can only turn over two cards. Which cards do you pick? If you are like most people, you'll have to think about this a bit and might even make a mistake. You should turn over the second and third cards. But the main point is that it seems the brain can *simulate* logical deduction, in the same way a computer can simulate predicate calculus, but it is unlikely to use logic as a primitive construct—because if it did, presumably the card problem would be trivial.

And just to drive another point home, see how easy this becomes if we translate this problem into a familiar experience. Four people have met at a bar, two teenagers and two older people. Two of them have left for the restroom, so your view is of a teenager and an old person and the drinks of the other two—a beer and a Coke. Whose identity cards should you check? The fact that you can do this easily suggests that your knowledge base is constructed from explicit exemplars of real-world situations. We can learn to abstract these into formal logical statements, but only with effort.

An Attempt at Framing the Debate
In a truly remarkable book, the physicist Erwin Schrödinger tackled the question "What is life?"[19] His primary theme was that the increasing abstractions of physical systems are needed to tame the chaos of the atomic levels. Thus, one can view the increasing abstractions in the brain as introducing successive levels of stability and order that finally become sufficiently regular to support the demands of computation to work with very long strings

of repeatable symbolic sequences. From that perspective, computation is a superb theory, one that provides a fundamental account of what the brain is up to that has, to date, no good competitors. For this perspective, the theory is emergent and depends critically on the brain's ability to insulate itself from the myriad of variations at lower abstraction levels.

The opposing view has been articulated most forcefully by Penrose. He has argued from many perspectives, but one way to understand this view is from the perspective of chaotic systems. *Chaos* is a technical term used in science to denote the ability of simple dynamic systems to turn small perturbations into impossibly large deviations. Such systems are surprisingly common. A child's playground swing turns out to be such a system. Without a parent to regulate the oscillations, and with the generous assumption of no friction, one loses the ability to predict where the swing will be in its oscillatory cycle. You have probably heard this phenomenon explained in a manner similar to "A butterfly flaps its wings and a hurricane strikes miles away in the Caribbean." Applied to the brain, this would mean that the brain does not protect itself from the chaos of low-level abstractions, but rather depends on them for super-Turing powers that are used, say, to frame a theorem like Gödel's.

So does the brain depend on protecting itself from chaos or exploiting it? The reader can decide.

1.7 Summary

Hierarchical Brain Organization
Just like any complex system, the brain is composed of distinct subparts with specialized functions. To propose good computational models, it is essential to understand what these parts do. The brain's forebrain, which is the site of neural algorithm development and storage, draws heavily from primal abstractions discovered by evolutionarily earlier brain areas.

Computational Abstraction Levels
An essential take-home lesson from complex systems is that they are inevitably hierarchically organized. This means that to understand the brain, it is necessary to integrate algorithms at several different levels of abstraction. Abstraction levels in brain anatomy and in silicon computers are accepted, but abstraction levels for the brain's computational algorithms are still a novelty. The levels outlined here are provisional and will be revised as new experimental observations are made.

Specialized Algorithms

If you want to show that computation is a good model, then the way to
go about it is to show how the things that people do can be described by
recipes or programs that make use of the brain's neurons in a feasible way.
The brain's algorithms must be exotically different from those used by silicon computers in part because the brain's architecture is so different from
silicon.

What Is Computation?

Computation is described by what a Turing machine can do. If you want
to show that the brain is not a computer, the way to go about it is to demonstrate something that the brain does that cannot be reduced to a Turing
machine program. There have been several attempts to do this, but so far
they have not carried the day. The nature of Turing machines is that they
can only represent integers. Real numbers, random numbers, and predicate
logic must therefore be approximated.

2 Brain Overview

How the functional systems or circuitry can be arranged into a unified whole is a tantalizing, deep unsolved problem… [However as of 2012] the relationship between [central nervous system] macroregionalization and functional systems is not obvious.

—L. Squire et al., eds., *Fundamental Neuroscience* (Academic Press, 2012, p. 44)

The brain's 10^{11} neurons have common mechanisms of signaling and organization, but they are hardly distributed uniformly. Distinct circuits exhibit many different levels of specialization, their most abstract organization being their grouping into major subsystems. The introduction to the brain in the previous chapter highlighted the biggest picture, emphasizing the parts of the brain that have essential regulatory or life support functions. In this chapter, the focus is on the most recent evolutionary mammalian refinement, the forebrain, which includes the cortex, basal ganglia, hippocampus, and amygdala, as well as related essential components, the thalamus and hypothalamus. Such subsystems have distinct functions that can be interpreted from the vantage point of circuitry needed to develop and run the programs we use in thinking and everyday behavior.

Currently, a growing number and variety of noninvasive techniques are driving brain research, but it was not always this way. Much of the earliest knowledge of the function of these subsystems came from people who acquired selective damage of one subsystem or another either through disease or cerebral injury. Tragically, one way of damaging the brain is through battle injuries. The study of brain function advanced greatly during and after the Russo-Japanese War of 1904–1905 because the muzzle velocity of rifle bullets had increased to the point where bullets could pass through the skull. Consequently, wounded soldiers survived and could be examined for the effects of their brain injuries. In earlier wars such as the American Civil War, slower bullets typically lodged inside the skull, producing fatal infections. Nowadays, there is a wide variety of less destructive avenues that provide information on the differential functioning of the brain's different parts, but cerebral incidents that produce damage, such as strokes, continue to provide valuable information.

Another major source of knowledge about the brain's subsystems is studies of patients that have brain diseases or malformations. Epilepsy, a neurologic disorder that is treated more and more successfully with drugs, still manifests itself in severe cases that require surgery. Because epileptic seizures are primarily due to a runaway circuit overload owing to mutually excitatory neural connections, finding and removing a focal part of this circuitry often can alleviate the problem. The effect of seizures can be so debilitating—think of falling down suddenly while crossing a busy street—that the severely afflicted patient is happy to have the procedure done. Currently, there is a trend toward use of implanted electrodes to localize the seizure focus, which is difficult, but in one earlier variant of this procedure, the skullcap is removed, and the patient is shown a card with a common image on it while an area of the memory is stimulated with a small current. If the patient subsequently cannot name the item on the card, then this part of the memory is deemed important, and the surgeon will try and spare it. Parts that are apparently uninvolved in the correct naming are candidates for surgical removal. Figure 2.1 shows one of these earlier operations under way.

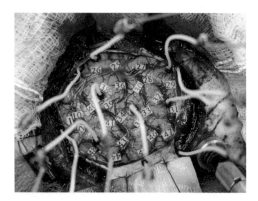

Figure 2.1
A patient undergoing surgery for epilepsy. An epileptic seizure is basically an instability in the network of cortical neurons that can be controlled or eliminated by removing portions of the cortex. The top of the skull is removed, and portions of the cortex that are important are determined by extensive testing and are labeled. These are spared during the final surgery. Reprinted with permission from Bob Wright.

Yet another font of information has resulted from huge advances in instrumentation for measuring brain signals. In particular, our understanding of the forebrain has been greatly improved by the development of noninvasive imaging techniques that allow the state of huge numbers of neurons to be studied remotely. One of the foremost of these is magnetic resonance imaging, or MRI. The MRI process creates images of pixel values that are related to the amount of resonance of atoms in a small region, and different atoms have different resonance values. Very important is *functional* magnetic resonance imaging, or fMRI, which measures the slight difference between the resonance of oxygenated and deoxygenated blood as a function of time. When neurons signal, they use lots of oxygen, so by subtracting the image for a baseline condition, one can locate the relatively more metabolically active neurons in the brain. Figure 2.2B shows fMRI data for a human subject, comparing neural responses for visual and auditory stimuli. Magnetoencephalography (MEG) data have a complementary focus. Bundles of nerve cells' signaling patterns create tiny magnetic field transients that can be measured and used to pinpoint neural events precisely in time. Figure 2.2A shows the different temporal responses to auditory, visual stimuli compared to motor responses.

As important as the fMRI and MEG studies are, they are just a small part of the total toolbox available for brain studies. If the information gained from all such investigations is combined, the amount is enormous, and the crucial question tackled in this chapter is how to make sense of this information at an abstract system level. The additional resource that we will draw on is computation, using its formal results, together with the breakthrough of the algorithm concept, and insights from the most successful complex device outside of life itself, the silicon computer.

One cannot stress often enough that a danger in making analogies with silicon is that mistakes can be made with hugely erroneous conclusions, but at the same time the key reason for going ahead is that, in the abstract, the computer has to solve many of the same problems as living things do, and thus the silicon solutions can point to novel and potentially helpful ways of interpreting the biology.

Our focus here is the forebrain, which makes and stores programs, and we will dive in with correspondences between the forebrain's specializations, and computational functions, as in fact they do line up in many places, but there is one important preparatory matter to attend to first. The mammalian forebrain is a late evolutionary invention, hence it gains great leverage in taking advantage of the exquisite organization of the neural systems that preceded it, and so one must understand the earlier structure

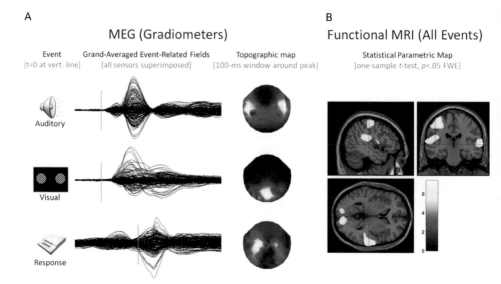

Figure 2.2
A multimodal approach characterizes the neural responses to simple sensory (brief auditory tones, brief visual checkerboard) and motor (right-hand button press) events. (A) Magnetoencephalography (MEG) data have a high temporal resolution (on the order of msec), which allows one to directly measure latency differences in these neural responses. (B) Functional magnetic resonance imaging (fMRI) measures have high spatial resolution (on the order of mm), which allows one to pinpoint the location of activity associated with a sensorimotor event. Reproduced by permission of Cambridge Centre for Ageing and Neuroscience (Cam-CAN), University of Cambridge and MRC Cognition and Brain Sciences Unit, Cambridge, UK, www.cam-can.com.

at least at a broad-brush level. Otherwise, one risks ascribing functions to the forebrain that have already been taken care of. So with this aim, we will revisit the systems introduced in the previous chapter.

2.1 Spinal Cord and Brainstem

Let's start the process of appreciating the forebrain's antecedent structures by focusing on an important system in the medulla oblongata that we neglected during the abbreviated pass in chapter 1; namely, the autonomic nervous system.

Autonomic Nervous System: Meta-states
The spinal cord was briefly introduced via a discussion of its reflex circuitry, but, in addition, it carries neural circuitry from the medulla oblongata to

regulate the body's organs. Strikingly, this neural circuitry is organized into two distinct systems of connections, designated sympathetic and parasympathetic (figure 2.3). The sympathetic system's job is to prepare the body for action (don't think "sympathy"!): heart rate is increased, breathing rate is increased, and so forth. The system is put in "fight-or-flight" mode. The parasympathetic system has the complementary job of slowing everything down. The system is put into a "rest and digest" mode where circuits facilitating digestion and so forth are given resources.

In thinking about the brain, one must always keep in mind the biggest picture, brilliantly characterized by Dawkins,[20] which is that the genes are directing the phenotype to reproduce successfully, and the human two-meta-state architecture somehow evolved to promote that success. Thus, one assumes that the two-state design has been much more successful than alternatives. Further along, we will discuss how the rest cycle is used by the forebrain, but for now this bipartite division should serve to emphasize that the human brain evolved from elemental functions and consequently all the brain's more elaborate programs ultimately must pay homage to these basic commitments. And in fact this set of constraints turns out to be a fundamental issue for the forebrain. From an evolutionary perspective, the programs that the brain is interested in are those that improve the chances of realizing the ultimate goals of its encompassing phenotype, and these are realized through the body's basic machinery.

For less complicated animals whose forebears got an earlier start, such as a praying mantis, a snake, or a frog, these programs can be deconstructed more easily, but for a mammal with a forebrain, the programs become increasingly elaborate, and the link between them and the phenotype's primary purpose is more difficult to calculate. The mammalian brain's response to this complexity has been to create and use abstractions, and with that in mind, let's revisit the brain's first elaborate abstractor, the cerebellum (figure 2.4).

Cerebellum: Sensorimotor Calibration

Both what the cerebellum does and how it does it are issues that are far from completely settled,[21] but a mainstream perspective is as follows. The cerebellum plays the major role in the memory for complicated sensorimotor experiences associated with actions. Catching a baseball in a glove requires associating the "thwack" sound of a successful catch with motor actions that control the glove's dynamics. Chasing a high fly ball to the outfield, professional baseball players use the sound made by the hitter's bat in determining which direction to run. The cerebellum handles these kinds of complex associations.

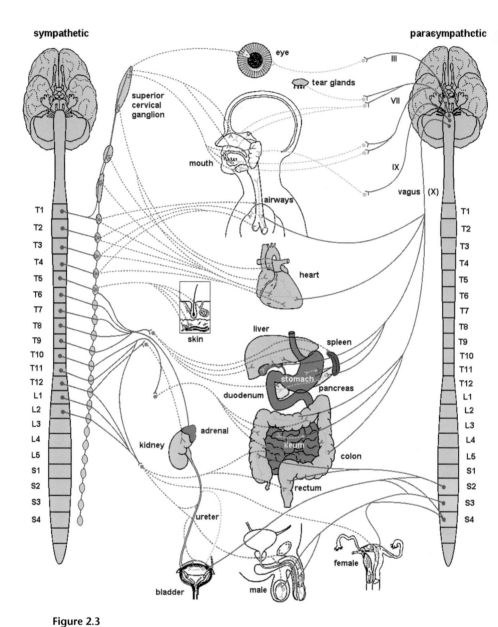

Figure 2.3
The basic neural wiring diagram for the body reveals two separate neural pathways:
the sympathetic system and the parasympathetic system. From Freeman, Scott, *Biological Science*, 2nd, © 2005. Printed and electronically reproduced by permission of
Pearson Education, Inc., Upper Saddle River, NJ.

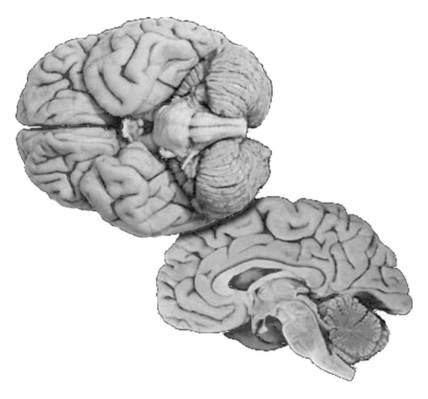

Figure 2.4
The cerebellum is a highly organized semi-independent nucleus that handles the coding and adaptation of sensory-motor experience. From Sylvius 4 Online, Sinauer Associates.

The coding of such experiences also requires constant adaptation. During development, the size of the body changes enormously, with the result that all its motor control circuits must be continually adjusted. This adaptation is handled by the cerebellum as well. Even in the moment, the sensory motor mappings may change, as when you try to balance a cup of tea while carrying a magazine under your arm. Yet another calibration job for the cerebellum.

If you are lucky enough to have access to a prism viewer, you can experience this all for yourself. Such a viewer looks like a pair of goggles, but it is special in that its lenses displace the visual field, typically horizontally to the left or right. When wearing the viewer, you can try and throw a wadded-up paper into a wastepaper basket set at a distance that makes the task fairly easy. You will find that you miss to one side initially, but over the course of three or four throws, you will adapt to be on target again. Now if you remove the viewer and try again, your first throw will miss on the opposite side, but again you will re-adapt in short order (figure 2.5). These adaptations have been studied extensively with interesting results. Walking with the goggles on will recalibrate reaching, for example.[22] A similar thing you can try that does not require prism glasses is running on a treadmill for about 20 minutes. Immediately after stepping off you will experience the effects of cerebellar visuomotor recalibration. Running on the treadmill, your ~5 meters/second speed sensed by your proprioceptive system adapts the visual system's report of 0 meters/second. Consequently, immediately after stepping off, the 0 meters/second proprioceptive signal causes a visual report of speed in the negative (backwards) direction for a few seconds.

An individual can still access his or her extensive library of movements without a cerebellum, but the movements will be laborious, and the individual will have lost the ability to adapt to new situations.

If you think about this process for a moment, you'll quickly realize what a profound ability it represents. As we get to in a moment—and spend a lot of time on in chapter 4—the cortex has the job of coding sensory-motor states in a vast table in a way that allows every response to be looked up. This coding process is also painstakingly laid out so that its computations can be done in real time. The adaptations to the table required by the prisms are huge. A movement that worked for the normal relationship between visual space and motor space is now off by a large margin. So there is the need for a device like the cerebellum that can quickly reestablish the new mapping between the sensory input and the correct motor response.

Up to this point, given our description, it would be easy to think that the cerebellum might work independently of the forebrain, but in fact they are intimately linked in that the forebrain is the complex that came up with the movement and thus created the expectation of what the sensory consequences should be. Thus, when the prism-glasses throw is way off, the reason we know this is that we have the expectation of what should have happened down to a fine level of detail. But in normal operation, the expectations set up in the cerebellum are met, and thus in a multistage movement the sequent stage can be engaged, and we can proceed on. Moreover, another advantageous feature is that, owing to the cerebellum's ability to check sensorimotor correspondences and compute corrections, the forebrain has less to do.

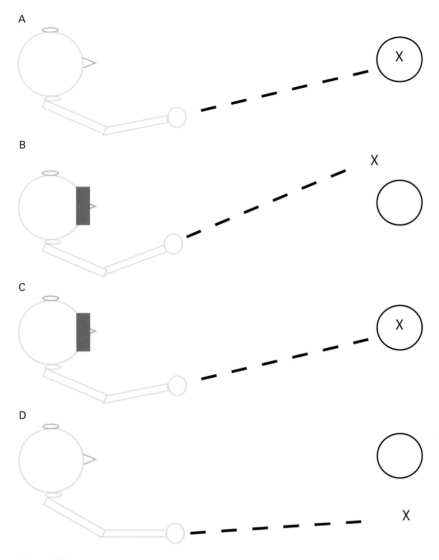

Figure 2.5

Adaptation of sensory-motor experience using prism goggles. (A) A person can throw a balled-up paper into a wastebasket. (B) When she wears prism goggles that shift vision to the left, she initially misses to the left. (C) After a few tries, she is able to recalibrate and hit the target. (D) Removing the goggles causes an initial miss to the right.

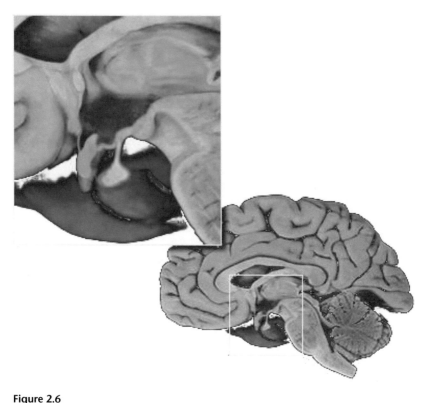

Figure 2.6
The hypothalamus is a relatively modest-sized portion of the brainstem, but it has
enormous influence as it controls our basic drives and is adjacent to reward centers.
The most abstract programs in the forebrain have to negotiate for approval with this
lower center. From Sylvius 4 Online, Sinauer Associates.

Hypothalamus: Rating What Is Really Important
If there was a candidate for central control in the brain, the hypothalamus
would be it (figure 2.6). This small region regulates visceral functions. The
hypothalamus can be thought of as mediating the four principal drives:
fighting, fleeing, feeding, and reproducing. Of these, the first two are associated with stimulating the body to vigorous action, for example by raising
the heart rate and increasing the production of adrenaline, whereas the
second two are associated with complementary acts such as stimulating
digestion and ancillary functions. This complementarity melds with the
brain's two different neural wiring systems: the sympathetic part of the
visceral nervous system is in charge of fight-or-flight decisions, and the
parasympathetic division is in charge of "rest and digest" activities.

The hypothalamus may have states that dictate the body's behavior, but it still has to have a way to get the message across. Thus, the hypothalamus is located right next to the brainstem nuclei that contain neurons that mediate neurotransmitters that modulate behavior, such as dopamine, serotonin, norepinephrine (adrenaline), and histamine. Dopamine is the major indication of reward in signaling the value of performing a function, but it is likely that the other transmitters can be seen as having behavior-related functions as well. The use of chemical rewards is of fundamental importance to computational models, and this will be taken up when we discuss reinforcement in chapter 5, but we can get started on the issue here.

Secondary Reward: The "Neuro"
In the course of using its programs, the brain has to have mechanisms for creating programs in the first place, and this is the job of the forebrain. Given that human programmers learn to program silicon computers only after a lengthy process of study, it seems nothing short of amazing that human brains can create complicated new behavioral programs naturally. Nonetheless, they do, and the process is understood in broad outline. In the first place, it is not that the forebrain has to create programs de novo, but rather that it can specify programs in terms of their similarities to existing forebrain programs. The forebrain just has to define the novel components. But even given that this can be done, there has to be a way of evaluating how well programs are performing. To do this, the brain uses a currency, neural "money" if you like, in the form of the neurotransmitter dopamine. Its use can be very complex, and it can mean different things in different brain regions, but for the main picture, the currency model is apt. In honor of the European euro, let us call it the "neuro." As a neural currency, it achieves the same goal as its monetary analog in that it makes many different programs commensurate. You have programs for reading a book, eating, and catching the bus. If you are faced with choosing one of these in a given moment, what could be the mechanism involved? The brain needs a way of reducing each of these options to a currency to compare values, and dopamine is that currency.

Like any currency, the neuro is prone to problems of inflation and deflation. What determines the value of running a program? If it is to get food, say an apple, then, in principle, its value can be ultimately reconciled after eating the apple. The body reports the apple's nutritional value in neural codes, and from that the cost of getting the apple in terms of physical effort expended, again translated into neuros, can be subtracted for the net. The system actually does a little better than this as it predicts the values of each in advance and only has to make adjustments if the expectations are not met.[23] But while the apple problem has a ground truth in that the nutritional value can be reported, the worth of more abstract behaviors (e.g., being a little more friendly than usual) can be much more difficult to evaluate.

An important additional insight as to reward management can be had by refocusing on the hypothalamus. The neural wiring to hand out the neuros begins in a small locus of neurons adjacent to it, so that the two centers work in concert.[24] The basic drives in the hypothalamus set up an agenda that the rest of the brain tries to satisfy. Over evolutionary time, the forebrain has become more and more elaborate, with the result that it can come up with more and more creative ideas about what should receive reward. But as its circuitry is wired to talk to the hypothalamus, no matter how abstract the proposed behavior, it has to satisfy this strategically located neural funding agency; the benefits of our most abstract plans have to be recast in terms of our most basic drives in order to be sanctioned.

If we combine the thoughts of the last two paragraphs, first the issue of calibrating the value of programs, and second the fact that reward is handed out by a master circuit in charge of our basic drives, you can see that the system is delicate. What if the forebrain is able to talk the hypothalamus into supporting programs that are not valuable, perhaps even destructive? The world of course provides helpful feedback, but as we are the ones that interpret those messages, there is plenty of room for mischief.

The Other Essential Neurotransmitters

Owing to its importance, we have highlighted dopamine, but dopamine is just one of four very important neurotransmitters that include serotonin, norepinephrine, and histamine. These each have a long-term evolutionary heritage and are used to modulate the state of the brain in its course of directing the body's actions. They have been studied intensively for decades, and their overall characterization is well understood both in neurochemical and anatomic terms. A summary of these functions appears in figure 2.7. Dopamine can be understood in terms of the neuro as just described, and evidence is accumulating that serotonin can be understood in terms of risk assessment as will be discussed later. Norepinephrine and histamine have well-characterized functions. Very crudely, the former can be thought of as turning up the gain in cognitive tasks. The latter is a wide-ranging modulator effecting at least 23 different physiological functions. To date there is no consensus on a satisfactory computational description of their roles, but their extensive individual anatomic circuitries suggest that

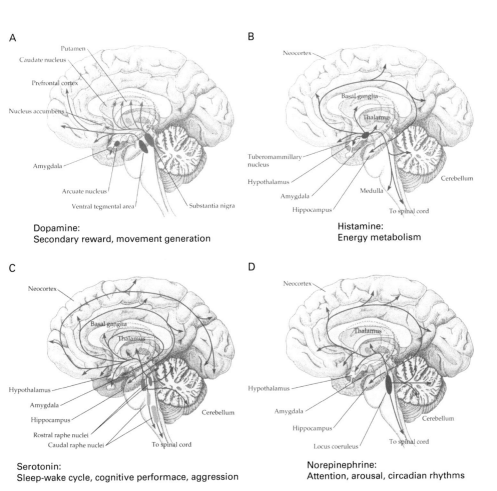

A
Putamen
Caudate nucleus
Prefrontal cortex
Nucleus accumbens
Amygdala
Arcuate nucleus
Ventral tegmental area
Substantia nigra

Dopamine:
Secondary reward, movement generation

B
Neocortex
Basal ganglia
Thalamus
Tuberomammillary nucleus
Hypothalamus
Amygdala
Hippocampus
Cerebellum
Medulla
To spinal cord

Histamine:
Energy metabolism

C
Neocortex
Basal ganglia
Thalamus
Hypothalamus
Amygdala
Hippocampus
Rostral raphe nuclei
Caudal raphe nuclei
Cerebellum
To spinal cord

Serotonin:
Sleep-wake cycle, cognitive performace, aggression

D
Neocortex
Thalamus
Hypothalamus
Amygdala
Hippocampus
Locus coeruleus
Cerebellum
To spinal cord

Norepinephrine:
Attention, arousal, circadian rhythms

Figure 2.7
Neurotransmitters are used in the midbrain to regulate the states of the body but in the forebrain are co-opted for more abstract purposes. In particular, dopamine has been characterized as the mediator of secondary reward. Reprinted with permission from Nicholls et al., *From Neuron to Brain* (5th edition, 2011). Copyright Sinauer Associates.

they too could be described computationally. These of course are not the only neurotransmitters. A reasonably complete list consists of about 50 neurotransmitters. However, the four neurotransmitters in figure 2.7 are distinguished by having their own pronounced cortical circuitry and being anatomically near the hypothalamus. Given our agenda of a compact overview, this discussion of essential neurotransmitters will have to suffice as a sketch of a workable context, and now we can delve into the mammalian invention, the program-developing forebrain.

2.2 The Forebrain: An Overview

An easy way to understand the forebrain's operation is to jump right in and compare its collective functions to those of a silicon computer. Despite our earlier caution, taking such a course cannot help but risk raising the hackles of those who are very aware of the silicon computer's vast dissimilarity to biology. Furthermore, as we still do not know exactly how the brain works, it is possible that even a very abstract comparison will throw us off track. Moreover, when we get to describe the details of the various components in later chapters, you will see that brain subsystems definitely are very different from conventional computers in the way that information is acquired and organized. Nonetheless, the brain can be described by computation; it's just that, from the standpoint of conventional computing, its architecture will be very alien. Those are the caveats. Having aired them again, we proceed with an introduction to the forebrain's components from the conventional computing vantage point. Let's start with figure 2.8, which compares functions of the major brain components to those of a conventional computer. You can appreciate that such a broad-brush correspondence is going to need considerable elaboration.

While it's not at all like a conventional memory, the *cortex* is an exotic memory that has an enormous storage capacity. The exact limits of cortical memory are completely unknown. And while it is not exactly like a

───▶

Figure 2.8

(A) The brain's major components to scale. (B, C) Comparing silicon computer functions with those of the cortex. The brain has functions for acquiring new programs that make use of the components for storing and executing its current programs. The amygdala represents a potential program's importance from the standpoint of novelty, the hippocampus represents a sketch of the program in terms of existing programs, and the hypothalamus scores it with respect to existing programs.

A

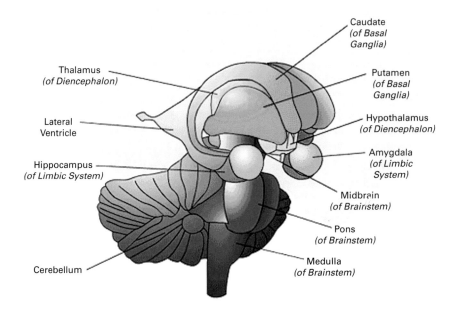

B

Basal Ganglia

Cortex

Thalamus

Amygdala

Hippocampus

Hypothalamus

C

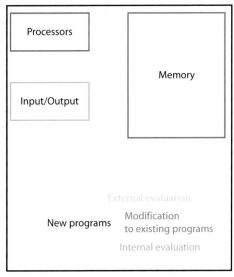

processor, the *basal ganglia* is the forebrain's main place for sequencing instructions. The cortex and basal ganglia need input and output and predominantly meet this need through the *thalamus*, a major subsystem that mediates the transmission of crucial information in and out of the forebrain. These three subsystems leave out the most formidable part: the brain's analog of the human programmer, which stitches together new programs that mediate new behaviors carried out by the body. Because programming in silicon by humans requires a substantial training period, it might seem daunting to characterize every person's brain as a very successful programmer in its own right, but nonetheless humans do acquire new behavioral programs, and one can sketch the role of the main subsystems that provide this functionality. In the brain-as-programmer model, it helps to eschew the goals of accounting for our lofty thought processes and to concentrate instead on essentials such as walking to a destination or preparing a meal.

Several organizational elements simplify the problem. In the first place, humans go through a lengthy developmental phase where behavioral programs are gradually and incrementally improved. Furthermore, as mentioned a moment ago, new programs are never coded completely from scratch but are extensions of the current repertoire. Thus, we are never creating brand new programs but always in the situation of modifying existing programs. The forebrain has a constellation of essential systems to exploit this gradualism. The brain has an open-for-business "awareness" state that, when activated, allows it to extend programs. In this state, the *hippocampus* parses attended inputs and outputs and codes them in terms of differences from a learned program. At the front end of the long tubular shape of each of the two hippocampi sits its *amygdala*, which rates these new codes in terms of importance. The more important variations are marked for encoding in long-term memory, and this consolidation processes happens during sleep. Harking back to the medulla oblongata's two-phase organization, you can appreciate how essential the sleep cycle is, as one of its premier features is that it sets up the permanent addition of new programs. Researchers are a long way from the actual specification of exactly how most of this happens, and proposals are constantly being tested and refined, but at our very abstract level of description, this is how it's done.

One very important area that this abstract first pass finesses is that of the timescale of this computation. So let's expand on this issue here. Conventional computing has become increasingly complicated with the use of more and more processors, but a still useful idealization is the random access machine, or RAM. The idea is that the program and data are stored in a memory. The program consists of a sequential list of instructions that a single processor knows how to execute. The execution of the program consists of a more or less sequential traverse of this instruction list, although some instructions can cause a jump to a more distal instruction location. Now, to situate this analogy in the cortex–basal ganglia loop requires choosing a spatiotemporal scale. It turns out that because many different kinds of computation are ubiquitous in the forebrain, at different scales there will be different kinds of computation, but at the most abstract level of our current focus, a particularly useful timescale is that of a few hundred milliseconds. Within this timescale, we can discuss the operation of the cortex and basal ganglia as memory and processor, respectively, within a RAM model.

Let's revisit the analogy with a Web search engine such as Google. To respond quickly to user queries, search engines "crawl" the Web continually, organizing its pages in ways that make them quickly accessible. Thus, Web pages are indexed, so that one can quickly navigate to any page and from one to another related page. The state of the cortex at any moment can be thought of as analogous to the right Web page for the user. There are lots of other pages accessible from that one, and by following a link, the user will transit to a new helpful page. The link transits may be thought of as analogous to what the basal ganglia does. So executing a mental program involves a series of cortex–basal ganglia transits and is abstractly like surfing the Web. Figure 2.9 summarizes these operations, showing the basic cortex–basal ganglia loop.

A processor and memory need to be able to access inputs and outputs. That role falls to the thalamus, an elaborate network of nerve cells that compose the interface between the cortex and sensorimotor circuits. Much of the detailed function of the thalamus is still a mystery, but, based on its position as an interface, we can speculate on the kinds of jobs that it should do. One of them is to mediate the consequences of a simulated action. Suppose that you need milk and imagine going to the store to buy some. You have not actually gone to the store, but given that the imagined program is executing in the cortex–basal ganglia loop, there is extra work to do. If you actually went on the journey, the sensory motor system would provide your forebrain with all kinds of data characterizing the journey's progression. However, given that you are just imagining the sequence, some vestige of the states that would have occurred have to be created in the brain without the helpful peripheral feedback. Enter the thalamus to help in this project. Current thinking is that the thalamus helps to create the result that would occur if the imagined plan went forward. This new state in turn sets up another cycle of the cortex–basal ganglia–thalamus loop.

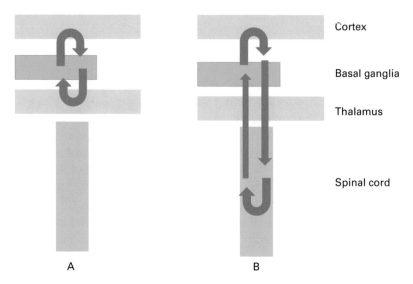

Cortex

Basal ganglia

Thalamus

Spinal cord

A   B

Figure 2.9
The general computation cycle in the brain can be seen as operating on a timescale of about a few hundred milliseconds. The cortex computes the state of affairs and a list of what to do next, and the basal ganglia takes one of the actions. The action might be (A) a mental action in simulating a process internally or (B) a physical action taken by the body in the world. In either case, the result is a new state for the cortex.

The other helpful thing the thalamus does is filter the input and outputs from and to the outside world. The best way to understand what is meant by this filtering is to take a quick inventory of the connections of its neural circuitry. It turns out that, using the visual sense as an example, only about 6% of the connections to the thalamus come from the eye's retinas. The other 94% are connections from the forebrain's programming core.[25] The implication of this is simply that "you see want you want to see." The interpretation of the outside world is based on a program's internal expectations. Of course it uses visual data, but expectations almost always get the upper hand. An easy way to appreciate this truth is to think of magicians' tricks. A magician performing for an audience throws a ball up in the air a few times, then quickly palms the ball and repeats the throwing action again. The programs in the brains of the members of the audience had evidence for the ball going up in previous throws and have additional recent evidence in the form of the throwing cadence in the final toss, so the absence of an actual ball is overruled in the mental programs, and the audience "sees" the ball thrown. The bottom line is that although inputs and output to and from the world are important, for the most part they are in service to the forebrain's agenda. So although we have covered some side issues, in the course of discussing the timescale of the cortex–basal ganglia–thalamus loop, the two most important are (1) the loop is organized into simulating (figure 2.9A) and acting (figure 2.9B) and (2) the transit time is significant, taking at least hundreds of milliseconds.

At this point, there is basic program sequencing and memory as well as input and output subsystems accounted for, but we need more discussion of how programs are created in the first place, which is the job of two more of the mammalian inventions: the hippocampus and amygdala. There are many hurdles to be overcome in making a program, but consider two of the most important ones. In the first place, given the seamless tapestry of experience, the hippocampus must somehow corral the important parts and organize them. And buried in this task is the second hurdle, and that is the hippocampus must be capable of scaling descriptions that have dramatically differing temporal extents. Consider the steps in pouring another cup of coffee. Do you have them ready? Now consider the steps in going out to buy a quart of milk. The first (pouring another cup of coffee) might take barely a minute or two, whereas the second (going out to buy a quart of milk) might take the better part of an hour. Somehow the hippocampus has to slice and dice these tasks so that they fit into the same apparatus for coding new programs. Of course, the coffee and milk programs are so routine that you might already have well-worn programs for obtaining them, so there would be little or nothing for the hippocampus to do. But suppose that for reasons unknown they were laced with novelty. And furthermore, suppose that they were each somehow important. Enter the amygdala. The amygdala is an extremely important subsystem that picks out what part of your experience is important enough to be remembered. Those chosen programs are marked for permanence and "downloaded" somehow into the cortex during the sleep cycle.

Thus with the combined hippocampus/amygdala system, there is a mechanism, albeit very sketchily presented, for making programs, but why does it do this? At this point, you have the information to elaborate on a lot of the details. First off, the programs are selected because they are good for the body's needs. The programs might be very long term and abstract, but that is what they are there for. Second, the mechanism for communicating these needs is established in broad outline: it is the essential neurotransmitters that summarize the state of the body. When they are used to organize the body's responses, they innervate the viscera, but in the forebrain they have been co-opted to signal various desiderata that can be sold to the hypothalamus as furthering its basic goals. In this sale, the brain's neuro, dopamine, is the best understood as it is a powerful reinforcer that is the medium of exchange. It is fairly well established that the programs that are useful are "paid" in dopamine at the start of their execution, and their payment is adjusted based on whether or not they meet expectations. Although we have lapsed into anthropomorphisms, do not be misled: The use of dopamine in this way can be completely explained by an algorithm once the requisite variables have been accessed. The other neurotransmitters do not have as firm an explanation as of this writing, but one new thought is that serotonin manages various kinds of risk.

By now, the reader should have a good idea of the overall functioning of the brain's subsystems. However, anticipating the identification of specific algorithms with these subsystems, we need take a more detailed pass through the list in order to tease out additional details that will be important later.

2.3   Cortex: Long-Term Memory

If you open up the skull, you would find that the brain is protected by rubbery encasements. But if you cut through these, you will be looking at the surface of the cortex, as you were in figure 2.1. The cortex is the main site of the brain's permanent memory. It has the structure of a thin, six-layer sheet of neurons that David Van Essen has usefully likened to a pizza in shape. To fit in the skull, the pizza has to be folded up, and hence the observer peering in sees folds, or *sulci*. Owing to their interconnectivity, neurons in areas of the cortex have very different characteristic properties, and these are best visualized if we have some way of unfolding the cortex. This is not so easy to do, but two popular ways are to flatten it surgically as shown in the upper part of figure 2.10 or to inflate its position data mathematically as shown in the lower part of the figure.

When we think of the concept of "memory," we typically conjure up rather elaborate sequences of events. We can recall whole conversations with friends sometimes verbatim along with their facial expressions. Or we can recall a scenic walk in a park with animal sounds and sights and smells. These exquisite continuous experiences of memory are very misleading in a discussion of cortical memory. For although the neurons in the cortex produce a major component of these experiences, they need help from other subsystems to do this. It is more accurate to think of the cortex as having major components of all these memories, but they are latent and coded. At any moment, the active part of the cortex is capturing a state in memory using a small instant in time to designate the referenced moment.

A
Flattened representation of monkey cerebral cortex

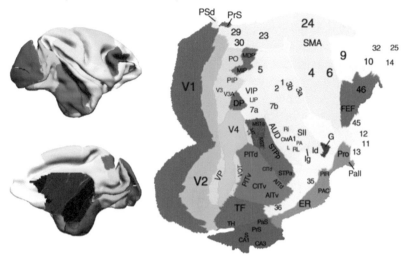

B
fMRI mapping of human visual cortex

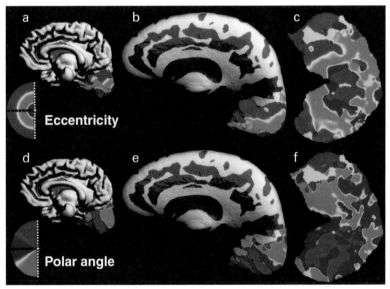

Figure 2.10

(A) The Van Essen laboratory's image of a macaque cortex that has been surgically flattened to easily visualize the different areas. (B) Martin Sereno's mathematically inflated human cortex being used to show the representation of the visual field obtained from fMRI data. a and d, original data; b and e, inflated views; c and f, expanded views of visual cortical area V1. (A) Adapted from Van Essen and Drury, 1998. (B) From M. I. Sereno et al., "Borders of multiple visual areas in humans revealed by functional magnetic resonance imaging," *Science*, vol. 268, p. 5212, 1995. Reprinted with permission from AAAS.

At this point, it is important to raise the distinction between computation time and the time represented by the running program. Of course, we can think of episodes that last weeks or years, but to do so we have to run them on our brain computer, and they all must share the basic cycling of the machine architecture. So the question is: How large can this temporal chunk be? It cannot be smaller than 1 millisecond, as that is the time it takes a neuron to create a spike to communicate with another neuron. And it cannot really be longer than 300 milliseconds, as that is the modal time that we hold our eyes still. Thus, the likely length for a cortical memory instant is about one third of a second or less, motivated by the modal time the eyes are fixed on a region of space, as well as the time needed to fit into a cycle of the cortex–basal ganglia–thalamus loop.[26] Researchers have long known that electroencephalographic (EEG) measurements exhibit a characteristic voltage transient at 300 milliseconds after stimulus onset (P300) associated with decision-making and recent more fine-grained measurements using MEG also supports the 100–300 millisecond constraint.[27]

Visual Fixations and Memory
Why should a cortical time constant be related to the stability of visual fixation? The human visual system is truly remarkable in that our continuous percept of the visual world is somehow created from these series of discrete instants lasting about 300 milliseconds, where during this time the gaze is held more or less stationary on a point in the world. We need to do this as the resolution of the eye is only very good for 1° of visual angle near the axis of gaze. To experience a visual degree, hold your thumb out at arm's length. After about 1 visual degree, the resolution drops rapidly by a factor of 100 at the periphery. In addition, high-resolution color vision is concentrated at the fovea. Nonetheless, for our discussion here, the most important point is this: The vast array of visually responsive neurons—estimated at one third of the cortical cells in a monkey brain—are *retinotopically indexed*. This means that their responses are sensitive to the position of the gaze points of the two eyes. Move the gaze point, and all these outputs of these neurons will change. As a consequence, what we would like to think of as a program state, at least for vision, is likely to be only stable when the eyes are stable, and thus this stability is unlikely to last more than 300 milliseconds.

That the human brain is dealing with the problem of having a very small visual area of good spatial resolution can be readily seen from human eye traces, as the brain is forced to choose special areas in the world to look at three times every second. These choices provide tremendous information as to the organization of running programs, as by recording eye movements we can watch a trace of running programs in action, as shown by Pelz's replication of some of the first eye-gaze traces obtained by Yarbus[28] in the 1960s (figure 2.11). We see the gaze vector rapidly scanning the image in very different patterns motivated by different questions asked of the viewer. In the figure, every time the trace stops at a location, indicated here by a small dot in the trace, it represents an instant when the cortex is carrying out some state-computing operation. We don't know exactly what operation it is except in very special experimental conditions, when there are strong hints. We will see some examples of this methodology in chapter 6.

Faced with all this data on the brain's interface with the visual world, you can now appreciate what a technical achievement our perception of a seamless stationary world is. Imagine the riot in a movie theater if the projectionist slowed the projector down from 24 frames per second to just three! Nonetheless, our brains somehow solve this problem for us in a very satisfactory fashion, as we are normally totally unaware that any kind of difficulty exists. Explanations of just how the perception of seamlessness happens is still being worked on, but one component of the answer is that the cortex uses its extremely compact memory encodings to predict what the image is likely to be.[29] In its extreme form, if the prediction of what would happen matches what actually occurs, then no explicit signal need be generated.[30] Only mismatches caused by the unexpected need be explicitly denoted. A bonus of this strategy is that the unpredicted is often what is of interest or importance.

To sum up the expanded introduction to this subsystem, the cortex is a vast memory of unknown capacity that contains literal and abstract representations of our life experience, as well as prescriptions of what to do about them. A part of the memory that is active, in the form of spiking neurons, which in computer science parlance is called the *state*, represents a small set of temporal instants from this enormous library. The job of stitching together successive instants falls on the basal ganglia (figure 2.12).

2.4   Basal Ganglia: The Program Sequencer

As shown in figure 2.13, the basal ganglia is a region of several interconnected regions in the center of the brain that play a major role in sequential actions that are central to complex behaviors.[31] Basal ganglia experts focus on deciphering the roles of the very characteristic connections between the regions. These connections are especially important for many reasons, one

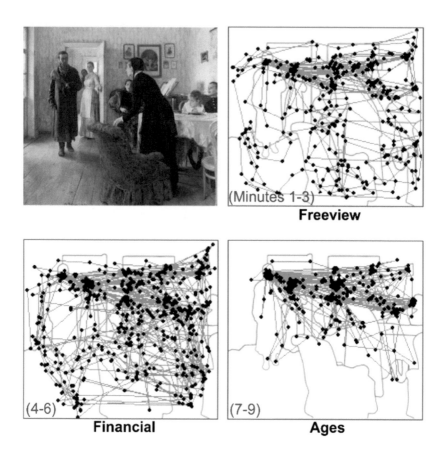

Figure 2.11

A human wearing a gaze-tracking device examines a famous Russian painting entitled *The Unexpected Visitor* for sessions of 3 minutes each. The different questions asked of the subject are as follows: (1) No question; free viewing. (2) What are the financial circumstances of the family? (3) What are the ages of the people? Note the very different scanning patterns for each of these questions, reflecting the need to extract very different information from the picture in each case. Reprinted with permission from Jeff Pelz.

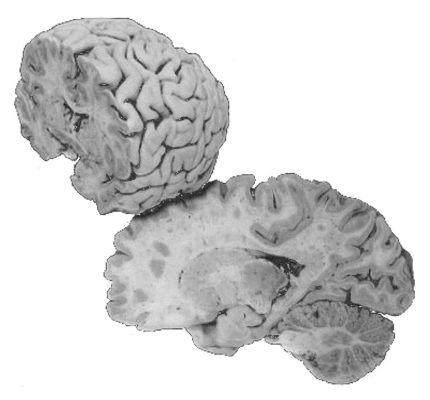

Figure 2.12
The basal ganglia is actually a collection of several interconnected areas implicated primarily in the control of movement but also secondarily in thought patterns that we can think of as simulated movement through an abstract domain. From Sylvius 4 Online, Sinauer Associates.

being that they play a major role in a disease such as Parkinson's disease, and the manipulation of connections between the regions can alleviate some of its symptoms. Nonetheless, for the points that we make here, we consider the entire subsystem as a unit.

The basal ganglia is also distinguished by having special concentrations of the dopaminergic chemical reward systems for rating the value of different action sequences. These are essential, as a fundamental problem is estimating the value of current actions that are done for future rewards.[a]

The basal ganglia guides a huge projection of neurons onto the spinal cord, and these govern body movements. Diseases that selectively attack the basal ganglia, such as Parkinson's disease, Huntington's disease, or

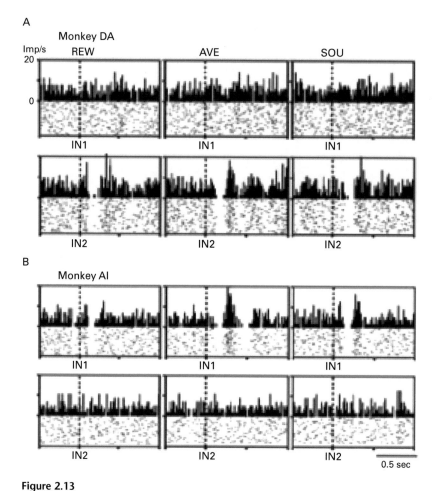

Figure 2.13

Tonically active neurons compose only 10% of the neurons in the basal ganglia, yet they play an important role in signaling task segments. Moreover, they do so by becoming silent. The rows of the figure show individual trails of neural recordings together with their histograms. The second and third sets of recordings clearly show the gaps, marking key points in a task. From Yamada et al., "Tonically active neurons in the primate caudate nucleus and putamen differentially encode instructed motivational outcomes of action," *Journal of Neuroscience*, vol. 24, p. 3503, 2004.

Tourette's syndrome, manifest themselves by producing movement disorders. Extensive clinical observations were a major part of the reason that for a long time, the basal ganglia was thought to be exclusively in charge of physical movement. For example, Parkinson's disease has sometimes been exclusively associated with the motor system, but as the following observation shows, this is an oversimplification. Patients that can be frozen when staring at a blank patternless floor can move forward when a very similar floor has patterned tiles. That is, the overall system is not exclusively motor but in this case needs to be triggered by the relative strength of visual input. The implication here is that, as described earlier, it is best to think in terms of sensory motor *programs* with sensory input interlinked with motor output. We can generalize this point with the eye movement system discussed earlier. Successive gaze points produce visual inputs, which in turn prompt the selection of new gaze points. Generalizing, we can think of visual and other sense data setting up basal ganglia–defined movements that in turn produce new data, and the process repeats.

This view of sequences is supported by data from special neurons in the basal ganglia called tonically active neurons, or TANs. These are a special class of cells that compose only 10% of the total number of cells in the basal ganglia, but nonetheless they play an important role. When a complicated sequence of movements is carried out, the TANs stop signaling vigorously precisely at the breakpoints in the task, as shown in figure 2.13. These data make the following important point: The basal ganglia does not define the details of the movement. Those details are handled by brainstem and spinal cord circuitry. Instead, the basal ganglia only has to specify the abstract components that are just detailed enough so that the more concrete circuitry can select what to do.

Finally, we can revisit the more general view of the basal ganglia, which is that of a subsystem that governs the generation of sequences. These sequences may be concrete motor movements, of course, but can also be abstract such as the steps in a mental program. Analogously to a motor program, we can think of a general-purpose mental program as one that has steps that produce new data just as motor steps produce new sensory data. To produce mental sequences, your brain co-opts the circuitry used to produce motor sequences. Of course, when you do this, you have to be able to make the distinction between what is real and what you have imagined. A grandmaster chess player can imagine what the board would look like 20 moves ahead for some situations, but she doesn't confuse that position with the current one. And even blindfolded chess players still can make the distinction!

The view of the basal ganglia as a general program sequencer is further supported by observations on an important brain feature termed *working memory*. Everyone has the experience of trying to remember a telephone number by rehearsing it either out loud or silently. It turns out that humans have a general property of only being able to keep a small number of things in mind at a time. This ability has been extensively studied, and we will have much more to say about it later, but we just touch on it here because it helps pin down the role of the basal ganglia and has a nice interpretation in terms of programming concepts.

Remember from the Turing machine description that an essential element of a program is the state. This is the information needed to keep track of where you are in the program. Think of making tea. You need to keep track of the state of the process. You have taken out the cup, added the tea bag and sugar, put the water on to boil, and are waiting for the water to heat up. For the next action, you need the information as to how you are going to measure when the water is boiling and the location of the cup. This information is what you need to keep track of in working memory. Of course, after the discussion of the role of the cortex, you know that the main site of the information in working memory is the cortex, but the basal ganglia needs to *refer* to that information to do its job.

Experiments[32] show that diseases that selectively damage the basal ganglia lower the capacity of working memory. Now here is why that data make sense. If the essential information in a program is held in its state (aka working memory), and if the basal ganglia is in charge of referring to that information in going from one state to the next, it is logical that damaging it would reduce the amount of state that one can refer to.

2.5 Thalamus: Input and Output

The *thalamus* is a major gateway that filters all sensorimotor input and output to the cortex (figure 2.14). However, unlike a conventional silicon-based "driver" program that shuffles input and output in standard computers, evidence suggests that the thalamus may use some kind of compressed code based on expectations.

To motivate the idea of expectations, let us examine some of the difficulties with an expectation-free account of an everyday activity. Imagine the job of reaching for your cup and picking it up. You could handle this in a Cartesian sense, which is the way robots would do it, by building a geometric model of the cup and calculating grasp points, moving the robot arm to the cup, and verifying that the grasp points were achieved. One special

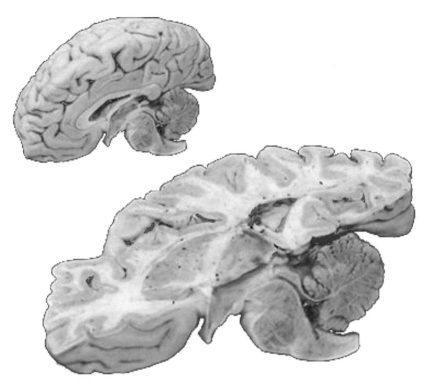

Figure 2.14
The thalamus is a large nucleus of cells that manages neural communication to and from the forebrain. It likely uses a compact expectation-based encoding, signaling the differences between what occurred and what the internal neural systems "thought" should occur. From Sylvius 4 Online, Sinauer Associates.

problem that robots and humans have to deal with is delays. If a finger touches something like a hot surface, then it has to be retracted quickly. In a robot, the heat sensor has to drive the motors that do the retracting, but this can take time owing to inertia. Robot systems do their best by calculating at rates of 10,000 calculations per second, but the rigidity of robot surfaces means that there are always problems. The brain has a much greater handicap because the neural circuitry that does the calculating is 10 to 100 times slower. The time for a signal to get from the hand's heat sensors to the cortex is of the order hundreds of milliseconds. To counter this difficulty, special reflexes are built in that do not go to the forebrain but instead connect more directly to the muscles via the spinal cord. An elaborate repertoire of reflex circuitry that handles these kinds of emergencies protects us, but this repertoire does not have anywhere near the range of responses that the forebrain is capable of. How can the slow cortical circuitry be made useful?

The way the forebrain can handle this is different and makes heavy use of prediction. It also exploits the body's fancy design, in the case of grasping with its elaborate sticky skin surface. Using these two attributes, prediction and stickiness, the grasp can be achieved with a more cavalier opening and closing of the five-finger system because many different configurations of the hand can be made to work. To see if in fact a grasp did work, the expectations in terms of specific haptic sensor readings can be calculated in advance. Thus, the detection of a successful grasp comes down to expressly *not* building and testing an entire and elaborate description of the cup configuration, but only the image of the parts that are relevant by being under the finger pads. The visual sensing can be handled in the same way, in that because we typically need very specific things from an image (e.g., where is the teapot?), we can use special-purpose visual filtering to just acquire the information that is going to help in this task. To make these ideas concrete, let's examine stages in the making of a peanut butter and jelly sandwich. Figure 2.15 shows moments in the sandwich construction process from the Hayhoe laboratory, which can pinpoint the location of eye fixations. When spreading peanut butter or jelly, the gaze follows the momentary tip of the knife, but when filling a cup with Coke, the gaze monitors the rising rim of dark liquid. And when affixing the Coke bottle lid, gaze aids in aligning bottle and cap to engage the screw threads. Each of these steps can be seen as running expectations that must be successfully met.

The idea that the brain might run on expectations is nicely developed by Baum[3] and Hawkins (see ref. 30), but still takes getting used to because the experience of seeing is so different from its neural implementation. We cannot escape the sensation of being immersed in an elaborate colorful, depth-filled, three-dimensional world. Yet the evidence suggests that the machinery that provides this experience is heavily coded and very unlike the literal sensation. You don't have to worry if this seems counterintuitive: Early theories of visual representation did not get this either and posited a "picture in the head," painted by eye movements. It was not until the realization that some inner "person," a homunculus, would have to look at this image—and thus no progress would have been made—that this idea was abandoned.

2.6   Hippocampus: Program Modifications

At this point, we have the image of the cortex defining a massive "state," and the basal ganglia being able to refer to that state in sequencing to new

A   B

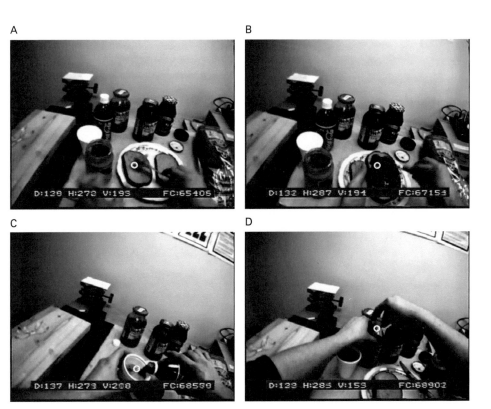

C   D

Figure 2.15
(A–D) In making a peanut butter and jelly sandwich, eye fixations are used to orchestrate sequential steps in the task. The momentary location of the fixation point
is indicated by the small white circle. Figure courtesy of Mary Hayhoe, University of
Texas at Austin.

states. So the basic elements of programs are defined. But how do we create
new programs? Or how do we run stored programs in novel environments?
We know that these questions are answered in most part by the brain subsystem termed the hippocampus (and the amygdala, which we will get to
in a moment), but let's elaborate. The hippocampus (figure 2.16) plays the
central role in the permanent parsing and recording of the sequences of
momentary experiences. Of the deluge of ongoing experiences that you
have every day, what is worth remembering? The hippocampus has mechanisms for choosing and temporarily saving current experiences until they
can be stored more permanently.

The effects of damage to the hippocampus and amygdala can be very
severe, as famously was discovered with patient Henry Molaison (H. M.),

A

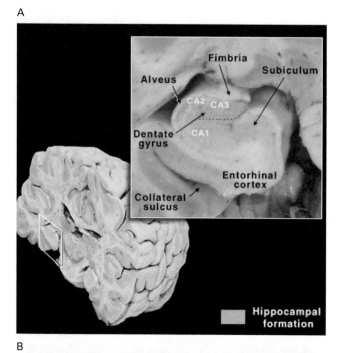

B

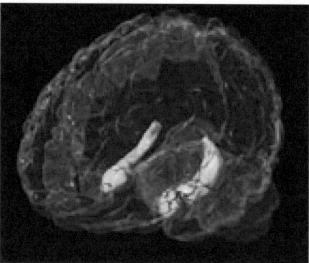

Figure 2.16
The hippocampus is an extension of the cortical sheet that reminds one of a jelly roll in cross section as shown here. The roll extends a good distance longitudinally to allow contact with all the cortical circuitry representing abstract descriptions. But its circuitry is very special and designed to extract and remember encodings of the crucial parts of everyday experience. From Sylvius 4 Online, Sinauer Associates.

who had both bilaterally removed to control severe epilepsy.[34] For people who have injured their hippocampus, like H. M., time stops at the point of injury: They can participate in conversations, behaviors, and the like but do not remember them. Their working memory is typically intact, so they can run programs that they have and deal with new information in the moment. But they cannot save this information for the next encounter. The movie *Groundhog Day* has this condition as its premise: All the citizens of a small town experience the same day over and over again. The hero knows this is happening and uses the information to comic effect, but everyone else is clueless. Their experience graphically suggests what it's like to do without a hippocampus. Every day is more or less the same new day. But like the foils in *Groundhog Day*, such patients are seemingly unaware of their lack of knowledge.

Even though you now know what the hippocampus does, you might be wondering why this function could not have been included in the cortex itself. After all, the cortex is the place where memory instants are stored. Why not just somehow add in the new ones? The problem is that the cortical memories are very compactly *coded*. Thus, when a new memory comes in, it cannot be added willy-nilly but must instead be filed near similar experiences. The task of doing this is delicate and takes time. (We'll describe some models of how it's done in chapters 4 and 5.) So the hippocampus has two main things to do: (1) It must remember the experiences that are going to be permanently saved and then (2) add them to the cortico–basal ganglia complex.

Conceptually, it would seem to be possible to have the permanent memory storage be an ongoing process. In computer jargon, this would be a "background job," something that can take place while the other important programs of the moment are directing the body's daily living activities. And very recent evidence suggests that some form of this may be possible,[35] but only in the sleeping state. Rats running mazes are using their hippocampus to remember sequences, and their hippocampal neural spikes can be recorded. However, when they stop for brief naps, it appears they play back the spike trains in reverse order (figure 2.17). While it is not known just how this information is used, the suggestion is that it is a part of the encoding process, as the rat is not doing anything else when this happens. However, it appears that encoding cannot be completely done without sleep. The most likely explanation is that some facets of the memory consolidation process interfere with the use of memory in these daily activities. There may be a possibility that auditory hallucinations are caused by this interference. The consolidation process accidentally turns on during waking hours. A person hears a voice playback, but the person

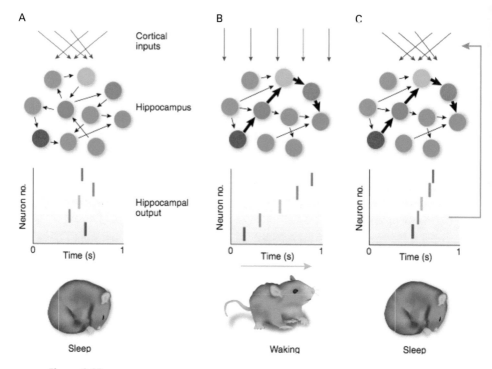

A

Cortical inputs

Hippocampus

Hippocampal output

Neuron no.

0   Time (s)   1

Sleep

B

Neuron no.

0   Time (s)   1

Waking

C

Neuron no.

0   Time (s)   1

Sleep

Figure 2.17

A synopsis of Wilson's discoveries in the encoding of programs by the rat's hippocampus. (A) Before learning a task, neural spikes are randomly organized. (B) After the task is learned, temporal sequences are strengthened and appear more frequently. (C) During sleep, the same sequences are played back during the encoding of the new sequence into the cortex's long-term storage. Reprinted by permission from Macmillan Publishers Ltd. M. R. Mehta, *Nature Neuroscience*, vol. 10, pp. 13–10, 2007.

"knows" it's not his or hers; ergo it must belong to someone else. Thus, in normal people the consolidation is postponed until sleep.

We know that human consolidation occurs in sleep because of experiments that were done by Karni et al.[36] In particular, their experiences showed that consolidation occurs during a particular portion of sleep termed REM sleep. REM stands for *rapid eye movement*, such as movements made by the saccadic system. During REM sleep, the eyes dart back and forth under the eyelids. As the eyeballs are imperfect spheres, this activity can easily be detected by characteristic wiggling of the eyelids. What Karni et al. did was have subjects learn a skill and then wake them up during the subsequent sleep period during REM sleep. A control group was awakened for the same amount of time but not during REM sleep. We can all be glad we weren't subjects in this experiment. The control group retained the skill when tested the next day, but those disturbed during REM sleep had impaired skill performance. In emphasizing REM sleep, we have glossed over the sleep cycle, which has several distinct phases, the exact purpose of which is still unknown. Nonetheless, it is extremely likely that they have important and related functions. What is clear is that the sleep cycle is essential for the hippocampus to do its encoding and downloading work.

It is hard not to appreciate what an amazing technical feat defining a program is. In conventional silicon computation, the way to get a program is to get a programmer. But the brain has to be its own programmer. Out of everyday experience, it has to try new programs and save the really good ones. It is impossible to save everything, and we just don't. Presumably, that would swamp the playback coding system. So a first task is to digest a day's experience into the important episodes worth saving. Saying that finesses another issue, though: That experience is a continuum of sights, sounds of the environment, and people you met. Furthermore the encoding, as reflected in REM sleep, can be quite literal. If we combine the insights from the rat and human data, it seems that a fairly literal slice, from a neural standpoint, is being selected for inclusion in the memory. But given the richness of a day's "fire hose of experience,"[37] how are the essential features for just a tiny subset of them selected? This is what the hippocampus has to do. It has to take a select few new descriptions of what happened and code them in a states and actions format.

The foregoing discussion has focused on the permanent storage of new programs, but another very important possibility is the use of novelty in the ongoing execution of programs. Presumably and perhaps especially if the new information is important, it would be useful for the ongoing programs that are concurrent with the novel additional structures. The elaboration of this point will need the development of later chapters of this book, but in terms of an overview, the stored programs that the brain uses make extensive use of statistics to achieve high compression. The careful management of probabilities results in the most useful sets of options being stored. However, besides this account of the augmentation of permanent storage, another extremely important issue is the use of the hippocampus in interacting with stimuli in the here and now.

This leaves a need to handle novel situations that are similar to a stored program but that have important differences. This case is handled by the hippocampus. If the novelty of a current stimulus can be detected[38, 39] it can be usefully melded into the stored program library and its existing protocol potentially can be adjusted. To elaborate, the stock response in a program at a given point might be action A, but if the current situation is slightly different, then the right action is B. The mainstream research view is that the novelty detection in the hippocampus contains the information that allows this adjustment to be made.

Such a capability is likely to be extensible and utile for the general cortex–thalamus–basal ganglia loop, which includes the hippocampus. This loop can be run in real time, in the course of directing behavior and steering the selection of actions, but also in "planning time," wherein it can use the predicted results of actions to explore their ancillary consequences. Experiments with rats exploring mazes show evidence that they use this capability.[40] It has long been recognized that explicitly recalled temporal sequences are maintained by the brain (see Tulving's comprehensive summary[41]). Whether the cortex–thalamus–basal ganglia loop can run backward in time is not completely settled, but its temporal sequence architecture is tailor-made for implementing the demands of episodic memories.

2.7 Amygdala: Rating What's Important

The *amygdala* is a smallish neural subsystem that is situated right at the front of the hippocampus's tubular structures as shown in figure 2.18. It plays a major role in arousal, orienting the brain to place emphasis on events that are especially important. In this task it is especially associated with fear. People who have had the misfortune of damaging their amygdalas do not get the adrenaline shock that you or I do when we see scary pictures. A particular patient with amygdala damage attempts to draw sketches of the various emotions. All are reproduced satisfactorily except for fear. This observation is just one of many. Rogan et al.[42] have shown that the amygdala is extensively activated when rats are conditioned to have a fear response to auditory stimuli. Subjects shown pictures of faces have enhanced fMRI responses to those that show fear. But in spite of all these results, newer research is showing that it is unlikely that only the emotional system per se is the issue; here instead the issue is danger. If you are picking out experiences to remember, then extreme danger is an obvious candidate to be handled specially.

Remember that when we discussed the cortex and hippocampus, one key point was its ability to encode continuous experiences and the associated difficulties in doing so. Another part of this difficulty results from the fact that the cortex is elaborately indexed, whereas the hippocampal coding is much more literal. Thus, between the cortex and hippocampus,

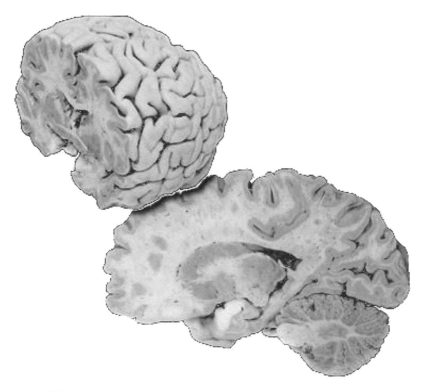

Figure 2.18
A relatively small area in the forebrain codes for the most important situations. The need to cram as much experience into the cortical memory as possible may have led to the elaborate cortico-hippocampal circuitry where new experiences are cataloged in terms of the memory's existing ways of parceling novel items. However, some extremely dangerous situations that may be cued by fearful companions may best be treated as one-offs that are remembered as is. In this case, evidence suggests that the amygdala circuitry is recruited to retain a more direct encoding that elaborates on the particulars of the near miss. Naturally, this strategy needs to be used judiciously given its relative expense. From Sylvius 4 Online, Sinauer Associates.

an elaborate electrical dance takes place to fit new experiences in the most efficient fashion.

But what if something really terrible happens to us, something so bad that it was life-threatening but still a near miss? It seems likely that in this case, the amygdala handles the coding in a straightforward way that bypasses or augments the cortico-hippocampal route. In this exceptional case, you get to burn-in neural circuitry that saves the details of this near disaster. Think of this mechanism as a "life insurance policy." If you had the good luck to deal with this successfully once and it was close to the dangerous edge, then you save the experience in a more verbatim form that preserves its gory details. Saving literal experiences literally is costly: If a young child is forced to code traumatic experiences, the damage, as reflected in extensive neural wiring changes, cannot readily be undone.

2.8   How the Brain Programs Itself

At this point, you have an idea of the function of the brain's major subsystems, but you are still missing a complete picture of how the brain solves the mystery of programming itself. This mystery is very much unsolved at this time, but enough is known to suggest a computational perspective in broad outline using figure 2.19 as a guide. Once a program is encoded, its execution is carried out via the cortex–basal ganglia–thalamus loop, discussed in section 2.2. But what of the creation of such programs?

To facilitate the ensuing description, it helps to introduce a distinction made in silicon program execution, and that is the difference between *interpreted* and *compiled* executions. When a programmer designs a program, he or she typically will use an abstract language, such as MATLAB, Python, or Java. However, before the computer carries out these instructions, they have to be translated into a much more basic language that the machine's hardware can understand. There are two ways to do this. When a program is *compiled*, all of its instructions are translated together. As a consequence, decisions can be made as to an efficient translation and consequently an efficient execution. In contrast, when a program is interpreted, the program's instructions are translated sequentially one at a time. The benefit of this is the programmer can easily and quickly modify the program. Instructions can be changed and the program re-executed in an efficient manner. Interpreting makes modification easy, whereas compilation makes execution efficient.

One can think of the job of the brain as to discover ever more programs in order to up its owner's chances of achieving his or her prime directives.

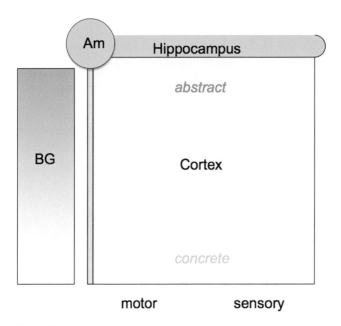

Figure 2.19
An overview of the brain's program development system. The cortex–basal ganglia (BG) system contains existing programs that can be either concrete and difficult to modify or abstract and accessible to modification. The instructions for modification are captured by the amygdala-hippocampal system. The hippocampus codes the difference between the current situation and remembered programs, and the amygdala scores this difference in terms of importance.

As it must perform in real time in the real world, these programs must be executed efficiently. So there would be a benefit to having some arrangement like compilation, so that the programs could "run" without being monitored. At the same time, it is possible for some unanticipated reason that the programs need to be monitored. Thus, there is need for some facility like interpretation, whereby a program can be improved by changing its instructions.

How should instructions be changed? This is handled by the amygdala-hippocampal system. This hippocampus is constantly slicing and dicing the interpretations of the current context into a format that can be matched against existing stored programs. But one doesn't want to modify just anything. Thus, the amygdala is constantly sorting potential changes in terms of their overall importance. Thus, to sketch the entire programming picture, the brain's end product is useful compiled programs. These programs are obtained by choosing important deviations from existing programs, selecting the right new program to encode, and then compiling it for efficient execution.

2.9   Summary

The site of behavioral programs is the forebrain. Different aspects of these programs can be associated with different forebrain regions:

• the cortex is the place where programs define their state, together with actions for changing state. This state can be extremely elaborate, involving multiple modalities.
• the basal ganglia is the place associated with sequences in programs. This sequencing can be for actual programs directing actions in the world or simulated sequences directing actions in the imagination.
• the hippocampus plays the role of creating the structure of new programs. New programs make heavy use of existing program descriptions.
• the amygdala codes for rare, life-threatening situations.
• the cerebellum controls the adaptation in sensory motor programs. Changes in the body from different loads or during child development are handled here.
• the thalamus handles input and output to the forebrain. To overcome the slow circuitry, heavy use is made of model predictions.
• the hypothalamus represents basic drives. Even the most abstract programs have to negotiate with this region, which is adjacent to the sources of regulatory neurotransmitters.
• the most important neurotransmitter is dopamine, which is a basic currency that allows different programs to be compared.

II Neurons, Circuits, and Subsystems

The basic goal of this book is to organize the principal computational issues the brain must face in terms of their natural place in an overall hierarchy. Part II focuses on the issues at the bottom of that hierarchy and, in particular, the computation that can be addressed while putting aside the brain's interaction with the outside world. We start with an exposition of neural firing codes. The most fundamental constraint that the brain must face is the slow firing rate of its neurons, which limits the speed of computation. To get around this limitation, the brain has evolved a number of ways of coding the spikes to speed things up. Some of these have been known for a long time, but some are very recent and have the status of provisional proposals that are in the process of being tested experimentally.

Whatever set of spike codes ultimately turns out to win the day, the issue at the next level of abstraction is the *table-lookup* organization of the cortex, which has vast amounts of information indexed so as to allow the moment-by-moment tracking of behavioral contingencies in terms of the immediate state of affairs and available actions. An enormous amount of information is known about the cortex, but even so, a huge amount of work remains to elucidate its exact structure and function. We review some recent experiments and computational proposals that tackle these issues.

Having a "program listing" in terms of states and actions does not get the program executed. This is the job of an elaborate system of loops, primarily involving the basal ganglia, which we focus on, but also the cerebellum for action monitoring. Basically, the basal ganglia is involved in picking the action, and the cerebellum is involved in monitoring its result. Recent results relate the role of the basal ganglia in learning and suggest roles for specific neurotransmitters.

3 Neurons and Circuits

The brain uses stereotyped electrical signals to process all the information it receives and analyzes. The signals are symbols that do not resemble in any way the external world they represent, and it is therefore an essential task to decode their significance.

—John G. Nicholls et al., *From Neuron to Brain: A Cellular and Molecular Approach to the Function of the Nervous System*, 3rd ed. (Sinauer Associates, 1992)

Like any organ in the body, the brain is composed of cells. But the brain's cells, termed nerve cells or *neurons*, are geometrically unlike any other cells in the body, being specially designed to send electrical signals over long distances. As Allman emphasizes,[43] this was a biological breakthrough that allowed sensory organs to direct distant motor systems. Consequently, the large-scale structure of animals, with their need to communicate across a very distant separation of sensors and effectors, became possible. In addition, the special geometry of cells allowed them to form intermediate circuits to analyze and condition the signals they were sending. In the forebrain especially, cells were able to take advantage of long-range connectivity to specialize complex circuits for each of its different subsystems.

In broad outline, nerve cells communicate by sending voltage spikes, which to a first approximation are all alike. However, the problem of figuring out what they do in detail is complex, and our comprehension at this point is still full of holes even though enormous amounts of information have been gained. To understand the magnitude of the problem, one can start with the issues of levels of abstraction. A neuron is characterized by a very complex chemistry that governs all aspects of its functioning. Molecules external and internal to the cell can change its electrical properties and thus make it function differently in different neurochemical contexts. At the next level of abstraction, there is the question of the electrical communication of the cells itself. The neuron's voltage spike can mean very different things in different contexts. And at the next level of abstraction above that, where circuits between neurons are formed, the many different functions of circuits have to be considered.

The goal of this chapter is to show how computational modeling of neurons in the cortex has enhanced our understanding of its circuitry and laid bare some of the more cutting-edge issues on spike signaling and to do all of this exposition in a compact format. To achieve this goal requires triaging the complexities of cell function, focusing on what we hope are useful abstractions that speak to its use in overriding computations. One way we try to achieve this goal is to relegate some of the details on spike generation that the reader might already be familiar with to an appendix at the end of this chapter, but even when this is done, a large amount of material has been sidelined. To obtain an idea of the functional context that is being glossed over, the reader is referred to three among many excellent sources. Rieke, van Steveninck, and Bialek[44] show that information-theoretic techniques can explain complex perceptual data, and Dayan and Abbott[45] provide a beautiful exposition of current analytical techniques. For modeling at the detailed level of a neuron's function, see Serratt et al.[46]

To understand the rationale for what does and does not get included in the models, we must first start with an overview of the most essential processes. The most basic we take to be the different ionic channels that regulate the passage of charge in and out of the cell. These channels guide the electrical functioning of the neuron itself. We will use a very basic model in which inputs add charge to or subtract charge from the cell multiplicatively through electrical contacts. The next level of abstraction is the circuit level, where different neurons are connected together for different functional purposes. The actual cells in such circuits are of many different types, but we will avoid that detail as well to concentrate on more central representational issues. How do cells come to represent the things they represent, and how do they signal this information to other cells? The representational answers find much consensus, but many different signaling models are possible, and which is best is unsettled. We will describe one of the more speculative to illustrate some of the crucial computational issues. Finally, there is the formative use of circuits in directing behavior. As a bridge to the very complicated sequences of behavior to come, we touch on a computational model of some amphibian behaviors, where pattern-recognizing circuits can be usefully and directly harnessed in generating motor commands.

3.1 Signaling Strategies

At the timescale of 100 milliseconds to 1 second, the action potential just looks like a single vertical line on an oscilloscope, so that it is compelling to think of the signal at this level as some kind of code, where a spike means a "one" and no spike means "zero." The sequence of ones and zeros over time is a code used by all spiking neurons to communicate. Although the sequence represents a compact code—as Nicholls et al. note in the introductory quote—it is far from being completely deciphered. One reason that the code has not been cracked is that there are many different ways that the brain uses spikes to do computation. To show off this variability, let us describe a few of the different schemes, which are depicted in figure 3.1.

Probably one of the easiest codes to comprehend is that used to control the vestibular-ocular reflex (VOR). This reflex stabilizes gaze on a target in the face of disturbances brought about by head movements. Try the following: Hold still and wave your finger back and forth in front of you all the while fixing your gaze on it. Now hold your finger in a fixed place and move your head back and forth. In the former case you engaged the pursuit neural circuit, while in the latter case you engaged the VOR circuit. It turns out that this circuit must respond so quickly that there is not time enough to sense the slip in the retinal image and correct for it, so it relies on the very fast vestibular system that can sense head accelerations and integrate them to obtain a velocity signal that can be used to counterrotate the head. Furthermore, and this would be the main point, the neurons that signal head velocity use a rate code. Their number of spikes per second can be directly interpreted as an analog quantity with about 300 spikes per second signaling the maximum velocity in one horizontal rotation direction, and 100 spikes per second signaling the maximum velocity in the other rotation direction. The VOR circuit's "rate" strategy is diagrammed in figure 3.1A, and an actual data trace is shown later in figure 3.3.

The next example comes from the hippocampus circuitry. Recalling the discussion of the hippocampus from the previous chapter, the hippocampus has the special job of slicing and dicing the continuity of experience into steps that can be seen as part of an existing program that directs behavior. A special problem it handles is that of coding spikes with respect to the part of the program that they are related to, and this can be done by using one of the forebrain's oscillations, the θ (theta) wave, which oscillates in the range 4 to 7 cycles per second. The convention is that every time the wave crosses zero and is increasing, this signifies the beginning of the program, and the end of a cycle, 360° later, signifies the end of the program. This

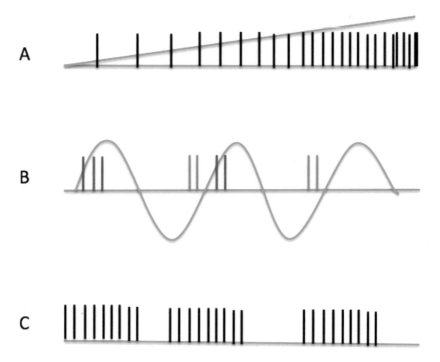

Figure 3.1

Spikes are used in different ways in different circuits. (A) In the brainstem, "spikes per second" denotes an analog quantity. (B) In the hippocampus, spikes are organized by the beginning and end of the current task of interest. The approximately 6- to 10-Hz Θ rhythm cycle is used for this phase code reference. Cells that code information at the beginning of the task send their information at 0° (red), whereas those at the end send their spikes at 360° (green). (C) A very different code is used by the tonically active cells (TANs) in the basal ganglia. They are tonically firing but silent at crucial breakpoints in a task.

provides a natural timing framework for spikes. Those at the beginning can signal that position by sending their spikes at a time near 0° and those at the end by sending their spikes at a time near 360°, as shown in figure 3.1B.

The final example comes from the basal ganglia. Signal processing there uses different spike codes, but, focusing on the important example introduced in the previous chapter, there are the tonically active neurons (TANs) that mark key moments in a complex task with silent periods, as shown in figure 3.1C. Silence can be of great moment in the basal ganglia because it is composed not of a single nucleus but of a succession of interrelated nuclei of cells wherein connections between nuclei can be inhibitory. Ergo, the lack of a signal can be an important message to signal that it is time to move on to the next step.

Signaling Spatial Position

The signaling of relative time by the hippocampus as a θ phase is exotic, but it has an advantage in that it allows time to be naturally scaled to be contained within the span of a behavior. If this was not done, an animal would have to keep track of elapsed time in some other fashion. However, even more exotically, recent experiments seem to point to the fact that an animal keeps track of space in a similar way. To record a location, one way would be to have neurons that are sensitive to a particular spatial location in a local map. It turns out that this happens, as shown by pioneering work by O'Keefe[47] and O'Keefe and Nadel.[48] However, it is also the case that the animal uses "grid" cells that fire at regularly spaced spatial intervals, as shown in figure 3.2A, which shows spikes sent by a single cell as a rat navigates. This work has been extensively modeled,[49] and the most recent model by Sreenivasan and Fiete[50] shows how the animal can take advantage of coding just one special spatial location (its own) in order to realize increased accuracy. Figure 3.2B shows how many grid cells of different relative phases produce a unique coding in the firing pattern for any one location. This strategy raises the possibility that animals may use a strategy where time is coded similarly, but this has yet to be tested.

Signaling Negative Numbers

In general, coding negative quantities is not a trivial matter for brains and computers, and it is handled differently in different situations, as shown in figure 3.3. In the vestibular-ocular reflex, "negative" numbers are signaled by lower firing rates. In silicon computers, negative numbers are signaled by a "twos-complement" representation: flip the binary code and subtract one. This makes addition hardware simpler. As we will see in a moment, the cortex exhibits complementary pairs, with only one of the pair typically having a response. Given that cortical neurons are sensitive to neurons and signal the presence of a particular pattern, you might wonder what happens when the complementary pattern arrives. For example, consider the motion-sensitive cell. What happens when the motion is in the *opposite* direction to the one for which it is most sensitive? If you didn't know already, you probably have already guessed that there is a neuron tuned to the opposite direction. In fact, this is a general coding strategy for the cortex. One way to think of this is as a strategy for signaling "negative" responses. For motion, one wants to impose a coordinate system wherein

A

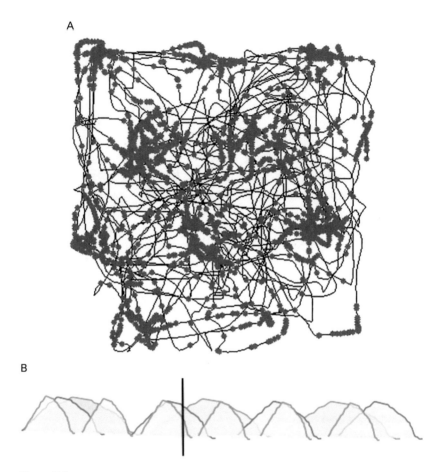

B

Figure 3.2
The special coding of position takes into account that the animal can only be at one place at a time. (A) Traces of a rat's path show that spikes (red) are signaled over a grid pattern. (B) A sketch of colored hypothetical grid cell responses reveals the methodology behind the coding strategy. The grid cells use several different frequencies resulting in phase offsets. Thus, for any particular position, for example the vertical line, the relative set probabilities of different grid cells sending a spike is unique. Grid cell responses courtesy of Torkel Hafting, CC Share-Alike 3.0; http://upload. wikimedia.

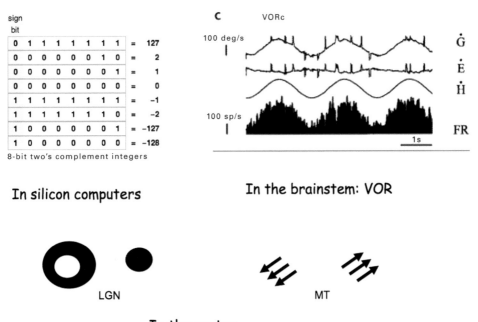

Figure 3.3
Signed quantities can be represented in several different ways. (Top left) Silicon computers use a "twos-complement" representation where in the binary code a number is negated by subtracting unity and flipping all the bits in the resultant binary representation. (Top right) In the VOR, negative quantities are signaled by spike firing rates below the average rate. (Bottom) LGN and cortical cells use pairs of cells with one cell representing the positive quantity and the other the negative quantity (e.g., motion coding in cortical area MT).

one direction is positive and the other has the opposite polarity. One reason why this convention might have been chosen is that the synaptic connections of any particular axon can have only one polarity. They can only be all positive (excitatory) or all negative (inhibitory). However, a better reason might be that the convention makes it easy to partition a signal into two halves and keep it calibrated.

3.2 Receptive Fields

Now that we've covered neuron basics, it's time to put these models to use and give their cells something to do. One central neuronal job is that of coding the external world into symbols, so we'll start here and see how that could be done. The initial model that will be used is very simple, and we'll have to resist the urge to overgeneralize it, and in fact it will need revisions here and later on. But it contains so many useful ideas that will stand us in good stead in discussions ahead that we'll take it very seriously here. The examples will be taken from studies of vision in monkeys: The reasons are that vision has been the most studied part of the cortex and that monkeys have neural wiring that is very similar to our own.

To make the idea of encoding concrete, consider how the eye handles an image. Light impinging on the eye contacts a matrix of discrete photo-receptors. These are specialized in terms of their wavelength sensitivities. *Rods* respond to all wavelengths of visible light and are very sensitive to low light levels. *Cones* come in three wavelength sensitivities peaked at red, green, and blue, respectively. Cones are concentrated near the center of the optical axis and become sparser toward the periphery where rods are dominant. We can see color in the periphery, but the spatial resolution of that "image" is drastically reduced.

After the retina, the image is sent to the thalamus, specifically to a small portion of the thalamus called the lateral geniculate nucleus, or LGN. The LGN is further distinguished as the place where inputs from both eyes are proximal. There are two lateral geniculate nuclei, and they each look after a hemifield of visual space. Looking down from the top, the left LGN takes care of inputs from the space to the right side of vertical. We know for a fact that the signal that arrives here is heavily coded, as the amount of cells sending axons to the LGN is about 100 times less than the rods and cones detecting light initially. However, the exact form of the transmitted code has yet to be determined in detail. The code used by the LGN is known in broad outline, though. It is in the form of cells that have a responsive center and an inhibitory surround, as shown by figure 3.4.

You can broadly think of the dots in a conventional dot-matrix printer, except that there are white dots and colored dots as well as black dots. You might think that a black dot could be signaled as the negative of a white dot, but the brain, as just discussed, has apparently not invented explicit negative numbers and instead resorts to a push-pull representation: For every value signaled, there is a cell dedicated to its logical opposite. By peculiar convention, cells that prefer a lighter center are termed ON cells, and those that prefer a darker center are termed OFF cells, but we shouldn't allow this idiosyncratic naming convention to confuse us. Both types have the same behavior in that they fire when the pattern they prefer occurs.

(a) ON-center ganglion cell

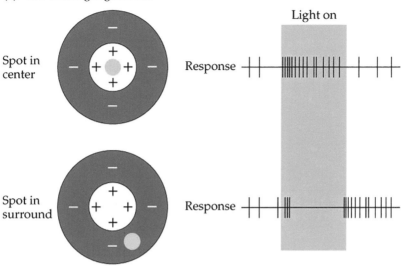

(b) OFF-center ganglion cell

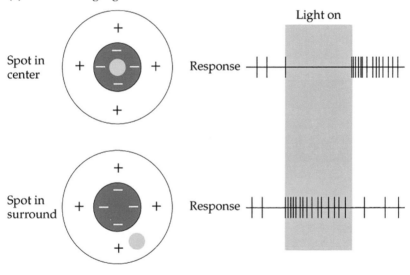

Figure 3.4
The high-resolution image of the retina is summarized by more coarsely sampled cells in the lateral geniculate nucleus (LGN) of the thalamus. The cells that code for light intensity are of two principal kinds: those that respond best to a lighter-than-average center spot of light (ON cells) and those that respond to a darker-than-average center spot of light (OFF cells). Reprinted with permission from *Sensation and Perception*, by Wolfe et al., 2012. Copyright Sinauer Associates.

Coding Sensory Data

The coding at the LGN is just the beginning of an elaborate process that is continued in the cortex. What starts out as a collection of light and dark dots ends up, after a succession of intermediates, as neurons that are responsive to an enormous library of very abstract visual properties such as the emotional expressions of faces. We will have much more to say about how this process operates in the next chapter. For the moment, let's just follow it for one more step to make a number of points about neural circuitry and its models.

The "dot-matrix" representation is unsatisfactory from the perspective of a code because it contains the ability to code many images that never are generated by the world with any significant probability. Pause for a moment and take a look at your surroundings. What you are likely to notice is a scene of textured surfaces circumscribed by sharp boundaries where the intensity changes quickly over a small spatial extent. What you won't see is anything that looks even vaguely like television "snow." However, the LGN neurons potentially can code the scenes from the real world and snow with equal facility. But as television snow almost never comes up for any significant interval and contains no useful information, a system that could code it with the same fidelity as real-world structured imagery would be very inefficient.

A coding scheme geared to the world would be especially organized to encode selectively images that are more likely to happen and discount those images that are very unlikely to occur. We say "unlikely" because once in a while we do look at television snow (even though we might have better things to do). Very satisfyingly, when experimenters look at the functional connections from the LGN to the striate cortex, or V1, the first stage of the visual cortical memory, V1 cortical cells seem to be doing just this: picking a coding scheme that emphasizes the structure in naturally occurring images at the expense of random noise.

Let's take a look at an encoding scheme that does the job for encoding small image patches. An experimental test of such codes conducted by Reid and Alonso is shown in figure 3.5B.[51] Suppose you would like to make a neuron that was responsive to such images. If you don't know the answer already and take a minute to think about this problem, you probably will come up with the scheme first suggested by Nobel laureates Hubel and Weisel using the neurons in the LGN. Where you want your neuron to be responsive to a relatively light pattern, you would find an ON cell at that location and connect it to your cell. You would do the same for the darker areas using OFF cells. As shown in figure 3.5, which shows some of the

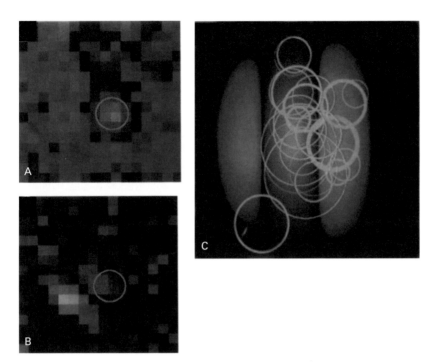

Figure 3.5
Two ways of sampling visual space with neurons tested with reverse correlation methods. (A) An LGN cell in the thalamus samples the visual field with the ON/OFF coding introduced in figure 3.4. (B) A V1 cell in the cortex samples the visual field in a different way, using oriented bar detectors. Hubel and Weisel, the discoverers of this code, posited that to make a white line detector, take the right set of ON cells from the LGN and connect them to the soma. Next, take flanking OFF cells from the LGN and connect them to the soma. You do not have to maintain the pattern geometrically. Just choosing the right sets ensures the desired response. The circle's overlap in B shows that this is indeed the case. (C) A summery of measurements made by Alonso and Reid coding the ON LGN cells as red circles and OFF LGN cells as blue circles. Their interrelation to a single cortical cell, whose line response field is shown by the red and blue ovals, is revealed by plotting the measured positions with respect to external space. Reprinted by permission from Macmillan Publishers Ltd: R. Clay Reid and Jose-Manuel Alonso, "Specificity of monosynaptic connections from thalamus to visual cortex," *Nature*, vol. 378, p. 6554, copyright 1995.

crucial experimental evidence for these connections, in fact you can do this by taking the axons of the requisite cells and connecting them with synapses to the coding cell.

To emphasize a particular value at a particular point, you can make the strength of the connection large. The simplest model used multiplies the input by the strength of the connection to obtain its overall effect. Thus, when the coded pattern appears in the dot-coding cells, it has the maximal effect on the coding cell. Remember that the cortex cannot solve this problem in the obvious way using negative numbers for the inputs x if required because there aren't any! But as reverse engineers after the essentials, we can take a shortcut.

In case a primer would be helpful, take a moment to glance at box 3.1, which introduces necessary notation. The upshot is that to get the response r of the neuron, multiply the image x by its synapses as follows (figure 3.6):

$$r = x_1 w_1 + x_2 w_2 + \ldots x_n w_n.$$

You can see that when the image x is a scaled version of the synapses, the response will be highest. Besides finessing the issue of negative numbers, another significant shortcut is to use a scalar value for the response. Keep in mind that the typical response for a cortical neuron is very close to random. The shortcut assumes that we can usefully summarize the number of spikes in some chosen interval as a "rate." Experimentalists take a

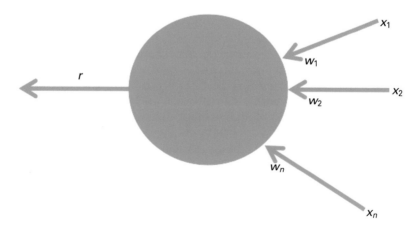

Figure 3.6
The simplest neuron model uses real numbers that can be positive or negative for inputs and outputs. Its linear output r can be expressed as $r = x_1 w_1 + x_2 w_2 + \ldots x_n w_n$.

similar shortcut by reporting the average of multiple trials, so that the rate approximation is a good model for trial-averaged data.

If we want to add extra detail, we can make this work with the positive-synapse convention, as the "negative" x is in fact positive—it is the output of an OFF cell and it pairs up with an excitatory synapse counterpart. Thus, looking at the connections, the only way we would know to draw the connections in the light and dark format is to know where they came from. This is, very famously, the idea of the *labeled line*.

3.3 Modeling Receptive Field Formation

Up to this point, the analysis has taken the perspective of the reverse engineer in that, given the data of cell responses, we try and guess the structure of their circuitry. Starting from the retina, information can be abstracted from the "raw" sensory data in successive stages. Only two—the LGN and the first stage of visual cortex V1—have been discussed, but later stages embody the same kind of thinking. It is just that the codes become more and more abstract and relevant to behavioral goals.

However, this approach has finessed the process that causes the receptive fields to get coded in the first place. For this to happen, there must be a normal developmental path working together with experience with natural images. One way this could occur, and the way we will elaborate, would be to use this experience with feedback circuits to set up the appropriate synapses. Cortical circuits make extensive use of feedback. Keep in mind that the cortical feedback paths typically have 10 times the connections as the feedforward path. The first stage of circuitry between the LGN and the cortex is no exception.

Figure 3.7 shows the anatomy and basic circuitry. Panel B shows a detail that we have not discussed so far. The left LGN integrates the connections from the right side of both the left and right eyes. Furthermore, although we will gloss over this significant aspect in the modeling that will follow, there are two principal kinds of cells: *magnocells*, which have coarse spatial resolution and relatively good temporal resolution, and *parvocells*, which signal relatively good spatial resolution and relatively poor temporal resolution. There are two layers of parvocells for each eye and one layer of magnocells, for six layers in all. All these cells send their connections to the input layer of V1, layer IV. In return, V1 sends feedback connections to the LGN from its upper and lower layers.

Imagine that a small image patch is present and that neurons have to signal its presence. We have seen how the LGN does this: It basically signals

Box 3.1
ARRAYS AND VECTORS

Much of the thinking about neurons and their receptive fields rests on geometric concepts. For instance, an image is an array of brightness values. To acknowledge the discrete nature of this representation, the image samples that they represent are called *pixels* and the associated values *pixel values*. Below is an example of a small, 16-pixel image:

3   7   15   18
6   1   20   22
2   5   8   17
10 3   19   13

What we would like to do is have this image input to a circuit that has other neurons and their associated synapses and keep track of what happens. To do this, we unravel the two-dimensional structure using *vector* notation, which lists all the values. To cue that it is a list, the convention is to use a boldface symbol; that is,

$\mathbf{x}$ = (3, 7, 15, 18, 6, 1, 20, 22, 2, 5, 8, 17, 10, 3, 19, 13).

Now if that image were represented by labeled lines, then each such line, arriving at a neuron, would make contact with a synapse. Supposing that there are also 16 of them, they can also be represented as a vector,

$\mathbf{w}$ = (w_1, w_2, w_3, ..., w_{16}),

where each individual w_k represents the strength of the synapse. In a simple model, the effect of such a contact is to multiply the image component x_k by the associated synapse strength w_k. Then the total effect of all the contacts is just the sum of each such interaction. Call this the response of the neuron, r. This can be written as

$r = w_1 x_1 + w_2 x_2 + ... + w_{16} x_{16}$.

Conventionally, we write this very tidily as $r = \mathbf{w} \cdot \mathbf{x}$, the *dot product* of two vectors.

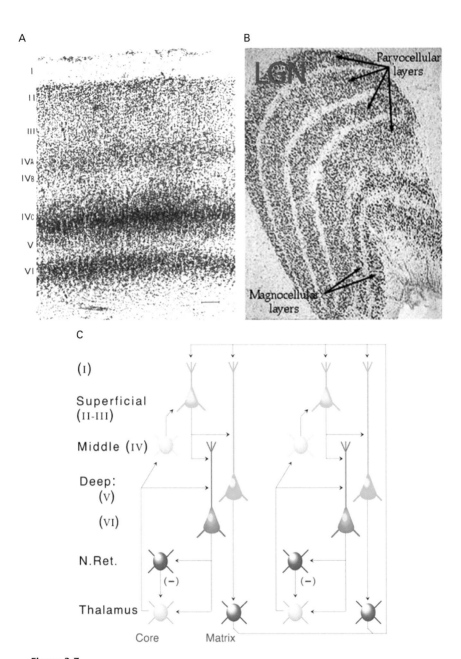

Figure 3.7
The distinct brain subsystems all have characteristic circuits. This example shows the circuitry that connects the cortex to the thalamus. (A) The cortex has a six-layer structure that consists of an input layer (IV), output layers (V and VI), and processing layers (I to III). (B) The output layers that feed back to the thalamus are 10 times as numerous as the input connections from the thalamus to cortical layer IV. Panel C shows a schematization of the connections by Granger. In this diagram, cortical cells are green and yellow, and LGN cells are blue. (C) Reprinted from A. Rodriquez et al., "Derivation and analysis of basic computation operations of thalamocortical circuits," *Journal of Cognitive Neuroscience*, vol. 16, pp. 856–877, 2004.

A   B

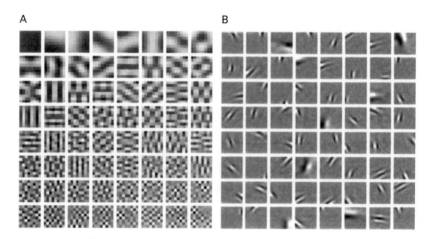

Figure 3.8

A portion of a set of basis functions obtained by Olshausen and Field's numerical simulations. (A) Functions found by counting the approximation error exclusively. (B) Functions found by trading off approximation error with the number of functions used. Reprinted by permission from Macmillan Publishers Ltd: Bruno A. Olshausen and David J. Field, "Emergence of simple-cell receptive field properties by learning a sparse code for natural images," *Nature*, vol. 381, p. 6583, copyright 1996.

the lightness or darkness of very local image areas. But, as we have seen, for natural images, there are ways to do this other than by very punctate local signals. One can have collections of neurons of the edge cells that are responsive to distinct edge orientations in the image. Figure 3.8 shows a portion of two such possible sets of receptive fields, learned from exposure to natural images in a breakthrough simulation by Olshausen and Field.[52] The two sets of 64 neurons have their synaptic weights arrayed in individual two-dimensional grids that are commensurate with the image. Each set can be used to reconstruct an approximation of the input image. To do this, weight each synaptic pattern appropriately to produce a subimage and add all of them up. It turns out that for almost all natural images, it is possible to choose the weights to approximate the original image. In terms of the earlier linear algebra notation, we can approximate the original image I in terms of N patterns as

$$I \approx \sum_{k=1}^{N} r_k \mathbf{w}_k .$$ (3.1)

The neurons' weights ($\mathbf{w}_k$) in such a representation can be described math-ematically as *basis functions*, and the neurons' corresponding responses (r_k) are termed *coefficients*. Here again, you need the concept of the labeled line. The coding neurons send their information on in terms of the responses. The particular axons, or labeled lines, denote the source or meaning associ-ated with the particular inputs. Of course, using an LGN-like strategy, for 64 image points you would need 64 neurons. It turns out that if you want to represent the image patch exactly, even with basis functions, you would still need 64 of those.

However, Olshausen and Field demonstrated that things change radi-cally if you have more than the full amount of basis functions and are will-ing to approximate the image, accepting small errors. In that case, certain pattern sets turn out to be much more economical. Figure 3.8 compares two computer simulations with different approximation criteria. Figure 3.8A shows the patterns computed when fitting the image exactly is at a premium. In figure 3.8B is the pattern set computed with a cost function that penalizes the number of patterns used and accepts approximate fits. You immediately see by inspection that the patterns in figure 3.8A are very textured, whereas the patterns in figure 3.8B appear much closer to those measured experimentally.

Learning the Weights

How can the feedback circuit help to set up the receptive fields of the V1 cells? This process can be described in terms of a learning algorithm wherein images are applied to a model retina, the signals are transformed to simulate a set of LGN responses, and then an algorithm is used to adjust the synapses of V1 cells.[53] To explain the algorithm, we'll start with a ver-sion that is almost right and then refine it. It still won't be the best algo-rithm, but it has a nice geometric interpretation. For a more sophisticated version that better approximates subtle features of observed response data, see Rehn and Sommer.[54]

The gist of the algorithm for learning receptive fields is extraordinarily simple. When an input arrives, the neuron whose synapses are most similar to it are made even more similar. All that remains is to say how this is done mathematically. Suppose the synapses of V1 neurons were initially random values and let us just concern ourselves with a tiny 8×8 image patch, so that $n = 64$, and the image values are given by $\mathbf{x} = (x_1, x_2, ..., x_{64})$. Because there are now lots of neurons, denote the receptive field of the kth one by $\mathbf{w}_k$. For each one, compute its response, which we can do as

$r_k = \mathbf{x} \cdot \mathbf{w}_{k'}$

Now pick the neuron with the highest response, let's call its index k^*, and adjust its receptive field to make it a little more like the image that caused it to win the competition. That is, change its synapses by a small amount $\Delta\mathbf{w}_{k^*}$. The algorithm is given by

while Number_of_samples < Total
do
    $\mathbf{x}$ = GetImageSample   Get patch from an image
                                            Compute the similarities:
    foreach k
       $r_k = \mathbf{x} \cdot \mathbf{w}_k$ Compute the projection
       Pick the k^* that maximizes r_k Find the neuron most similar
                                            Move the neuron toward the
       $\Delta\mathbf{w}_{k^*} = \eta(\mathbf{x} - \mathbf{w}_{k^*})$ winner

In the algorithm, η is a small constant that the experimenter gets to pick. What this algorithm does is create single neurons that approximate as best they can the images that can arise in the natural world. The question is: For small image patches, what is the average patch like? The answer is that it looks like an "edge," a light-dark transition. These receptive field patterns were dubbed "simple cells" by Hubel and Weisel, who discovered them.

The algorithm is not complete for a couple of reasons. Firtoff, it continually picks the best neuron, whereas equation 3.1 prescribes that we pick the N best. The trick is to allow the subsequent neurons to approximate the leftover or *residual* after the best neuron has done its job. In other words, after fitting an image, subtract the fit and have the neurons compete to represent the information in the residual. This can be defined mathematically: let's let w_{k^n} denote the winner for competition round n. Then for $n = 0$,

$r_0 = \mathbf{x} \cdot \mathbf{w}_{k^0}$

as before. Now let's bring the other neurons that are going to work on the residual into play. For $n > 0$, the best approximation so far is

$$\mathbf{y}_n = \sum_{i=0}^{n-1} r_i \mathbf{w}_{k^i} \, ,$$

so you can subtract this from the image to get the leftover (i.e., $\mathbf{x} - \mathbf{y}_n$). Now just find the best neuron to represent this vector, which we can denote as

$$r_n = (\mathbf{x} - \mathbf{y}_n) \cdot \mathbf{w}_{k^n} \ .$$

The second modification to the algorithm corrects another problem. It turns out that in the competition for matching the image, some neurons never get picked and so they never develop receptive fields. This can be fixed by defining proximities and allowing the cells near the winner in response to adjust their receptive fields also. The more elegant fix is to allow all the neurons to participate in the selection process probabilistically, with their response magnitudes determining the odds of their being chosen., By doing this we can we can increase the library of receptive fields to get much closer approximations to the stimulus. As a bonus, this modification produces the kinds of sets seen in figure 3.8.

Figure 3.9 shows stages in these operations. The input I is a vector, as are all the weight vectors, one for each neuron. The responses are computed by projecting the input onto each neuron. The winning response, shown by the red projection, defines a vector that can be subtracted from the input, leaving a residual, shown in green. The residual then plays the role of a new input, and the process repeats. Winning neurons learn by having their weight vector rotated toward their input by a small amount.

How many neurons should one use? As the patch has 64 numbers, to be sure of approximating it perfectly would take 64 neurons, but it turns out that almost all natural images can be approximated with much fewer, say 8 to about 12 neurons. Furthermore, the simulations that use fewer patches are the ones that produce the biological-looking receptive fields. Why is this so?

While we do not know the answer to the question of why codes that use small numbers of neurons are biological, a very intriguing suggestion is that the brain is trading off the cost of coming up with the neurons against the accuracy of the approximation. Remember that the response is going to be signaled somehow with spikes, and signaling with spikes is very expensive; half the metabolic cost of the neuron. So if there are "good enough" approximations that are much cheaper than the very accurate ones, the pressure to choose them would be very great indeed. A grand way of thinking of this strategy is that the brain is engineered to come up with a theory of itself. It chooses theories (read: modifies representations) that work in that they account for what happens in the world, but an additional feature is that the cost of the theory itself, here in spikes, must be taken into account. The way this is done in our small example is by limiting the number of vectors that are used in the approximation.

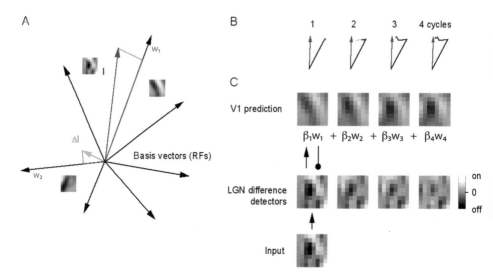

Figure 3.9

A geometrical explanation of the matching pursuit learning model. (A) An 8 × 8 filtered image patch can be represented as a blue vector with 128 coordinate values (in our notation). The neuron whose receptive field is most like the patch, in this case w_1, is chosen to represent the patch. Because there are also 128 synapse strengths, or weights, these can be represented as a vector also. The difference between them is termed the *residual* (green) and is sent back to the LGN as feedback, and the process repeats. A very small number of repetitions produces an accurate representation. (B) Four steps in the vector approximation. (C) The evolution of the approximation in pictorial terms. The green vector is also the basis for the learning algorithm. After each vector is chosen, it is moved closer to the input by adding the residual into its synaptic weight vector. The weight vectors are normalized to unity, reflecting a constraint that limits the total strength of the synapses.

The thrust of this section was to show that the receptive field of a cell need not be specified in the genes per se, but rather once a basic circuit connectivity is set up, it can learn these values using a coding principle. The algorithm is purely geometric, and as the circuitry in the cortex contains repeated units just like it, we might expect to find it used over and over again as a general strategy. But once the receptive fields are set up, they can be used in general decision making and in fact are, and that is the topic of section 3.5.

3.4   Spike Codes for Cortical Neurons

Given the introduction to the cortex in chapter 2 and the modeling studies here, you might suspect that its use of spikes would be well understood.

However, this is far from the case, particularly owing to a raft of recent data revealing timing effects between spikes. Here, we will just sketch some of the main issues; however, before we proceed, some cautionary words are required. In the previous development of the formation of receptive fields, the synapse strength and firing codes were numbers in the models. However, if we jump down one level of abstraction, we can see that the signals must be individual spikes, and the synaptic efficacy must be represented in terms of a number of synaptic channels. This lower level allows us to address the problem of interpreting its more basic signals and structure. Of course any such low-level model must include specifications as to how its representations can be translated into previously developed more abstract spike models, but the attraction of working at the lower level is that it allows many fundamental questions to be addressed. The particular spike level solution described here has some evidence in experimental findings but is one of many possibilities. It is very far from being generally accepted, but it has the virtue of addressing very important issues (particularly related to timing) that are finessed at the next level up.

The most used abstraction to characterize cortical spikes is that of a rate code. This idea is enormously compelling and has been a hallmark of thousands of neural recording experiments. Figure 3.10 shows why. In a typical observation from a sensory neuron, the more the input stimulus is matched to the pattern that the synapses are tuned for, the more spikes that are produced. This neuron is taken from an area in the cortex that is measuring motion in a smallish area of visual space. If the world in that small area is moving in a certain direction, the neuron signals this by sending lots of spikes, whereas if the world is moving in the opposite direction, the neuron sends just a few spikes. The fastest that neurons can spike is about 500 to 1,000 spikes per second, but most forebrain neurons, and especially cortical neurons, have rates that are much lower, spiking at rates of about 10 to 50 spikes per second.

Given its status as a ubiquitous observation, why should there be any doubts about the use of a rate code as the fundamental message protocol between neurons? There are several reasons to be skeptical, but let's focus on two important ones. First, the rate code is a unary code. To send a larger number, there must be more spikes, and the number of spikes is proportional to the number's magnitude. Note that this is hugely more expensive than a binary code used by silicon computers where the number of binary bits sent is proportional to the logarithm of the number that needs to be sent. This difference is very important as the main expense for a neuron is sending a spike[55] (as shown later in figure 3.23). And the brain itself is

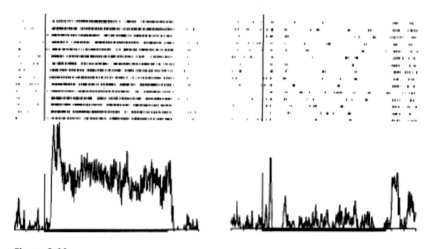

Figure 3.10
A visual motion–sensitive cell in the cortex studied by Newsome. Motion at a certain location in space in a preferred direction produces a large amount of spikes, in contrast to motion in the opposite direction, which produces few spikes. From Britten et al., "Responses of neurons in macaque MT to stochastic motion signals," *Visual Neuroscience*, 10:1159, 1993.

metabolically expensive: It only comprises only 2% of total body weight but accounts for approximately 20% of the body's metabolism.

The second reservation with rate codes is that the observed spike signal from a cortical neuron is very close to a random process. So there is the strange situation with the neuron trying to send a rate with high accuracy with a random spike train. Let's expand on the accuracy issue. Assume that a temporal interval can be adjusted so that in a Poisson model, the probabilities of having either zero or one spike dominate the probability of having more than one spike. In that case, an estimate of the probability of an input rate ϕ can be made by counting spikes. Where the input is a probabilistic variable Z that can be zero or one and has $P(Z = 1) = \phi$, the estimate $\hat{\phi}$, which is the parameter representing rate, is given by

$$\hat{\phi} = \frac{1}{m}\sum_i^m Z_i \, ,$$

where m is the number of intervals. You can imagine that as zeros and ones are equivalently heads and tails, this formula has been extensively analyzed. A straightforward application of the Hoeffding inequality shows that for a precision $\delta = |\phi - \hat{\phi}|$ known with confidence C, the number of inputs required—and consequently synapses—is given by

$$1 - C \leq e^{-2\delta^2 m} .$$

Plugging in the numbers, to just get within 10% with 95% confidence requires about 150 synapses. So for 100 independent inputs, the total synaptic budget necessary would be 150 synapses per input × 100 inputs, or 15,000 synapses total. Given that some estimates of the number of total synapses on a cortical neuron are as low as 6,000, you can see why there might be cause for concern. Of course, there are lots of tricks one can use to save synapses and/or neurons, but they all fall prey to the basic stance of trying to estimate the precision of number with coin flips. If a neural network is commited to counting Poisson spikes to obtain an accurate scalar estimates, the process is going to be costly in synapses, and either space or time.

Because sending a unary-code spike is so expensive, researchers have been exploring alternate coding strategies that would use the spikes more efficiently, but to date no satisfactory alternate scheme has been found. However there is reason to be hopeful. Numerous recent experiments show that the neurons exhibit many different kinds of oscillations, opening up the possibility of oscillation-based codes. One possibility is to use a latency-based code. This would be similar to that used by the hippocampus in the θ frequency band, but geared to the much higher frequency γ band. Anticipating this use, we will take a special look at the potential codings used by the salamander, a very primitive amphibian. Gollisch and Meister[56] were able to record retinal neural data for a flashed image stimulus, as shown in figure 3.11. They interpreted the resultant spikes in two ways: as latencies, with short latencies from the image onset signifying large responses, and as rates, with large spike counts signifying large responses. In this case, the spike latency interpretation seems to carry more information than the spike rate interpretation.

The problem with using latencies in the cortex is that neurons throughout a vast network of neurons all must have a latency reference. In the salamander retina experiment, a natural reference is the flashed stimulus. The cortex's problem is: latency with respect to what? Here is where the γ frequency potentially could be utilized. Spikes that are sent near a phase reference of 0° could represent large quantities, and spikes near 180° could represent small numbers. This would be a γ frequency latency code. Thus, neurons all over a large network could potentially communicate numerical

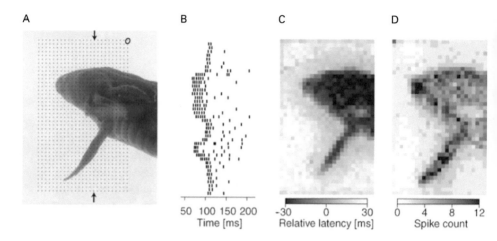

A   B   C   D

50 100 150 200   -30   0   30   0   4   8   12
Time [ms]   Relative latency [ms]   Spike count

Figure 3.11
Experiments by Gollisch and Meister showing the impact of different coding interpretations on the decoded image. (A) Original image with overlaid retinal recording
grid, showing one column of recording electrodes (arrows). (B) Spike response for
a flashed version of the image for that column. (C) Image reconstructed for all recording columns assuming the code is specified by the *latency* of the first spike. (D)
Image reconstructed assuming that the code is the total number of spikes. From Tim
Gollisch and Markus Meister, "Rapid neural coding in the retina with relative spike
latencies," *Science*, vol. 319, p. 5866, 2008. Reprinted with permission from AAAS.

quantities by using a common γ phase reference. With just a small amount
of modification, we can adjust the simulation in figure 3.9 to use such a
code. Figure 3.12 shows the main steps. Figure 3.12A shows that the phase
latency with respect to zero phase is interpreted as a numerical scale. Figure 3.12B shows the modification we alluded to earlier: The candidate is
chosen probabilistically from nearby neurons with odds proportional to
the projections.

Figure 3.13 shows the result of simulating the coding of the same small
image patch 200 times using the γ latency protocol. Owing to the probabilistic selection, different basis functions are chosen at every iteration. Furthermore, the latency code allows the spike delays to be indicated with
color code. This very different coding strategy shows that, even though
the signaling is very different than a rate code, it can first of all do the
job and also appear like a rate code under many natural conditions.[57] The
analysis here assumes that exemplars from the g-frequency range of 40-90
Hz are available, and postpones the issue of how they are created. However
another group led by Koppell is addressing this question.[58]

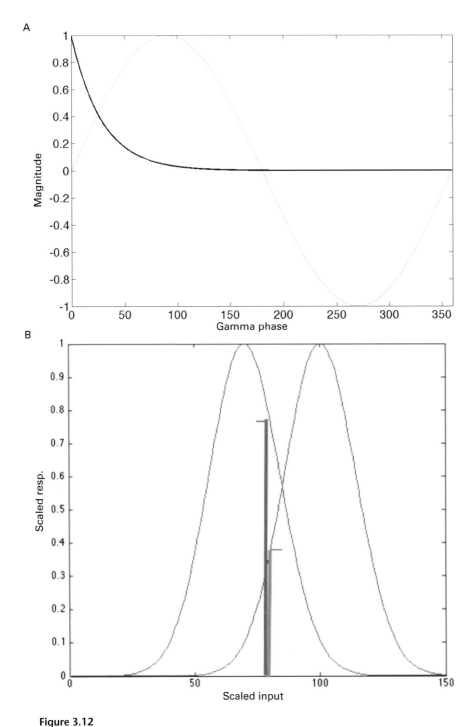

Figure 3.12

(A) A γ latency code uses the high-frequency γ range (40 to 90 Hz) to increase the communication rate. Individual spikes can signal a numerical quantity by coding it as an analog delay from the zero phase points of a γ frequency cycle. (B) To make the learning of receptive fields work, neurons with overlapping receptive fields have to be selected probabilistically with odds determined by the ratios of their overlap at the input coordinate.

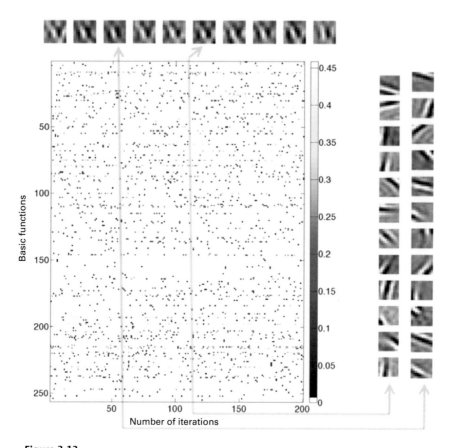

Figure 3.13

Responses of 256 model neurons when recoding a single 10×10 image repeatedly for 200 γ cycles. (Top) Reconstructions for 10 instances spaced 20 cycles apart show that they differ only slightly. (Center) Individual spikes communicate numerical values by a latency to a γ phase. Owing to the probabilistic nature of cell selection, at each cycle different neurons are selected and send a spike. The latency of each spike is denoted by a color. (Right) Basis functions for two coding cycles show that their sets of 12 neurons sending spikes vary.

This way of coding information has increasing evidence[59–61] but at this point remains speculative. Nonetheless, it has the advantages that:

- it is 1,000 times more compact than rate coding,
- it explains why randomness is seen in a spike train,
- it is compatible with spike timing dependent learning,
- it is significantly faster than rate coding, and
- it allows different computations to use the same neuron as long as they use different γ-frequency references

Alternative methods would have to address all these issues.

3.5 Reflexive Behaviors

The simplest behavior we have is the reflex circuit that links sensation to action. Tapping the knee in the vicinity of the patella will cause the knee to jerk forward suddenly. This response does not need the forebrain at all. The sensors that detect the mild tap are connected almost directly to neurons in the spinal cord that innervate the muscles that do the retracting. Sherrington's discovery of the reflex arc was enormously influential and provided a way of thinking about more complicated circuitry that did involve the forebrain. Sensory neurons analyzed the stimulus and extracted features. These neurons in turn were connected to motor neurons that triggered an appropriate response. The feature neurons in this line of thinking were anticipated in the experiments on frogs by neuroscience's pioneer, Adrian:

I had arranged electrodes on the optic nerve of a toad in connexion with some experiments on the retina. The room was nearly dark and I was puzzled to hear repeated noises in the loudspeaker attached to the amplifier, noises indicating that a great deal of impulse activity was going on. It was not until I compared the noises with my own movements around the room that I realized I was in the field of vision of the toad's eye and that it was signalling what I was doing ...[62]

Adrian's observation famously was made more concrete by Lettvin's experiments on the frog's retina that revealed neurons that responded selectively to dark, moving spots.[63] The spots were just the right feature to describe moving insects. These and other experiments have shown that a frog has rather stereotypical behaviors that use features that are directly computed in the retinas. Such behaviors are so regular that they can be effectively modeled. Figure 3.14 shows the flow diagram for Cobas and Arbib's model.[64] The early sensory neurons provide an estimate of the bearing of potential prey in terms of a distance and heading. If the distance is too

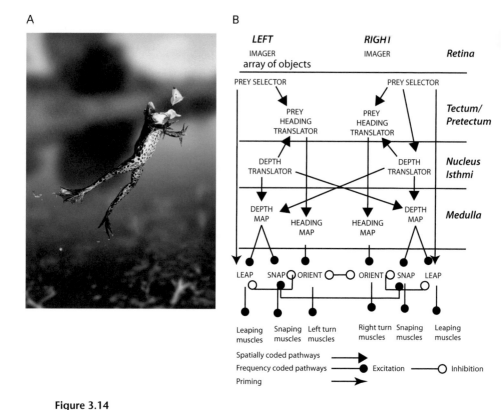

Figure 3.14

A frog has a regimented prey-finding strategy that is modeled in this circuit diagram. The centerpiece is a reflex: If neurons in the visual field signal a moving object at the right size and depth, the frog snaps its tongue out to get it. If the prey is too far away, the frog will approach it. If the prey is out of reach of the tongue, the frog will orient toward it. (Left) Kim Taylor/Caters News. (Right) Redrawn from Alberto Cobas and Michael Arbib, "Prey-catching and predator-avoidance in frog and toad: Defining the schemas," *Journal of Theoretical Biology*, vol. 157, Copyright 1992, with permission from Elsevier.

far, the frog will advance. If the heading is not appropriate for the frog's tongue-snapping reflex, the frog will orient itself, and finally, if both those conditions are met, a snap will occur.

The computation of features at the beginning of the circuitry occurs in other animals, too. A rabbit's retinas contain neurons that compute features that measure three-dimensional rotation. These circuits also include inputs from the rabbit's vestibular system, which is measuring head accelerations. The objective of the circuit is to make sure that the rabbit keeps very still under certain circumstances so as to be less detectable by predators. Note that these circuits are an advance over the more basic reflexes such as the patella response. In that kind of reflex, the response is hardwired in the spinal cord, and there is no sophisticated differentiation of the stimulus. In contrast, the frog behaviors, although still primitive, reveal a more advanced level of processing. In snapping, the coded feature neurons can select a candidate target stimulus that might be food compared to similar but less effective stimuli. But even though there is an increase in sophistication, there are many steps to go before approaching the repertoire of mammalian forebrain behaviors. The next step up is to introduce some modifiability in the circuitry.

Learnable Behaviors

When we study primates, we find that the neurons that are representing these kinds of features are computed by more central circuitry in the cortex. Why is this so? One idea is that in the cortex, the features can be used in many different ways by combining them flexibly in programs, whereas when they are found in the retinas as in other species, they cannot be flexibly combined. But at any rate, they can still be used in the role of feature detectors in decision making. A spectacular demonstration of this was done by Newsome, who trained monkeys to make decisions based on moving dot stimuli. Neurons that do the frog's job of detecting motion are found several synapses into the cortex in a cortical region called the medial temporal cortex, or simply MT. Monkeys were trained to signal with eye movements whether the dots in a pattern of moving dots were moving up or down. Newsome laced the stimulus with randomly moving dots and varied the percentage of them to make the problem successively more difficult. His data showing the different spike rates were shown in figure 3.10.

Because spike rates are low, experimenters use *histograms* to sum up the results of many different trials and thus emphasize small differences. Thus, the upper two panels in figure 3.10 show individual trials with black dots indicating spikes and then the sum of these trials immediately below.

Strong motion differences elicit very pronounced differences in spike rates, but very weak differences still show up as well.

The monkeys made successively more errors with the more difficult patterns. Amazingly, when Newsome recorded the responses of neurons in a motion-sensitive part of the cortex, the individual neurons had firing rates that increased in a way that paralleled the success of the monkey on patterns in the task. In a coup de grâce of experimental artistry, stimulating a "down motion" neuron when the pattern was up motion so that it sent out more spikes made the monkey reverse its normal decision.

While on the surface, it appears that the motion neurons in area MT are performing the same kind of function as those in the frog's retina, there is a very important difference in that the cortical neurons were trained to do this job. The circuitry for making decisions about the particular representation of motion on the stimulus screen was not prewired but instead was formed, presumably by changing synaptic strengths, into one that would do the job. Given the vast array of different neurons in the cortex, this experiment hints at the huge flexibility the cortex has in creating different states that can be used by programs. How that might happen is the subject of the next chapter.

3.6 Summary

1. Neuron spike codes in the forebrain are specialized in different systems:
• θ-phase codes in the hippocampus delineate task spans,
• pauses in tonically active cells in the basal ganglia delineate sub-task boundaries,
• low spike rates in the cortex are puzzling but they may allow multi-tasking whereby neurons can participate in different active memories.
2. Neuron receptive fields in the cortex can be learned from experience and are sensitive to the statistics of natural stimuli.
3. For the most important part of the forebrain, the cortical memory, the exact signaling strategy is still under study. The rate code is a model that has good explanatory properties, but timing codes are receiving intensive study. One such avenue explores the use of a phase code in the g-frequency range to allow a spike to signal a scale quantity.
4. The feedforward connections form the input to internal cells, such as those in the cortex, construct abstract patterns from the inputs. These patterns can be thought of as codes that allow the specification of the input

very concisely in the cortical memory, so that only a very limited number of cells can express any particular pattern.

5. The low number of neurons can be taken as evidence that the cortical memory places a premium on compact codes. This emphasis on compactness may reflect metabolic costs of which the transmission of spikes is the major fraction.

6. Without a forebrain behaviors can still be sophisticated as in that of a frog leaping to catch a flying prey, but the repertoire and adaptability of such behaviors is limited.

7. In contrast, the mammalian cortex is distinguished by exhibiting enormous plasticity in the learning of arbitrary behaviors. Moreover the cortex stands by having an enormous capacity to remember vast numbers of such behaviors.

3.7   Appendix: Neuron Basics

A nerve cell has three main functional components, its *dendrites*, its *soma* or body, and its *axon*. The dendrites are a tree-like structure that form the main area where axons of other cells make connections. The connections are anatomically specialized structures termed synapses (figure 3.15). The synapse is the junction that allows one cell to affect another. The connecting cells alter the charge balance on the dendrites of a cell that they connect to. If this change is significant, the soma will send a voltage signal down the length of the cell's axon to be communicated with the cells with which it connects. The axonal distances can be enormous when measured by the scale of a standard cell's central component, its soma. The soma is about 10 μm (millionths of a meter) across, but if you imagined it to be the size of a marble, then the axon of a typical pyramidal cell would extend for 10 m, or about 30 feet. Furthermore, the axon bifurcates repeatedly until it has approximately 10,000 terminals, or *butons*, which connect to other neurons by making synapses with dendritic *spines*.

As figure 3.16 shows, the combination of axons and dendrites makes up about 60% of the cellular material by volume.[65] Taking just the axons alone, if you think of each axonal segment as a small "wire," then you have so much of such wire in your head that if you attached all the separate segments end on end, their combined length would be sufficient to get to the Moon and back. This remarkable observation is reinforced by figure 3.17, which shows the main axonal pathways obtained with diffusion tensor imaging. Clearly, the connections between cells are key.

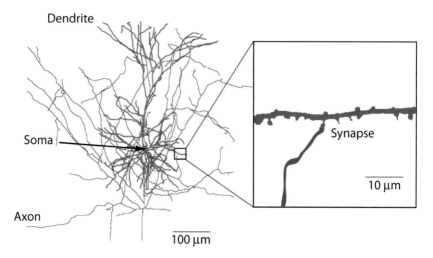

Figure 3.15
When the axon (blue) arrives at a target cell (red), its terminal buton encounters one of the cell's dendrites where it makes a connection. The connection is made at the dendrite's spine, a microextrusion that provides a site for the connection. As shown, the collection of dendrites is peppered with synaptic spines to accommodate the thousands of incoming signals. From D. B. Chklovskii, "Synaptic connectivity and neuronal morphology: Two sides of the same coin," *Neuron*, vol. 43, p. 5, 2004.

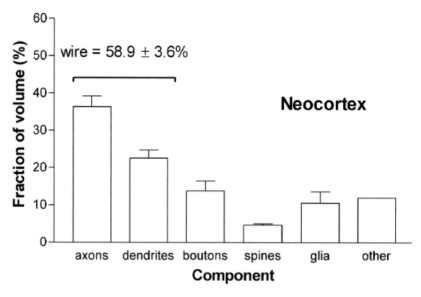

Figure 3.16
If we add up the volume of the neuron that is extruded in the form of the axon and the dendrites, then the brain is mostly "wire." From D. B. Chklovskii et al.,"Wiring optimization in cortical circuits," *Neuron*, vol. 34, p. 3, Copyright 2002, with permission from Elsevier.

Figure 3.17
Diffusion tensor imaging exploits the fact that small molecules can move slightly under magnetic fields to track the orientation on axonal fibers. Red-blue shows vertical pathways. Green-yellow shows horizontal pathways. These primitives represent neural "buses." From the NIH Human Connectome Project.

Total wire length is just one of several informative neuronal features, which are summarized in table 3.1. The total number of neurons in your brain is estimated to be of the order 10^{11}. Taking this number together with the figure of 10,000 connections per cell gives us 10^{15} connections total. While this number is formidable, it is an extremely small fraction in terms of the 10^{22} synapses it would take to have every neuron connected to every other neuron. Thus, the nerve cells working together form a very sparsely connected system, and the connections are the major component that dictates how the brain will process information. So much so, that scientists working on large-scale models of neural networks are characterized as *connectionists*. The other shocking feature in the table is the slow speed of axonal propagation, a far cry from silicon circuits where the speed of propagation is of the order 3×10^7 m/s.

Table 3.1
Important neuron features

Feature	Value
Number of neurons	10^{11}
Average connections per neuron	10^4
Total connections	10^{15}
Speed of communication	1~100 m/s
Average signaling rate	10 spikes/s
Total axonal length	10^6 km

Note: The large range of spike propagation speeds results from the fact that a neuron's axon may be unmyelinated, or myelinated, which greatly improves the propagation speed. In addition, the diameter of the axon may vary, and larger diameters result in faster speeds.

Our focus is the neuron and its interconnections in circuits, each of which in itself is also a complex venue. In the first place, all the different forebrain subsystems that were introduced earlier have their own very characteristic circuit. Typically, the circuit in each individual area is like a signature for that area. And each characteristic circuit is replicated millions of times within its area. Thus, if the basic circuit is understood, then enormous progress could be made. However, these circuits are complicated, and each one implements a number of different functions. So they are not just doing one thing but several things. Just to guess what some of these functions might be requires a lot of basic reasoning on the part of the reverse engineer. An indication that it's a very hard problem is that none of the circuits in the subsystems has been satisfactorily deciphered yet even though a lot is known about them. However, there has been recent promising work done in the cortex that we have focused on to give you an introduction to the sleuthing process.

Neuron Outputs: The Action Potential

Nerve cells in the brain communicate via voltage spikes. The generation of those spikes is a complicated process that was worth a Nobel Prize for Hodgkin and Huxley, who described the first generally useful model.[66] Here, we will present it in caricature. To set things up, the membrane of the cell has special pumps for ions that move sodium ions, Na^+, out of the cell and potassium ions, K^+, into the cell. When they equilibrate, there is a charge imbalance, so the net result is that the membrane potential is at −72 millivolts (mV). The ions would like to diffuse across

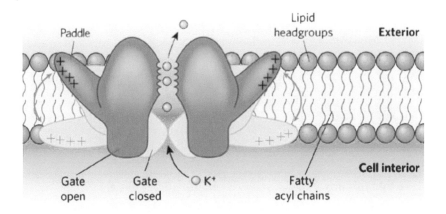

Figure 3.18
An abstract characterization of a voltage-gated potassium channel that sketches the channel protein embedded in the membrane, showing that the voltage levels can manipulate the transition between open and closed positions. Reprinted by permission from Macmillan Publishers Ltd: Anthony G. Lee, "Ion channels: A paddle in oil," *Nature*, vol. 444, p. 7120, copyright 2006.

the cell membrane to restore the balance, and they can, potentially, but only through orifices defined by folded proteins, termed *channels*. Figure 3.18 shows the structure of a channel designed to modulate potassium levels.

The spike generation channels have a special property in that they are voltage gated, meaning that they can be open or closed depending on the differential voltage across the cell membrane. At –72 mV, they are closed, but if enough other cells that are connected to a cell send excitatory spikes to its synapses, they can raise the potential to –55 mV, a rise that will open the cell's sodium channels, allowing the positively charged sodium ions to rush in. This initiates a runaway process that rapidly increases in membrane potential, causing more and more gates to open. The result is a rapid increase in voltage. This increase is very short-lived, as at more positive membrane potentials of greater than +40 mV, voltage-gated potassium channels open, allowing positively charged potassium ions to rush out, consequently rapidly lowering the potential and shutting off the process (figure 3.19). The process of converting ions to changes in potential is very efficient, so much so that a cell can send hundreds of spikes before taxing the pumps' abilities to restore ion concentration balances.

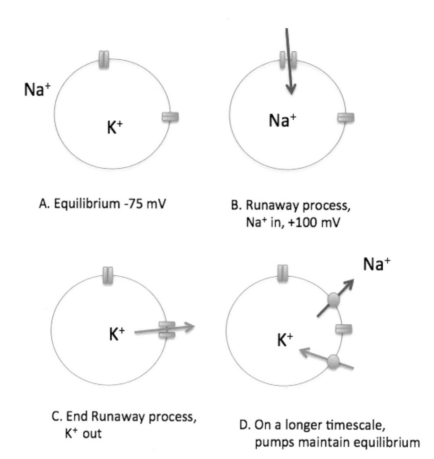

A. Equilibrium -75 mV

B. Runaway process,
Na+ in, +100 mV

C. End Runaway process,
K+ out

D. On a longer timescale,
pumps maintain equilibrium

Figure 3.19
A cell with a single representative channel illustrates the delicate process of spike
generation. (A) At equilibrium, charge repulsion and diffusion forces balance the
concentrations of ions in the cell, resulting in a net voltage potential of −72.5 mV.
(B) A spike arriving at a synapse opens up sodium gates on the recipient cell, causing
ions to flow in and raise the voltage potential. The rise causes more gates to open so
that the voltage increases rapidly. (C) The rising voltage opens potassium gates, causing positive potassium ions to rush out and lowering the voltage potential. (D) Ionic
pumps restore the concentrations of ions to their equilibrium state.

The rapid increase and decrease of the action potential, or voltage spike, is a neuron's fundamental way of signaling and is shown in figure 3.20. Although this profile's shape can vary slightly for different cell types, and this can be useful for identifying the type, given an individual potential, it is believed that recipient cells cannot distinguish these differences. Consequently, it is believed that all spikes are functionally identical so that in crude computational terms, a "one" is sent at a specific time.

Via mechanically gated channels in synapses, each axon connecting to another cell's dendrite is able to cause electrical charge to build up in the target cell. This triggers the runaway process just described that allows the target cell to produce a voltage pulse that has a typical amplitude of about 100 mV and a pulse width of about 1 millisecond. The pulse travels down the target cell's axon at about 10 m/s. When it comes to axonal branches, the pulse is divided and proceeds along each branch. The special construction of the axon allows it to propagate along all of the 10^4 branches. You could not do this with plain electrical wire as the capacitance of the wire would attenuate the signal too drastically, but biology does not have this problem as it is specially designed to propagate the spike actively. The speed of 10 m/s might not sound like much, but when you consider that most of the brain is reachable in 10 cm, it will only take a millisecond to get the pulse to any particular new target neuron. Once it gets there, it triggers a chemical process (that we discuss next), and the process repeats. For a long time it was thought that this process of dumping charge on the target neuron was deterministic, but some recent research suggests that it may be a probabilistic process.

Neuron Inputs: Synapses
Now for the cell's input, which is handled at synapses. The basic abstract model of a synapse is simple, in that it provides an electrical connection between cells. However, the chemical machinery that creates, maintains, and manages it is very complex. As you know now, the axons of a cell may have to go on extraordinary journeys to find their targets. They do this during a developmental period by following chemical gradients like bloodhounds following a scent. They can use a mixture of molecules for complicated steering and then stop when they identify markers on the surface of a target cell. Once they stop, there are specialized processes including *glial cells* to guide the manufacture of a synapse. They do this by using a combination of a special protein together with a measure of the electrical activity between the incoming axon and its potential target cell. It's a bit of a chicken and egg problem, but if the electrical activity of the two are

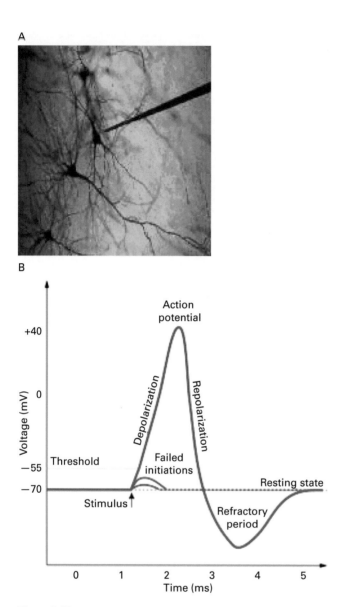

Figure 3.20
(A) Single neuron recording techniques allow the measurement of the basic signaling method of a neuron—the action potential. If a small wire can be advanced sufficiently near the soma of a neuron, the voltage changes in the cell body can be measured. (B) A schematized version of the voltage spike, or action potential, created by adding charge to a neuron. At threshold, the soma suddenly depolarizes or increases in voltage and then just as suddenly reverses itself. From Wikimedia Commons/Creative Commons Share Alike.

related, then the glial cell can build a more permanent connection between them in the form of a synapse.

Synapses can be of two forms: *excitatory* where positive charge is added or *inhibitory* where, depending on the type of synapse, negative charge is added or positive charge is taken away. A single neuron can only have synapses of one or the other type.

Figure 3.21A is a very-high-magnification electron micrograph that shows a synapse and some glial cells. What you see is a beautiful picture of the cross section of a dendrite (de) that has a raised spine (sp) to make contact with an incoming axon terminal (bo). To the side is a glial cell (gl). When the spike arrives at the terminal, it triggers the release of chemical neurotransmitters in the synaptic cleft (indicated by the dark patch between the terminal and spine): these migrate to the spine and mechanically open channels to make small holes in the spine. The outside of the neuron is relatively positively charged, so that when these holes open, charged ions rush in and raise the soma's voltage potential. The rise in the potential, if it's some combination of fast enough and large enough, will trigger a spike via the runaway mechanism just described.

How Neurons Learn: The Hebb Rule

At this point, you have a crude picture of a neuron's basic operation. Inputs at synapses result in charge flowing into the cell; this raises the membrane potential, and if the increase is sufficiently large, a runaway process is triggered that results in an action potential, or colloquially, a spike. What has not been discussed so far is how the circuit can be modified. One straightforward way to change a circuit would be to make its firing patterns more likely. This could be done if the effect of an incoming axonal spike could be modulated. In 1949, Donald Hebb proposed that if the input spike preceded the output spike by a small amount, the cell might be able to increase the strength of that synapse in a way that would make the effect of future spikes larger. This is the famous Hebb rule, taken for granted in many neural computational models by having the effect of the input spike and synaptic strength be multiplicative. In addition, over the past few decades more and more evidence has accumulated that the timing between the incoming and outgoing spike is vitally important.[67]

The importance of timing is elegantly illustrated in an experiment by Bi and Poo[68] working with the hippocampus. The hippocampus is special in that an excised circuit can be kept functional in a dish, which facilitates complex recording strategies. Thus, they were able to simultaneously monitor the input to a synapse together with the cell's spike generations. This

A

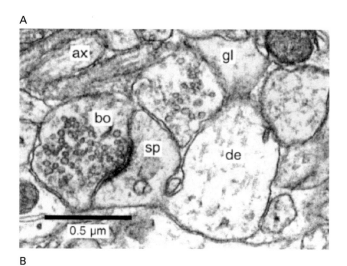

B

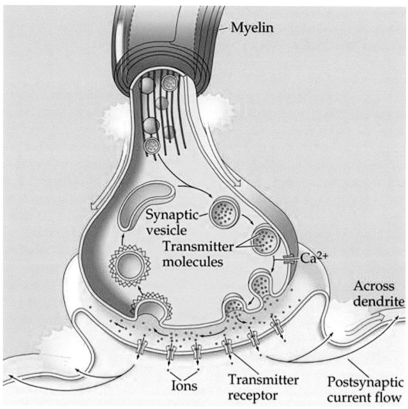

Figure 3.21
(A) A colorized electron micrograph shows dramatically that neural circuitry is tightly packed. The different components are as follows: axon (ax), buton (bo), spine (sp), dendrite (de), glial cell (gl). (B) Schematic of the operation of a synapse. A spike arriving at the end of an axon causes vesicles containing a neurotransmitter molecule such as acetylcholine to be released into the tiny space between the cells. It quickly migrates across and binds with an ion channel. The binding process changes the shape of the channel protein, literally prying it open so that ions can get inside the cell, raising the voltage potential of a synapse. (A) from Dmitri B. Chklovskii, Thomas Schikorski, and Charles F. Stevens, "Wiring optimization in cortical circuits," *Neuron*, vol. 34, p. 3, Copyright 2002, with permission from Elsevier. (B) Reprinted with permission from *Neuroscience*, 3rd edition by Purves et al., 2004. Copyright Sinauer Associates.

allowed them to characterize in exquisite detail just how a synapse's efficacy depended crucially on the relative timing between the input and output spikes. If the input spike preceded the output spike by a small amount, the effect was large and positive. But if the input spike arrived behind the output spike, the effect was large and negative. Figure 3.22 shows their result.

It turns out, however, that just spike timing is not enough of an explanation. There needs to be a mechanism for strengthening synapses. This is done through a fancy ion channel termed an N-methyl-D-aspartate (NMDA) receptor, which is both ligand gated and voltage sensitive.[a] At low levels of excitation, the ligand glutamate pries it open but cannot remove a magnesium plug. But at high levels of excitation, the voltage is significant to pop the magnesium ion "cork" and let calcium ions (Ca^{2+}) into the cell. These initiate a complicated chain of events that results in more conventional (AMPA) channels being formed, thus increasing the efficacy of the synapse.

At this point, the major neurotransmitters have been introduced, and their functions are summarized in table 3.2. The transmission of the spike, both its generation at the soma and propagation down the axonal tree, is handled by voltage-gated sodium and potassium channels. The spike generation process is triggered by ligand-gated channels that let positive ions into the cell, but it can be blocked by a complementary set of channels that we did not focus on, which work against this process by letting negatively charged chloride ions into the cell. The learning process whereby additional channels are created at the synapse is controlled by special NMDA channels that work by letting calcium into the cell. Finally, the overall modulation of the circuits that the cell is part of is handled by another

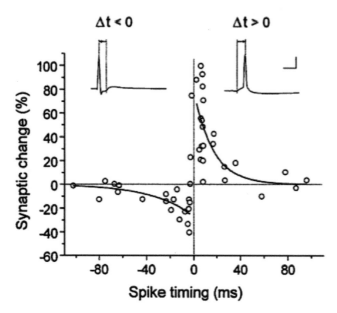

Spike timing (ms)

Figure 3.22

Effects of spike timing on synaptic efficacy. In this ingenious experiment, the change of the synaptic strength in response to variations in timing between the input spike to a synapse and a resultant output spike was measured. If the output spike almost immediately followed the input spike, then the change was large and positive. If the output spike preceded the input spike, the change was negative. From Guo-Qiang Bi and Mu-ming Poo, "Synaptic modifications in cultured hippocampal neurons: Dependence on spike timing, synaptic strength, and postsynaptic cell type," *Journal of Neuroscience*, vol. 18, p. 10470, 1988.

Table 3.2

Neurotransmitter summary

Function	Neurotransmitters
Spike transmission	Sodium ions, potassium ions
Spike excitatory input processing	Ligand gates—glutamate
Spike inhibitory input processing	Ligand gates—gamma-aminobutyric acid
Learning	Calcium, NMDA channels
Reward	Monoamine channels (e.g., dopamine)

chemical set of channels, the monoamines, which include norepinephrine, epinephrine, histamine, dopamine, and serotonin. Of the monoamines, dopamine would be the most important for regulating large-scale behavior, as discussed in the previous chapter.

Without understanding what computation the spikes are used for, one can still paint a very abstract picture. Incoming spikes to a cell may trigger outgoing spikes, and, in all likelihood, timing relationships among spikes, as well as selective neurotransmitter modulations, are used to steer the neural circuits into useful configurations. This whole process takes up a huge portion of the cell's metabolic energy, as shown by figure 3.23, and as the brain's portion of the total body weight is approximately 3% and can use 20% of the body's metabolism, it is likely that the circuits are adjusted to use fewer spikes where they can.

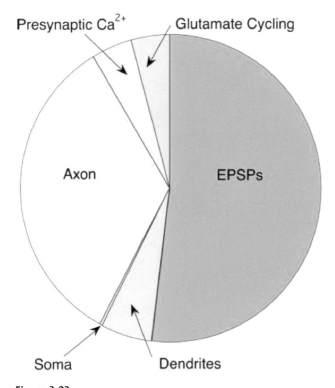

Figure 3.23
Sending a spike is very expensive for a cell (see ref. 54). On the right, we see that the excitatory postsynaptic potentials (EPSPs)—read spikes—consume more than half a neuron's energy. Reprinted from Peter Lennie, "The cost of cortical computation," *Current Biology*, vol. 13, p. 6, copyright 2003, with permission from Elsevier.

Appendix Summary

1. Cell ionic transport is increasingly well understood. Different neurotransmitters are used for a neuron's different functions. Voltage-gated sodium and potassium channels govern spike generation. Ligand channels manage spike information transmission across a synapse. Dopamine, serotonin, and other neurotransmitters modulate the state of a neuron. Calcium and N-methyl-D-aspartate (NMDA) modulate the consolidation of synapse modification in learning.

2. Synapses are constructed by the action of glial cells. The course of learning is a very active process in which synaptic connections are made and broken.

3. Synapses are the key to programs. As modifiable connections between cells, they are the basis for programming.

4. Spikes are the primary long-distance signaling method, and they can be used by different circuits in different ways.

4 Cortical Memory

At any given time in the awake organism, a widely distributed and changing representational network is active in the neocortex. That is active memory. Its extent and topography are determined by multiple factors, including the present and recent sensory inputs, the neuronal assemblies that, by association (as a result of prior experience), are activated by those inputs, and the drive or motivation prevalent at the time.

—Joaquín M. Fuster, *Memory in the Cerebral Cortex* (MIT Press, 1995, pp. 4–5)

One of the most difficult things to come to grips with is that the basic machinery of the brain seems to work in a way that is nothing like what we might like to infer from our conscious experience. This is particularly true for the cortex, which we posit is the site for remembered programs. Later when we develop the learning of programs that use cortical memory as a component, it will become easier to relate cortical memory to our everyday ideas about memory, but here the task is to dissect the cortex's functionality to expose its special structure.

Now is a good time to revisit the important analogy that will start us on the right track for this job, and that is that of a search engine such as Google. Just as billions of Web pages are indexed by a search engine based on importance as measured in a composite scoring, in the same way the cortex can be thought of as containing codes for representing billions of useful microexperiences that are categorized according to what next action the body should take. A particular momentary experience is the analogy of a Web query in that the moment situates the brain's internal state with its external surround. It is used to "look up" the right internal description, which is indexed with appropriate courses of action for that venue. At this level of abstraction, then, a brain "program" can be thought of as following a set of "links" from the cortical "Web pages."

Following links from a Web page is an excellent metaphor as it conveys the right image in two important ways. In the first, the memory has to have a description of the spatiotemporal state of affairs, which must include both the current state, the page, and the actions that can be taken from the state—the links. In the second, the memory has to have a time constant associated with its retrieval of this information, which is necessarily short, but not too short. The latter idea needs some elaboration. The brain's body has to interact with the world, and that interaction is conceptually not continuous, but divided into discrete chunks of experience. In reaching and grasping an object, there is the reaching stage where the arm is extended toward the target object followed by the grasp. In looking for the object to grasp, discrete fixations examine the scene. While these actions are not clocked, nonetheless they have a characteristic time constant in the hundreds of milliseconds range. Eye fixations are very representative as their modal distribution centers around 200 to 300 milliseconds. Just to be concrete, we will use 300 milliseconds as a representative time, acknowledging that the actual time is a distribution about this value.

In concluding this introduction, let's recall that the cortex is part of a cortex–basal ganglia–thalamus loop. Thus, to play its role in this loop successfully, it has to retrieve program fragments repeatedly using just a few hundred milliseconds for each retrieval. While how it is able to do this is far from settled, an enormous amount of information is known about its structure and constraints. Let's examine these elements from a computational perspective.

4.1   Table Lookup Strategies

The cortical memory has a huge job to do in its Web page–finding instant. During its 300-millisecond main cycle, it may have to read an emotional expression, size up a scene for danger, or retrieve the motor commands for clasping a door handle. From the perspective of its millisecond spike generation circuitry, these operations must be done extraordinarily fast. Consider that each neuron in the cortex typically communicates at the rate of about 10 to 50 spikes per second. This means that in 300 milliseconds, there is only time to send 3 to 15 spikes. From this cursory analysis, it's obvious that the answer cannot be computed from scratch, but instead has to be already in the memory in some form and looked up. For this reason, we say that the cortical memory uses a *table lookup* strategy. Moment-to-moment experience is coded in a table at a microlevel.

It is useful to compare the cortical table lookup strategy to one with which you are already familiar. Imagine looking up a telephone number in an old-style telephone book. In the phone book, all the numbers are precomputed and indexed by surnames. If you have done this before, no doubt you will use a divide-and-conquer strategy. You'll divide the book into two parts, determine which of the two your surname is in, and repeat the process on that portion of the book. With this strategy, formally the number of operations needed is proportional to the logarithm of the number of names. With a 1,000-page book, it takes an average of about 5 tries to get to the right page. Plus, it's important that you have the surname handy. You will be at a loss if you only have the street address.

The cortex's lookup strategy differs from the phone book in two important ways. First of all, it has to be even faster that your divide-and-conquer strategy. It has to take a few tries independently of the size of the entries. In computer science formal jargon, this is termed *constant time*. The other improvement the cortex incorporates is that it will work for any index or fragment of an index. So notionally if you only remember part of the street address and something about the surname, it may be able to complete the entry for you. This property is termed *content addressable memory*, or *CAM*, and is completely unlike silicon memory. Conventional silicon memory is divided into two parts: addresses and contents. For example, if a program line was "x:= 3" (read: "set the variable *x* to value 3"), one would expect to find an address allocated for the variable "x" and its contents would contain the binary code for "3." Content addressable memory has no such distinction. Of course, the silicon structure has some advantages, one being that if one wants to refer abstractly to a body of data, as in "the bee that is chasing me," that is easy to do. For a CAM, additional machinery is required. But we will postpone dealing with this issue until later chapters. For now, let us see what a CAM does well.

Figure 4.1 shows one of the first simulations of a CAM model from Hertz and colleagues.[68] There have been a lot of additional developments since this formulation, but it remains a beautiful example for highlighting the essential concepts and placing subsequent developments in relief. In this model, the image pixels are abstract model neurons that can be in one of two states: −1 (black) or +1 (white). Their values start in some random initial state where only some values are correct and then the connections between the neurons drive the states to recover a stored pattern that has the most in common with the initial state pattern. In this model, the connections are extensive; every neuron makes a connection with every other neuron. This is a very oversimplified and abstract model of what the cortex can do and is incorrect because the cortex does not store pictures in any literal sense. In fact, it is better to think of the image simply as a pattern of binary bits that could stand for anything. Nonetheless, the figure still depicts an important

Figure 4.1
The fundamental idea behind a content addressable memory illustrated with three examples that evolve in time horizontally from left to right. Each image is conceptually composed of a large array of neurons whose states are illustrated by pixels and are either firing (+1 or white) or off (–1 or black). Given a set of patterns, each neuron can connect to all the others with synaptic weights that are pattern dependent. The initial condition causes some neurons to fire and others to remain off. This in turn causes the neurons to change state at the subsequent time step. Owing to the way the synapse strengths have been picked, the patterns that were coded via synapses are recovered. Copyright © 1991 John A. Hertz, Anders S. Krogh, Richard G. Palmer. Reproduced by permission of the Perseus Books Group.

aspect of the CAM simulation as it illustrates the course of the dynamics involved in a table look-up. The cortex has vast numbers of these bit patterns precategorized, and if the input does not correspond to any one of these exactly, the dynamics can quickly recover the closest match.

Content addressable memory is a wonderful concept. If you have a stored pattern, you can retrieve it by specifying some of the parts of the pattern. The wonderful part is that the exact parts do not matter, you just have to get reasonably close to your pattern. You smell a fruit pie and it reminds you of your grandmother's home. The idea would be that the whole complex pie-grandmother-home pattern was stored as a unit and can be thus recovered by specifying some arbitrary bits and pieces. The secret is in the connections between different neurons. A particular pattern is stored by adding small increments or decrements to the strength values of all the synapses. The effect of this storage strategy is to allow mental "pictures" to be retrieved just by specifying some of the correct values initially. Once this is done, connections in the memory neurons are able to complete the pattern quickly. The values that are correct influence other neurons and are able to turn them on to their correct values. This process repeats until all the model neurons have correct values.

Content Addressable Memory

Content addressable memory is such an important concept that we will enflesh the mechanics behind the example. To do this, we will use an abstraction of a neuron as representing a memory bit. This abstraction needs some additional structure to set the value of the bit. The model of a neuron that is used for this purpose is a slight variant of the one from the previous chapter and is still much simpler than a real neuron, but we'll use the term *neuron* anyway for simplicity. A large number of interconnected units is a *network*. The state of a unit will be discrete, either 1 or –1. In other chapters (e.g., chapter 3), an analog state of a real number between 0 and 1 can be very important, but here the ±1 state simplifies the analysis. Recall that real neurons can represent negative numbers in at least two different ways. One is to have a baseline firing rate mean "zero," but the method used by cortical neurons, as previously mentioned, is to use two neurons to represent the signal, one for the positive part and one for the negative part).

The key parameters that define the memories are the weights that model synaptic connections between units. Biological synapses are very complicated, so modeling them with a single real number is a gross simplification. The goal is to specify the weights so that the changes in the state vector can function as a memory. The weight w_{ij} between unit j and unit i will be a signed real number.

Another helpful thing to do is to have a notation to summarize the entire collection of neurons' momentary firing patterns. As you know, we can do this by describing the state of the network in terms of a state *vector* **x**. This just has the states of individual units as components; that is, for N units, $\mathbf{x} = (x_1, x_2, x_3, ..., x_N)$. All we have done is designate a single symbol to cover the entire state of the network. This coding scheme is illustrated in figure 4.1 where each panel is composed of a collection of N neurons that are either OFF (–1, black) or ON (+1, white). As time evolves, the panels in figure 4.1 have a different state vector for each of the time steps. You just viewed one of the classical simulations of this in figure 4.1. By setting the individual pixel values appropriately to one of two values, we can create all nine of the images shown.

Creating the time evolution of those three examples requires a way of changing the state vector at each point in discrete time; that is, $\mathbf{x}(t)$. The first equation one might think of using is to make the state dependent on the product of the input and synaptic weights as we have done before:

$$x_i(t + 1) = \mathbf{w}_i \cdot \mathbf{x}(t). \tag{4.1}$$

The problem with this equation is that x_i in this case could grow arbitrarily large. To restrict it to be either ±1, let us use a limiting function:

$$g(u) = \begin{cases} +1 & u \geq 0 \\ -1 & \text{otherwise} \end{cases}$$

$$x_i(t + 1) = g(\mathbf{w}_i \cdot \mathbf{x}(t)). \tag{4.2}$$

The remaining step is to specify a formula for the weights. This CAM formulation is due to Hopfield, whose key insights were that every unit should be connected to all the others,[70] and the sets of weights connecting the units should be symmetric; that is, $w_{ij} = w_{ji}$. In fact, symmetrically weighted, totally connected networks are known as Hopfield nets. It turns out that, for any given set of patterns to be stored, it is relatively easy to pick w_{ij} so that the network will act like a memory. Given a set of P patterns to be stored $\mathbf{x}_p$, $p = 1, ..., P$, the appropriate setting for the weights is given by

$$w_{ij} = \sum_{p=1}^{P} x_i^p x_j^p . \tag{4.3}$$

This is a variant of the Hebb rule that makes the weight strength proportional to the product of the firing rates of the two interconnected units. As introduced in the previous chapter, these types of rules are named after Donald Hebb, who first suggested that synaptic strengths might be determined by the correlation between presynaptic and postsynaptic neural firing patterns.[71]

To see the weight calculation details, let's pick three patterns to be stored in a memory. In this case $P = 3$, and let

$$\mathbf{x}^1 = (-1, 1, 1, -1, \ldots)^T$$

$$\mathbf{x}^2 = (1, 1, -1, -1, \ldots)^T$$

$$\mathbf{x}^3 = (-1, 1, -1, 1, \ldots)^T.$$

Now calculate one of the weights, for example, w_{23}:

$$w_{23} = x_2^1 x_3^1 + x_2^2 x_3^2 + x_2^3 x_3^3$$
$$= (1 \times 1) + (1 \times -1) + (1 \times -1) = -1.$$

Once the weights have been computed, the memory is ready for operation. The input layer is initialized to some state. The dynamics equation, equation (4.2), determines the state vector at subsequent time steps.

Is this going to work? Suppose that you had just one pattern and consider connection w_{23} as before. That connection's value is now just $x_2 x_3$. So if we started with the pattern, it would have its product $w_{23}x_3$ equal to $x_2 x_3 x_3$. But this is just x_2 because $x_3 x_3$ is always 1 no matter whether the value of x_3 is plus or minus. So the effect of the pattern is to deposit the sum of N copies of x_2 as "raw" input. And when the function g is finished with it, it will of course be whatever value x_2 was in the first place. All this is a long-winded way of saying that if the pattern starts out with each neuron in the correct state, it will stay put.

What about the case when there is noise in the pattern? Because noise is random, then sometimes they will accidentally have the correct value and sometimes the opposite. For a large value of N, the accidentally correct values will cancel the accidentally incorrect values, and the correct input will still result in the pattern being correct, usually after one step. You can now see what other patterns might do. They will have the effect of raising the chance that a particular neuron will have the incorrect value. Mathematically it can be shown, although we will not do it here, that as long as the number of patterns is small compared to N, even though the

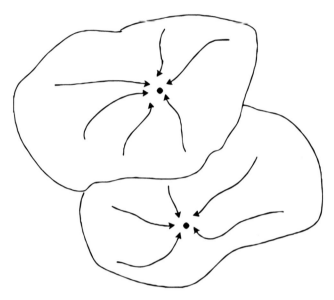

Figure 4.2
The abstract picture of content addressable memory depicts stored patterns as points where each point stands for an entire pattern. Nearby patterns are also points, but the neural dynamics of equation (4.2) is such that they will gravitate in discrete steps toward a stored point in the state space. The region defining the points that will move toward a particular point is referred to as that point's *basin of attraction*. The figure shows two such basins, but there will be many others, one for each pattern.

original pattern has lots of mistakes, as the examples in figure 4.1 do, they will converge to the correct pattern as the examples do. Mathematics has a succinct way of describing this: Remember that we can represent the entire remembered image as a point. Such a particular pattern vector, say $\mathbf{x}_0$, is called an *attractor*. Now imagine all the possible $\mathbf{x}$'s. There will be ones that are close enough to $\mathbf{x}_0$ so that when the equations are used, they will each gravitate to the stored pattern. The region that contains these $\mathbf{x}$ values is called the *basin of attraction* for $\mathbf{x}_0$ (figure 4.2).

The basin of attraction concept is a useful abstraction, but it is very important to draw the right lessons from it. For the cortex, a pattern will not be an image. If it were, then some mechanism would have to look at it, and we would be right back at the retina where we started! Each pattern will be some coding of a state. The state of course might refer to some aspects of an image, but in general it will have lots of other information— perhaps the food value of something in the image if you are hungry or a logical assessment of the prospects of catching it—at this point we cannot say exactly what might be represented, but in a moment as we examine the principal details of the cortical architecture, there will be lots of suggestions. However, it is important not to lose track of the CAM's main attractive property: its neural dynamics that can quickly compute a state from a vast number of nearby starting points. This computation has to be done in just a few hundreds of milliseconds as it is the platform from which the brain can then compute actions.

A lot has happened since Hertz et al.'s simulations of attractors, but the basic system is such a powerful illustration of the ability of an enormously high-dimensional system to be governed by a low-dimensional state that understanding its structure proves an enormously useful way of conceptualizing the operation of very large cortical networks. In addition, it serves as a springboard to appreciating the newer developments that we will touch on later in this chapter. One is the concept mentioned in the first chapter, and that is probabilistic computation. About the same time as Hopfield's CAM result, Pearl developed fast algorithms for probabilistic inference in networks representing conditional probabilities.[72] More recent work has extended his ideas to neural networks where the elements are individual neurons. Another development has tackled specific kinds of networks that have cascades of layers of units. These learning algorithms for these architectures suffered from being impractically slow and inaccurate. However, recent progress has successfully addressed these issues.[73] As the cortex has such a layered architecture, this is a welcome development, and it is to the cortex's elaborate layered architecture that we now turn.

4.2   The Cortical Map Concept

Chapter 3 showed examples of how the information in a small image patch could be coded. Owing to the need to save spikes, it seems to be important to the brain to sink its effort into coding the useful patterns that come up in natural environments, and these are a very small fraction of the possible patterns owing to the structure of the world. Implementing this strategy algorithmically, the first set of features that appear are small linear discontinuities such as those shown in figures 3.8B and 3.9 in chapter 3. This result represented an economical coding in terms of spikes, but only for a tiny visual area spanned by a small image patch. For the entire visual field, it turns out that the area of the cortex that receives the connections from the lateral geniculate nucleus (LGN), V1 (or striate cortex), is laid out in a *retinotopic map*. The map concept was introduced earlier, and here we will

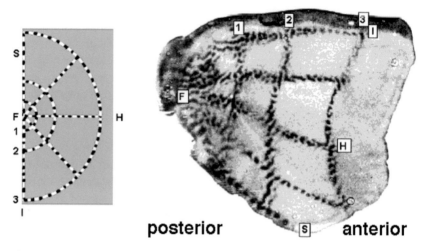

posterior   **anterior**

Figure 4.3

Tootell's classic demonstration of retinopathy in cortical area V1. An anesthetized monkey gazes fixedly at the blinking target shown in the left panel while being infused with a radioactive glucose tracer that marks active cells. Subsequently, an overlaying photographic plate is exposed to V1 and reveals the organization of the visual field as a retinotopic visual map on the cortex. The exaggerated sampling of the central foveal area is clearly seen (see ref. 74). From R. B. Tootell et al., "Deoxyglucose analysis of retinotopic organization in primate striate cortex," *Science*, vol. 218, p. 4575, 1982. Reprinted with permission from AAAS.

need to develop it in much greater detail. For vision, the visual field of the retinas is initially mapped onto a cortex with roughly each hemifield taking a slice of Van Essen's "pizza" that comprises visual area V1.

The reason for thinking of it as a map is as one traverses it while the animal is fixating, the responses of cells at a particular point on the map are sensitive to the structure arising from the corresponding portion of visual space. Figure 4.3 shows the results of Tootell's classic experiment that illustrates this retinotopic property.[74]

A monkey stares at a screen with a flickering radial pattern shown in the top segment. The monkey is anesthetized so that the eyes are fixed and centered on the target. At the same time, the monkey receives an injection of radioactively labeled 2-deoxyglucose that is rapidly taken up by spiking neurons. The monkey is sacrificed, and its V1 is put on a photographic plate. The right panel of figure 4.3 shows the developed image, which reveals bands of cells that sent large amounts of spikes. To gain an idea of the scale, it is estimated that there are 150,000 cells per cubic millimeter, so one is looking at a lot of cells in each darkened area.

A more detailed inspection of the map reveals another important feature of vision, and that is that the central area of the pattern is exaggerated in the cortical area. This of course reflects the fact that the central area of the visual field is sampled in the retinas by their high-resolution foveas. Thus, for vision, the brain allocates neural circuitry proportionately with image samples. It has been estimated that if we had brains that had foveal resolution over the whole visual field instead of the central 1°, the brain would have to weigh 300 pounds. It is easy to appreciate why it's a lot cheaper just to move the eyes!

Figure 4.3 shows the spatial layout of the neurons but does not reveal the characteristics of individual cells. However, we know from the previous chapter that the photometric patterns are coded. One such code is that of photometric orientation, and we saw how to build a circuit that would convert cells that responded to points of light into those that would respond to light and dark bars. Given that there could be many such possible orientations in a particular region of visual space, one might wonder just how these patterns are laid out on the map. This has been studied in great detail using various kinds of imaging techniques. These techniques allow one to reconcile two sets of information: the properties of cells' receptive fields and their spatial layout.

Figure 4.4A is made by using an optical imaging technique that takes advantage of the difference in reflectance between oxygenated and deoxygenated blood. By choosing oriented stimuli and recording the light coming from the surface of the cortex's area V1, the locations of cells responding to that orientation can be selectively mapped. This technique shows that (1) the properties vary across retinotopic space and (2) for each local area there is a collection of cells coding for different orientations. It turns out that there is a compromise between grouping the cells together based on orientation and based on position. This compromise is shown in figure 4.4A, which shows that cells of similar orientation tend to clump together, where at a much smaller scale the cells of similar orientation are color coded. So for small spatial areas indicated by the insets, all the orientations are represented. Figure 4.4B shows a much smaller scale testing *direction selectivity*, a property related to orientation. Oriented cells can be sensitive to motion perpendicular to their orientation in one of the two possible directions. The very large magnification shows that, in the cat, the line of demarcation between populations of cells that are sensitive to the different directions is very pronounced.[75]

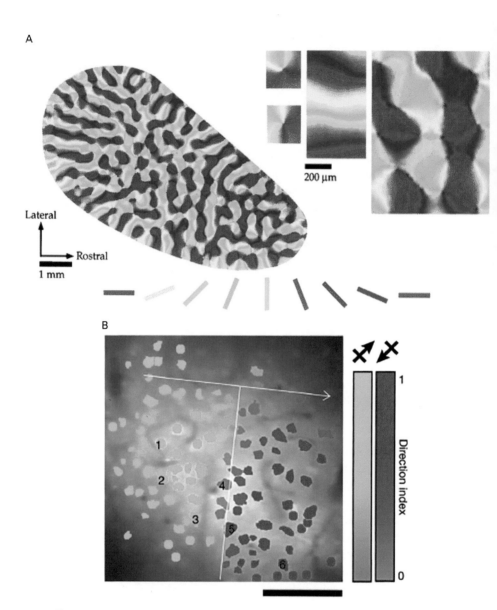

Figure 4.4

(A) The large-scale layout of cells sensitive to different orientations revealed by the voltage-sensitive dye technique. Inserts show various geometric features of the spatial microstructure. (B) A related feature to orientation is direction selectivity, which is the sensitivity of a cell to one of two directions of motion measured with respect to its orientation sensitivity. Here, Clay Reid's group has used a calcium imaging technique to delimit the individual somas of cells that have different selectivities. Reprinted by permission from Macmillan Publishers Ltd: K. Ohki et al., "Functional imaging with cellular resolution reveals precise microarchitecture in visual cortex," *Nature* vol. 433, p. 7026, copyright 2005.

Orientation and direction selectivity are just two of a handful of properties that are represented. The short list includes *ocular dominance*[76] (which
eye has the strongest connection) and *color*. Just by counting, you can
quickly see that there is a limit to the number of properties that can be
represented, because if there are too many, there would not be enough cells
to represent all the combinations.

4.3 Hierarchies of Maps

In the elementary CAM whose output was shown in figure 4.1, the different "neurons" representing the image all have the same status in that
they all receive direct image input. However, nothing could be further from
the way the real cortical neurons are organized. The tour of V1 provides a
motivation for what is coming next, as, via the explicit property enumera-
tion scheme, only a small number of such properties can be represented in
a fixed population of neurons. Many other properties are of interest, but
luckily it turns out that they can be seen as abstractions of these very basic
properties. Thus, it is important to refine the CAM concept to include the
notion of abstraction.

Motion Maps

Let's start with the visual measurement of movement. How can one detect
that an object in an image is in motion? The easiest way is to detect changes
in the individual image elements, termed *pixels*, as pixel values changing
over time can signal motion. The difficulty is that pixel changes do not
always signal motion but can also signal illumination changes, like the
lights turning on and off. It turns out that image motion can be separated
from illumination changes, but only after a certain amount of computation.

The first step in that process is the computation of *optic flow*. Optic flow
is in the form of an image: At every point in the visual field there is the
velocity of that point in space, but we only see part of the three-dimensional motion. The component along the ray of projection is lost. Nonetheless, the collection of remaining two-dimensional velocities form an image.
In the cortex, midway up the hierarchy, the collection of neurons that are
sampling this image form a retinotopic map. What this means is that if we
could stimulate the retina at some specific location with a small patch of
motion, then there would be a corresponding small group of cells in the
optic flow map that would respond (two examples of which are shown in
figure 3.10). For this reason the pure motion image, termed *optic flow*, is
an abstraction of the photometric change image in that temporal changes

have been factored out. This abstraction process can be repeated. Consider that the optic flow can represent many different kinds of motion from all the motion going in one direction to all the motion elements signaling a random direction. But now think of the optic flow fields that you might experience as you walk through the world. These are a very special subset of all possible flows. For example, if you look straight ahead and move straight ahead, all the flow vectors will point radially out of the center of your image. This is given vivid demonstration in movie fiction when the *Star Wars* spaceships jump to light speed. The particular patterns in the flow images that result from self-motion are useful in navigation, but apparently are extracted with a computational cost; registering their presence appears in a separate map.

The elaboration of the different forms of representation that are computable from the movement in an image is an abstraction hierarchy. Just to emphasize this point, the hierarchy is summarized in table 4.1. It turns out that this hierarchy is computed by successive retinotopic maps in the cortical memory and that these maps are interconnected in just the way that allows this to be accomplished.

In the same way, color has a similar hierarchy. If we take a look around indoors in daylight, we see a vivid impression of the colors of things, and it is easy to name their colors. Now suppose that we go and get a green light-bulb and illuminate the scene with the new bulb. We should still not only be able to name the colors correctly, but also we would perceive them more or less to be their normal colors. This is a remarkable achievement because now most of the light coming from the objects is green! The unraveling of the contributions of the differential reflection of the light spectrum by the surface and the spectrum of the illumination can be done but requires computation. Once the color of the surfaces of the object have been identified,

Table 4.1
A portion of the maps in the motion hierarchy

Self-motion	Cells respond to full visual field translations or rotations.
↕	
Optic flow	Cells respond to motion stimuli in a particular direction and speed range. Response can be modulated by motion in the surround.
↕	
Temporal change	Cells respond to temporal illumination change, such as switching the lights on and off, and to the illumination changes due to object motion.

Table 4.2
A portion of the maps in a possible color hierarchy

Color object labels	Cells respond to colored object features.
↕	
Color constancy	Cells respond to surface reflectance color and are insensitive to illumination color.
↕	
Opponent color image	Cells code color using an opponent color scheme; e.g., Red Center–Green Surround, Blue Center–Yellow Surround.

they can serve as a label for that object. The labeling level has not been pinned down in the cortex in the same way that the motion representations have, but it has been shown computationally that this level of representation can be particularly effective for multicolored objects.[77] But in any case, to agglomerate the amount of different colors of a multicolored object requires some computational bookkeeping. So the upshot is another albeit provisional hierarchy, shown in table 4.2. Like the motion hierarchy, these properties are to be computed in successive cortical retinotopic maps, although the descriptions of the color maps is not quite as settled.

The overall hierarchical organization of the cortex has been characterized by Van Essen's laboratory and others via a huge amount of painstaking anatomic neural tracings. To appreciate these findings, it helps to recall figure 3.6 in chapter 3, which introduced the layers of the cortex. Rather beautifully, these layers are grossly organized into layers I to III (processing), IV (input) and V to VI (output). It turns out that the cortical abstraction hierarchy can be intuited from the layer information in conjunction with the anatomic connection patterns between maps. In the more abstract of two connected maps, neurons in the processing layers and output layers communicate with neurons in a less abstract map by sending their axon terminals to that layer's levels I to II and V to VI. The less abstract map's I to III neurons communicate with the more abstract map's layer IV neurons (figure 4.5). This principle was a remarkable discovery: Just by studying patterns of connections between maps, the relative abstractions of their information could be ranked with respect to each other.

Figure 4.6 shows a flattened anatomic picture of the first two visual maps in cortex, V1 and V2, revealed by a cytochrome oxidase stain. Area V1 shows up as spotted and area V2 as striped. Each of these maps covers one hemifield of the visual space. The neurons in a specific location in one

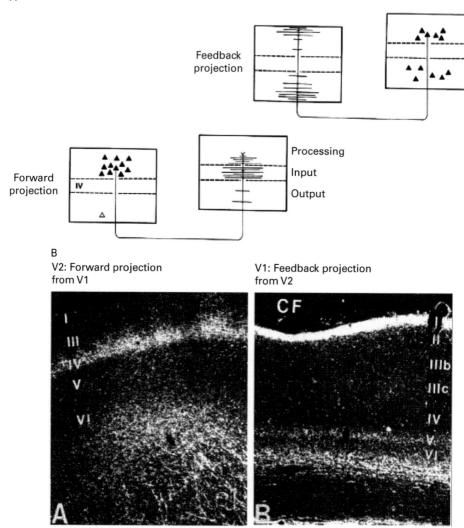

Figure 4.5

Slices of the cortical thickness show the anatomic rules that define cortical areas. (Bottom left) A schematic of the principle that shows how areas are related hierarchically. Layer II to III neurons in an area lower down in the abstraction hierarchy send their terminals to the upper area's layer IV. Going the other way, neurons in II to III and V to VI connect to the corresponding layers in the lower area. (Bottom right) These connections can be revealed by anatomic staining. Left: Reprinted from J. H. R. Maunsell and D. C. Van Essen, "The connections of the middle temporal visual area (MT) and their relationship to a cortical hierarchy in the macaque monkey," *Journal of Neuroscience*, vol. 3, no. 12, 1983, with permission from the author.

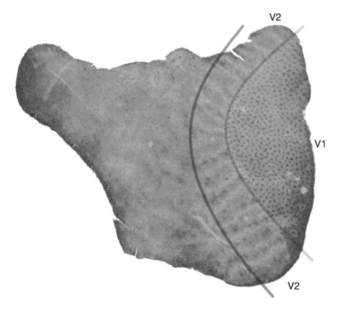

Figure 4.6
The first two visual maps in monkey visual cortex, stained with cytochrome oxidase, clearly indicate differences between areas V1 and V2. From R. B. Tootell et al., "Functional organization of the second cortical visual area in primates," *Science*, vol. 220, p. 4598, 1983. Reprinted with permission from AAAS.

map are reciprocally connected (via intermediate neurons) to neurons in the other map.

The general principle is that all of the cortex is organized into hierarchical maps: The two examples from vision that we described, color and motion, are the kinds of computational chains that are repeated for different sensory modalities such as audition and touch. One way of remembering this organization is to think of an old roll-down window shade. Such shades were on spring-loaded rollers and could be pulled down to block sunlight from entering a window. In this analogy, the hippocampus is the roller, and the cortex is the shade material itself. The overall hippocampus circuit organization is quite close to a rolled up sheet, so it is very cooperative in this analogy! At any rate, as depicted earlier in figure 2.19 of chapter 2, neurons near the roller represent very abstract concepts. The closer one moves to the lower edge, the more we approach the sensory-motor periphery and the more concrete the cell codes become. A detailed summary of the connections and areas for the macaque is shown in figure 4.7. Figure

A   B

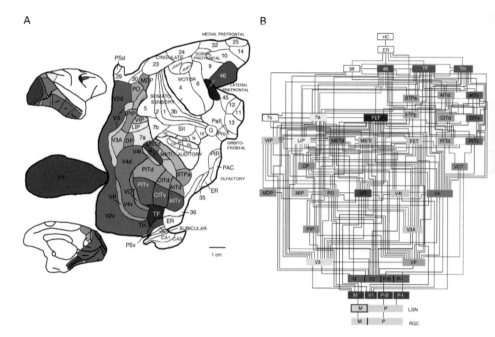

Figure 4.7
The Van Essen laboratory's tour de force summary of hierarchical connections
between visual areas in the macaque monkey cortex. Symbols in colored boxes
denote complete visual hemifield maps. Each line between boxes denotes a point-
to-point retinotopic connection. At the top is the hippocampus (HC). The bottom
areas track the magnocellular (M) and parvocellular (P) cells. Reprinted from D. J.
Felleman and D. C. Van Essen, "Distributed hierarchical processing in the primate
cerebral cortex," *Cerebral Cortex*, vol. 1, p.1, 1991, with permission from Oxford University Press.

4.7A shows the flattened map of monkey cortex revealing the extent of
the different retinotopic maps. Summarizing each of these as a rectangle,
figure 4.7B shows the map hierarchical organization. To get an idea of
the data compression in the figure, each colored box is a separate map,
and each line between boxes represents a two-way series of neural connections that are roughly point to point with respect to visual space as rendered in retinal coordinates. To expand on these points, each labeled box
is a retinotopically organized map of neurons responding to parts of visual
space. Thus a line on the figure from V2:M to MT represents a complete set
of two-way connections between corresponding points in the two visual
maps.

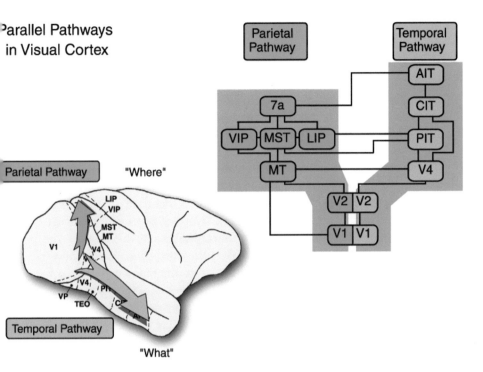

Figure 4.8
A portion of the visual maps in monkey cortex illustrating the *what-where* distinction. The calculation of motion and self-motion is done in the V1-V2-MT-MST hierarchy, and the calculation of color properties is done in the V1-V2-V4 hierarchy. Figure courtesy of John Maunsell.

Large-Scale Map Organization

Like computer graphics, the cortex distinguishes between the transformations that describe the relationships between objects (they may be in motion); that is, *where* they are from their properties such as their color and texture or *what* they are; massively so as there are two huge streams of two-way connections between maps. One for the connections that go over the top or *dorsal/parietal* part of the cortex, and another for the connections that go along the side or *temporal/ventral* part of the cortex. Maunsell's figure 4.8 summarizes some of the visual maps, again for a monkey, emphasizing this dichotomy. The graphics perspective suggests some obvious intuitions for this organization. When we would like to manipulate an object, many of the motor commands can be generic to a large object class (e.g., small cylindrical) and do not reference all the individual elements of the referenced object. As a consequence, connections between the dorsal and ventral areas can be more spartan than they would have to be for a naive approach.

Our Google metaphor clarifies another aspect of this organization. The pioneering work of Mishkin and Ungerleider showed via lesion studies in monkeys that the functionalities of object properties and object relationships could be seen as localized to ventral and dorsal areas respectively leading to the conclusion that vision could be parsed into separate functions (i.e., what and where).[78] Goodale and Milner have argued persuasively for a what/*how* characterization on the basis of two patients that have selective damage to these areas.[79] The common ground is that both groups describe the dichotomy in terms of visual streams. The Google metaphor suggests a rapprochement between the two characterizations in that a stored program will need data and instructions. Here the instructions specify how to bring about geomechanical transformations in different situations. So we can envision the dorsal stream as focusing on "Web links," or actions, and the ventral cortex as focusing on the state of affairs, or the data on the Web page. Naturally, because we are primates and place a huge premium on visually directed behavior, visual data will be an important part of computing the two sets of information, but from the stored program viewpoint, it is the state-action division that is central.

4.4   What Does the Cortex Represent?

While we know the *kinds of things* that are represented in the cortical memory, we are far from knowing *exactly* what gets represented. One general computational way of thinking of what is being represented is in terms of very complex codes. Paul Revere's revolutionary war audience just had to know how the opposing British forces were going to attack; hence, the famous light code, "one if by land, two if by sea." In the same way, we have to respond to everyday circumstances by triaging them quickly. Is it safe to make a right turn now? You have to analyze a complex traffic pattern and come up with a go/no-go decision. Detailed coding models have been difficult to come up with for the most abstract maps, but we saw that some success has been achieved at the earliest map, V1, where Olshausen and Field showed that the groups of cells there can be thought of as trying to code image patches with the smallest number of firing neurons (see figure 3.9 in chapter 3). The receptive fields of those neurons result because most of the time the local photometric variations look like linear edges separat-

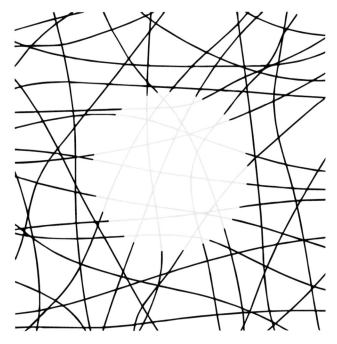

Figure 4.9
An illustration of the idea of pattern completion. There is no translucent blue disk!
Seeing it is an illusion created by having all the blue lines just the right lengths so as
to form the boundary of a disk. From blelb—The Visual Perception Lab (http://www.
blelb.com; info@blelb.com).

ing two areas, and the number of neurons that could represent this was
constrained to be small.

The smallest number of neurons cannot be the whole story though.
Consider figure 4.9, which shows sets of colored lines. You should be able
to have the impression of a translucent blue disk on top of a set of black
lines. This is an illusion. If you compare small white areas either side of
the disk by covering everything else up, you will see that they are the same
shade of white. Debates have raged over just how this experience happens.
At one extreme, researchers argue that the early retinotopic areas of cortex
fill in, that is, make the neurons that are in charge of the empty space
here actually send more spikes, and have developed extensive models to
show that this is possible.[80] At the other extreme, researchers maintain
that the "blue disk" neural code does not have to be literal. Just turning
on the abstract representation, whatever it is, is enough to experience the

bluc disk image. The CAM model might lead us also to allow for a middle ground. The neurons that are receiving unambiguous input try and complete a pattern. That pattern will involve both abstract and concrete parts.

There *is* evidence that the lower cortical areas are modulated. In a related experimental setup, von der Heydt has shown that a neuron's firing rates are sensitive to the figure that edges belong to. If an edge is part of a complete figure that "owns" it, it will fire more than if it is part of the background (figure 4.10). What this means is that, to work this out, the neuron has to be communicating with other neurons in the map. The fact that the firing measure represents an important property of the world—closed boundaries—means that the computations are very sophisticated.

Summarizing the representational discussion so far, the evidence shows that the cortex represents a coded version of the world, but also that that code can be adjusted to reflect not only the low-level pixel correlations needed to produce edge-like cells in V1, but also complex figural properties. Thus, the correlations that the CAM is sensitive to can be very high order.

Finally, we will introduce another important and related representational idea: that of a *distributed representation*. For that we turn to Tanifuji's studies of object coding in monkey inferotemporal cortex.[82] Because infero cortex is at the end of the temporal (*what*) neural hierarchy, it is the logical place to look for the representation of abstract object properties. That portion of the cortex is relatively flat in a rhesus monkey, so it can be studied with a special optical technique. When neurons fire they use oxygen and so remove it from blood. Relatively deoxygenated blood is darker than oxygenated blood and so can be measured with an optical microscope backed by image differencing techniques. The result is that such areas can be detected and displayed on top of the conventional optical image as is done in figure 4.11. While these measurements do not specify exactly what is represented, they do indicate that large cabals of neurons in disparate cortical locations are responding to the single stimulus. And as the nature of the most abstract areas is to represent very distinct properties, it is odds-on that the image contents somehow have been partitioned into useful components. This view is supported by studies of inferotemporal areas exhibiting responses to abstract properties of faces by Tsao. Her work implies that the perception of faces is mediated by the interactions between on the order of five separate areas that cooperate in the computation of different components of a facial recognition system.[83,84]

The view of the cortex as having distributed representations has been greatly refined by Gallant's group at Berkeley, who had subjects watch movies of natural scenes while in the functional magnetic resonance imaging

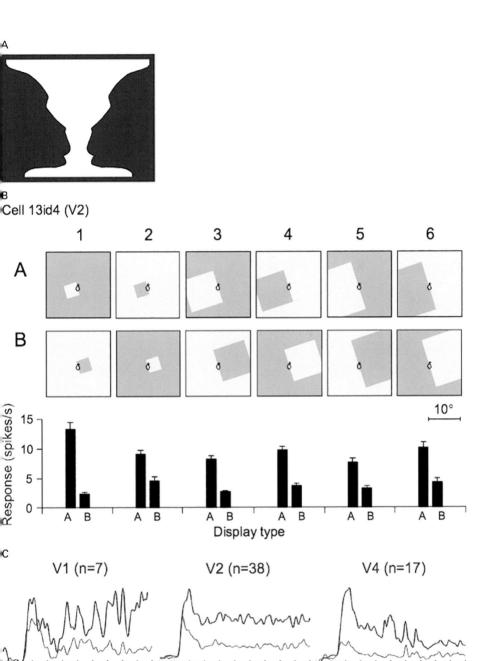

Figure 4.10

(A) Reversible figure depends on the "boundary ownership." If the boundary belongs to black, you should see two monkey faces, otherwise a vase. (B) Recordings in V2 show that edge cells are sensitive to boundary ownership constraints. (C) Spike histograms reveal that this ownership question is resolved extremely quickly, only 25 milliseconds after stimulus onset.[81] From H. Zhou et al., "Coding of border ownership in monkey visual cortex," *Journal of Neuroscience*, vol. 20, pp. 6594–6611, 2000.

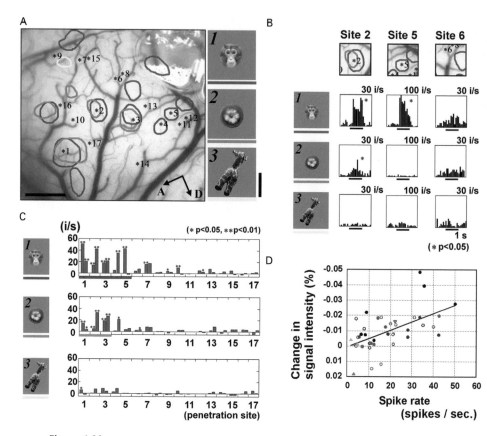

Figure 4.11

A view of a portion of temporal cortex that represents object properties. The figure shows the results of two kinds of experiments. (A) The blood oxygenation technique uses optics to measure small blood flow opacities produced by firing neurons. Colored marked areas show regions of the cortex that responded to the different stimuli shown on the immediate right panels. (B, C) The other experiment is the familiar single-spike recording method. (D) This panel shows that the two methods are correlated. The important point of the figure is that the representation of objects is distributed across this cortical map, with a single object causing modulated firing in disparate parts of the map. Reprinted by permission from Macmillan Publishers Ltd: K. Tsunoda et al., "Complex objects are represented in macaque inferotemporal cortex by the combination of feature columns," *Nature Neuroscience*, vol. 4, p. 8, copyright 2001.

(fMRI) scanner. The team painstakingly annotated the content of the movie frames at 1-second intervals, building a vector of categories marked as one (present) or zero (absent) on each frame. The categories they chose were a subset of a vocabulary graphical data structure termed WordNet, which links words with categories. This allowed them to extend their hand classifications by inference to approximately 1,700 total. As implied by figure 4.12, the relationship between the measured responses at the voxels can be linearly related to the categories at each interval of time. Given this scheme, they used a standard linear regression method to map the values of the order 5,000 small volumes (voxels) onto the categories. Thus, a set of fMRI voxels for a small time instant can be translated into WordNet coordinates. The set of all such data for a 2-hour movie was then analyzed using principal components to describe the data in terms of its four principal components with the greatest magnitude. Let's summarize for clarity. Using the regression methodology the approximately 5,000 voxel measurements are mapped onto WordNet coordinates. Next WordNet coordinates are analyzed to find their principal components. Two interesting data depictions can now be done. Using a clever color coordinate scheme, the WordNet items can each be given a specific color, which helps us visualize how the categories are grouped on the basis of a subject's data. Figure 4.13 shows this result. The other way of utilizing the analysis is to project the principal components back onto the voxel data. Figure 4.14 characterizes the results of this step. Like Tanifuji's result, the data show that concepts are distributed over the cortical maps, but the additional important insight is that they are organized into map regions of semantic significance. This result is also important as the analysis shows that the results can account for approximately 22% of the BOLD signal, a huge improvement over other methods. (A typical fMRI experiment reports differences between the desired manipulation and a similar control stimulus, and these differences can be of the order 1% of the signal.[a])

The principal components coordinate system in the WordNet data space provides a way of grouping the data. Simply put, the individual coordinates show groupings that can be semantically interpreted. For example, examining panel (C) of figure 4.13 shows that the third principal component is correlated with animals, especially humans. This different way of looking at the neural activation correlations implies that, at least in the context of movie watching, vast disparate areas have correlated activations that, from the WordNet principal coordinates perspective, somehow are involved in processing related to the same concept. You might be thinking how these studies can be reconciled with the inferotemporal data that show distributed but still relatively more focused representations. Two things to keep

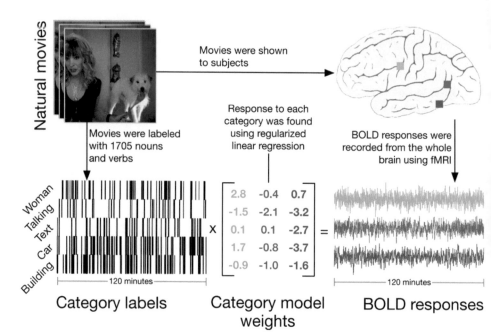

Category labels   Category model   BOLD responses
                              weights

Figure 4.12

To map fMRI BOLD responses onto categories, Huth et al.[85] recorded the objects on the frames of an approximately 2-hour movie. A total of about 1,700 categories were used, taken from the WordNet vocabulary. Once this was done, the BOLD responses of a subject watching the movie were registered. The three-dimensional signal over the cortex was then divided into small three-dimensional voxels. About 5,000 useful voxels were obtained. Assuming a linear relationship between voxels and categories allowed these measurements to be expressed as a set of linear equations with unknown category weights that could be estimated with linear regression techniques. The net result was that the BOLD signal at each instant could be mapped into the WordNet coordinate system. Reprinted from A. G. Huth et al., "A continuous semantic space describes the representation of thousands of object and action categories across the human brain," *Neuron*, vol. 76, p. 6, copyright 2012, with permission from Elsevier.

Figure 4.13

The data set from a subject coded in the WordNet coordinate system can be analyzed to find the coordinate axes of most variance, termed principal coordinates. The panels show how the WordNet categories are related to the principal coordinates using a color code. The bottom panel shows how the individual colors map into the principal coordinate system. Reprinted from A. G. Huth et al., "A continuous semantic space describes the representation of thousands of object and action categories across the human brain," *Neuron*, vol. 76, p. 6, copyright 2012, with permission from Elsevier.

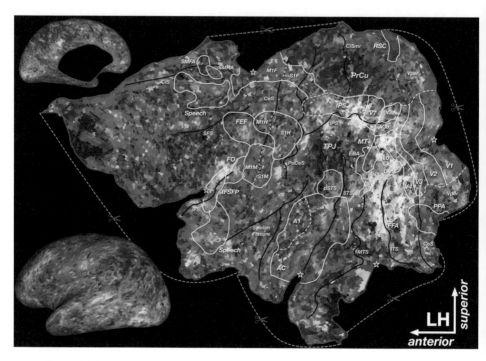

Figure 4.14
The principal coordinates of the WordNet coordinate system also can be re-projected back onto the voxel data. The figure shows how the individual voxels map into the cortical map. Unexpectedly, rather than being tightly confined to cortical map boundaries, the data are distributed over large areas over the flattened cortical surface. Reprinted from A. G. Huth et al., "A continuous semantic space describes the representation of thousands of object and action categories across the human brain," *Neuron*, vol. 76, p. 6, copyright 2012, with permission from Elsevier.

in mind are that the Word Net study takes advantage of a longer temporal interval and also the categories of stimuli are also more general. With this distributed focus in mind, we now switch to a computational models venue.

4.5   Computational Models

The description of the cortical memory started in the previous chapter with the structure of simple cell visual receptive fields. Here, the cortical memory's structure has been elaborated in a number of very important ways. One is the organization of visual receptive fields in retinotopic maps. Another is the elaboration of the maps in abstraction hierarchies. One can focus on the map hierarchies' feedforward connections in describing the progression of the abstraction of the cell's properties, but as we have just seen, there is also the complementary direction wherein the abstract cells can influence the responses of cells that are lower in the hierarchy. Since these feedback connections outnumber the feedforward connections by a ratio of 10:1 or more, it is unlikely that the cortex can be fully understood without accounting for their purpose.

To introduce the computation that the cortex might be doing, we used the concept of the content addressable memory, but the example of memorizing image patterns was simplistic, in that all the model cells were directly connected to image input. From the hierarchical structure of the cortex, you know that this example is greatly oversimplified, as successive maps of neurons are not directly connected to input, but abstract the derived properties of the layers below. So there is a need for computational algorithms that can handle this case. Given the basic circuit that was used for the simple cells, and the fact that the circuits in each map have hugely similar features, it is easy to hypothesize that all of these circuits can be analyzed computationally in a similar manner. It is odds-on that they can, but the mathematics of the hierarchies turns out to be more delicate, so that it is still under construction, with several different possibilities in play.

One class of models that has proved extraordinarily influential is a probabilistic one using Bayes.[86,87] Suppose that the task is to estimate a quantity y that takes on a range of numerical values. For each of these values, let us assume that it has a probability estimate $P(y)$. Now further assume that you have done an experiment and measured a value for the quantity x. So now the problem can be stated as the combined probability $P(x, y)$. If we were to estimate what the quantity might be, a logical thing to do would be to pick the value that makes $P(x, y)$ the largest. Here is where Bayes comes in. It turns out that $P(x, y)$ can be factored into two parts as follows:

$$P(x, y) = P(x|y) \, P(y),$$

where $P(y)$ can be thought of as the distribution of prior beliefs about the quantity, and $P(x|y)$—read "the probability of x given a value of y"—is the estimate for the value of x given a particular value for the prior belief. You can see how this might work in the context of the boundary ownership example of figure 4.10. The local edge cell receives an estimate of the response from cells that are lower down in the hierarchy, but those measurements, in isolation, are not enough to decide on whether the figure is to the left or right. But in the context of the cells higher up, additional information is potentially available to decide the issue. Thus, in this case the measurements from below might have estimates for $P(x|y)$, where let's say x and y can both take on values "LEFT" or "RIGHT," and the higher values would be modeled as $P(y)$. Because the higher values have a larger perspective, they can tip the balance, potentially resolving the issue.

The previous discussion has been a thought experiment on how the Bayesian framework could potentially resolve the issue, but now let us turn to a more exacting model that explains a visual illusion of motion. Consider the thin rhombus shown in figure 4.15. If the figure moves horizontally to the right and has a high contrast with respect to the background, the perception of its motion will be veridical. But the situation changes if the contrast is lowered, when it looks as if it is moving downward on a slanted course. How could this be? Weiss, Simoncelli, and Adelson showed that this effect could be explained via a Bayesian model whose computations are also summarized in the figure. To understand the explanation, you have to appreciate that when a local edge moves, only the velocity perpendicular to the edge is measurable. To see this, cut a circle in an opaque piece of paper and move the edge of another piece of the paper across the hole from behind. When you move the paper parallel to the edge, the motion will be very hard to perceive. This sets up the explanation. Let the motion estimates from each of the edges be $P(x)$ and $P(z)$. So extending our formula, the Bayes estimate for the motion is given by

$$P(x, y, z) = P(x|y)\, P(z|y)\, P(y).$$

In this formula, note that x and z are independent of each other if we assume nothing about the figure's shape.

Figure 4.15A shows the high image contrast case. The top of the figure shows local measurements of the motion, which we have denoted x and z, and their loci in the space of possible velocities are plotted in panels "Likelihood 1" and "Likelihood 2." Because only the perpendicular component of a velocity is measurable, the parallel component is arbitrary, with the consequence that the actual velocity could be anywhere along a line. The prior belief y is here crucially assumed to be a distribution that favors low velocities, hence the fuzzy ball around zero velocity in the leftmost mid-panel. In the high-contrast case, multiplying all these probabilities together produces the veridical velocity shown in the "Posterior" panel, but in the low-contrast case, noise introduced into the likelihood measurements shown in figure 4.15B results in the downward estimate. This has been a small example, but the reader is asked to imagine that its implication is that vast

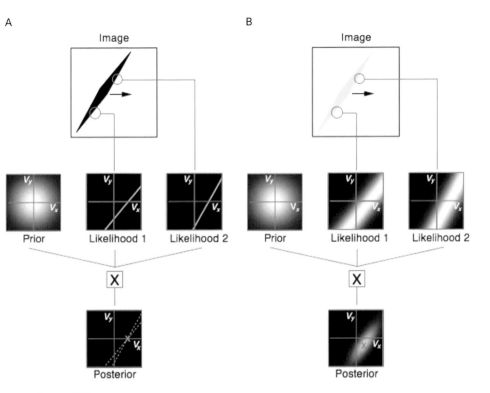

Figure 4.15
When the rhombus moves, local cues at a moving edge can only measure the component of the motion perpendicular to the edge. The Bayesian estimate for the direction of motion is the maximum after multiplying the estimate for prior motion with the two estimates at the edges. (A) At high contrast, the edge motion estimates are accurate leading to a veridical result of horizontal motion. (B) At low contrast, the edge estimates are noisy leading to a final estimate that has a downwards component.[88] Reprinted by permission from Macmillan Publishers Ltd: Y. Weiss et al., "Motion illusions as optimal percepts," *Nature Neuroscience*, vol. 5, p. 6, copyright 2002.

amounts of probabilistic information could be arranged into a very large network of probabilities (without loops), perhaps on the scale of all of cortex. Under such circumstances, fast algorithms can propagate information added to any part of the network to form a globally consistent interpretation. Furthermore, it could easily be the case that distant peripheral data, when combined, could lead to very confident central abstract conclusions.

The reader may have noted that the Bayesian model of motion perception does not have explicit neurons, but rather represents mathematical constructs that a more detailed model that had explicit cells with more exacting cell properties such as spike generation would have to implement somehow. Such more detailed models are in play, but their details would take us too far afield, so there will be just an introduction of one of them here to provide a flavor for the modeling effort and to discuss some of the key computational issues.

The discussion of the Bayes nets assumes that the networks have been learned and their connectivity is in place. However, a complementary problem is to learn these networks. An important modeling issue is to allow nonlinear transformations of the code between levels. Without this arrangement, mathematically one could always reduce a multilevel system to an equivalent single level. Thus, the case of purely linear transformations cannot really add anything important to the representation. Figure 4.16 shows a result from Lee et al. of Ng's lab.[89] A three-level neural network is trained exclusively on images of faces. At the bottom level, the V1-like edge responsive cells appear, but at the two subsequent levels, face parts appear at level 2, and then whole face receptive fields appear at level 3. This network uses a big shortcut of sharing the information across the unit "cells" on its image. This means that when a cell's receptive field is updated, the update is shared across copies of that cell that span the image. In the mammalian system, it is not obvious how this sharing could be done, but the thought is that this is not a burning computational issue, as the biological system can compute in parallel across the image locations.

Three levels work well for a data set that has only faces, but you can see why the much richer set of images from the real world requires additional levels. The cortex of a macaque monkey uses 10 such levels. When training networks that have additional levels beyond three, there have been technical problems as the algorithms depend on passing error measures through the networks, and these measures tend to become unreliable as the depth of the network increases. However, recent advances such as Hinton's "deep learning" have found ways to circumvent this difficulty, opening up the study of learning to cortical-scale architectures (see ref. 73).These

A

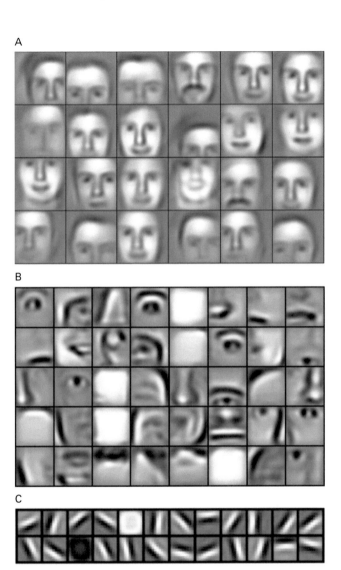

B

C

Figure 4.16
Lee et al.'s multilevel model learns receptive fields from data using a probabilistic model. When trained exclusively on images of faces, the higher levels of the network form features that capture face essentials. Reprinted with permission from H. Lee et al., "Convolutional deep belief networks for scalable unsupervised learning of hierarchical representations," *Proceedings of the ICML,* 2009.

model architectures are undergoing a rapid development in terms of reconciling them with existing cortical features, notably in the Poggio[90,91] and DiCarlo[92,93] labs at MIT.

At this point an outstanding issue is all the models referred to invoke purely image-based coding criteria of creating compact image codes. There is no way the images per se can anticipate what is especially valuable content from the animal's perspective. So another vital component is to allow the animal's needs to influence what must to be done. The next chapter discusses ways of doing this.

4.6 Summary

The cortex is a complex CAM indeed, as it does a number of things simultaneously:

1. It must compactly code sensed data. Most of what we know about how this is done comes from work with vision and image coding. In V1, a model that measures the statistics of images in the natural world produces similar receptive fields to those observed experimentally. As the main feature of this model is adding a cost to spikes, there is hope that this might be a general principle.

2. The interpretation of images that we see reflects complex ways in which the world can present itself. Thus, an image might reflect a translucent object that is occluding another object further in the background. Illusions can be created that play on the brain's ability to compute these kinds of interpretations. Such illusions suggest sophisticated neural codes, and these have been detected experimentally.

3. At the most abstract sites in the cortex, the coding of object properties is distributed, with properties being represented by neurons in disparate areas of cortical maps. This may reflect the fact that we typically only need to access the properties of one or a small number of objects at a time. Thus, the coding of objects can be shared without producing confusion as to which one is being referred to.

4. All the machinery in the chapter has focused on learning of abstractions from the bottom up, that is, from the data nearest its inputs, but, as Baum has argued (see ref. 33) and Wagner et al. have shown,[93a] basic abstractions may also be genetically driven. The search for formal models of these is the focus of Tenenbaum et al. and other groups.[93b,93c]

All these aspects are very important in understanding the cortex, yet perhaps the most important one is still missing. This is the use of the information in a program. Recall the example of the previous chapter where monkeys learned to discriminate motion patterns. While the cortical areas can have all the properties just discussed, they still need to be capable of being configured to solve particular problems. How does the monkey's cortex "know" that a particular motion-sensitive neuron is the one to pay attention to in making a decision? In the monkey, we can reach for the answer of training. The monkey learns the task arduously over many months. So perhaps there is time to "burn in" a circuit that does exactly what is needed by adjusting synapses. However, humans can do tasks like this with only the barest of instructions. So there is yet a final thing to explain, and that is how the cortex can be reconfigured quickly to test the representations that it has in the course of computing a program's crucial state. This issue is extremely important but to discuss its complexities, we will have to jump abstraction levels and take it up in part III.

5   Programs via Reinforcement

[A] habit is a behavior that we do often, almost without thinking. Some habits we strive for, and work hard to make part of our general behavior. And still other habits are burdensome behaviors that we want to abolish but often cannot, so powerfully do they control our behavior. Viewed from this broad and intuitive perspective, habits can be evaluated as relatively neutral, or as good (desirable) or as bad (undesirable). Yet during much of our waking lives, we act according to our habits, from the time we rise and go through our morning routines until we fall asleep after evening routines.

—Ann Graybiel, "Habits, rituals, and the evaluative brain,"*Annual Review of Neuroscience*, vol. 31, p. 359, 2008

Everyday behaviors might seem simple, but they require that the brain have the ability to create and execute complex programs. These are coded sequences of abstract actions to achieve goals. Most of these involve physical behaviors in which the body is directed to do something. Consider making a cup of tea. The program shown in box 5.1 should work.

But how do we come up with these action sequences? For many such things, we can be taught by getting verbal instructions or having someone else demonstrate. But for a vast amount of behaviors, we are left to our own devices and have to generate the recipes alone. Ultimately, we are programmed to propagate our genes, but primate life is long and complicated, and it is not always clear how the behaviors of the moment are related to this ultimate goal. Nonetheless, this is the problem brains have to solve.

In modeling this solution process computationally, a basic first step requires describing the brain's programs in the abstract, using the states and actions formulation. We already introduced the former in the previous chapter as one of the principle jobs of the cortex. Suppose each specific stage in a recipe can be instantiated in one of the dynamic cortical "Web pages." The new addition here is the *execution* of actions—instructions for

Box 5.1
PROGRAM TO MAKE A CUP OF TEA

Get the kettle.
Fill it with water.
Put the kettle on to boil.
Get a cup.
Get a tea bag.
Put the tea bag in the cup.
If the water is boiling, fill cup.
Discard the tea bag.
Add milk.

the body to do something. The mainstream view is that this is the job of the basal ganglia, but the exposition here is not new: it rather interprets versions of this idea made by Houck, by Haber, and by Graybiel.[94–98] The basal ganglia's interpretations of codes for actions may be carried out physically or, in the case of mental simulation, they may put the cortex in a new state that would reflect the result of carrying them out. The particular emphasis in this chapter is the basal ganglia's focal role in learning the value of programs by reinforcement.

This notion of state is so crucial we must risk belaboring it. Figure 5.1A shows the point stressed in the discussion of the cortex. Millions of neurons, perhaps billions, in the course of their spike activations produce an attractor that carries the summary of the state of affairs. All this work is reduced to a single state symbol. Thus, the next step of complexity is to have the capability of transiting from one state to another. Such an idea famously emerges in Pavlov's experiments on dogs. Dogs that would salivate at the prospect of an immediate food reward had learned that a preceding bell sound promised that reward and salivated at the bell sound. Figure 5.1B shows this case. The direct presentation of food is designated the conditioned stimulus and the bell sound the unconditioned stimulus. An early computational model of this is due to Rescorla and Wagner.[99] With this background, it can be seen that the tea-making program adds levels of complexity. In the first place, as shown in figure 5.1C, from any state, there may be several possibilities. And after one transition we are not done; we have to string together the succession of remaining steps. The all-important notion of state is that it does not have to be exhaustive but just carry the information that signifies that the next step can be carried out. Once

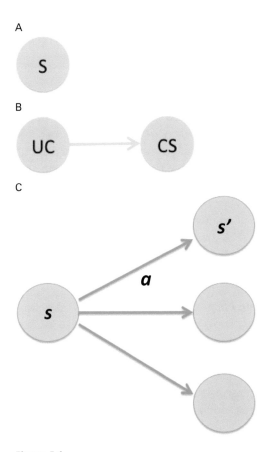

Figure 5.1
(A) Basic state abstraction where the state of huge numbers of neurons is symbolized by a single state (S). (B) Pavlovian conditioning. (C) The basic vocabulary for programs. The useful conditions of the world are represented by abstract states, denoted by circles that can be transited by actions, denoted by arrows. Each action is denoted by a transition function $T(s, s')$ that results in the brain being in a resultant state s' after starting from state s.

the kettle is filled with water, it's ready for the next step. The color of the kettle is (almost always) inconsequential.

This notation allows us to specify potential programs in terms of a *graph* of possible state transitions. Figure 5.2A shows the situation in the abstract with a graph of states and actions that represents a model of possible program versions. Each circle represents a certain state that you could be in. In terms of brain codes, we do not know what these states are yet, but, say for the tea-making example, a state might code enough information to capture the moment that the boiling water has been placed in the cup and it's ready for the tea bag. The action might be to put the tea bag in the cup. This graph is the source of particular programs that are found by searching possible paths in this space of alternative programs to come up with a specific program indicated in red in figure 5.2B. How the brain organizes the details is still a mystery, but in broad terms the program of states and actions would naturally be encoded in the cortex and the execution of the program handled in the basal ganglia.

However the states and actions are partitioned, the problem of developing programs would be much easier to solve if we had some way of rating the value of doing things. If a scoring mechanism were in place, then programs could be scored and the best ones picked. However, getting the information needed for this rating system is not straightforward at all. Consider eating an apple. It is easy enough to design an internal system that measures the caloric value of food and reduces appetite when "full." So conceptually at least, we can design a system for linking the value of the program to its execution. But there is a problem here, and that is the fact that the reward is delayed. We have to eat the apple first to know its value exactly. In this particular case, we can imagine that the relatively small delay between eating and scoring is handled somehow, but for longer delays this turns out to be a big problem.

Consider the value of foraging to find an apple. One could of course set off and just hope that apples are on the route, but what if there are multiple possible routes with probabilities of success that depend on current conditions? Certainly if the effort that would need to be expended exceeds the value of the apple, then one shouldn't set off in the first place. Unlike the first calculation for eating an apple already in hand, this harder calculation has to be done much more abstractly. We have to evaluate a plan for getting the apple that involves evaluating these possibly risky uncertain steps. At least for the apple, the end product is something of concrete value. Very abstract programs with uncertain consequences are much harder to evaluate.

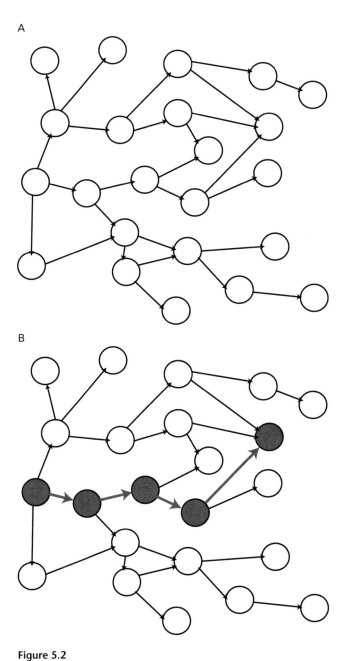

Figure 5.2
(A) The problem of finding a behavioral program can be cast abstractly as searching through a maze of states and actions. (B) A successful program can be represented as a specific path through the space.

5.1   Evaluating a Program

Finding the right program involves two very different but important steps: generating a plan from a repertoire of possible steps and scoring the plan. As far as our current understanding goes, the first part, plan generation, is the most difficult part, and the details of how it is done contain many open questions. The second step, scoring a plan, is better understood.

Searching for the Right Program

We know how to search through possibilities when there is a modest number, but when we have to choose from a vast number of plausible alternatives, there is no general agreement on how to narrow down the possibilities. It could be the case that we are stuck with thinking of plans that are "close" in some way to previous successful plans. This would allow them to be discovered by brute-force searching methods. In this case, the alternative ways of doing the task are enumerable in some way, and we systematically try them out.

In fact, there is a very good way to conduct this search in the case where there is a deterministic model of the space of states and actions. A deterministic model allows you to go back and forth between states easily. If you are in a state and take an action, you know the state you will end up in. Just as important, if you are in a state and know the action you just took, you know the state you were just in. In this idealized case when we can depend on the actions from a given state taking us to another specific state, then you can exploit this property by starting from the goal and working backward as shown in figure 5.3. (We will elaborate on how to do this in a moment.) However, the overwhelmingly common situation is much more complex, owing to the uncertainties in the world. Just consider rolling dice in a game. If the action is to roll the dice, before rolling them you won't know what state you will end up in owing to the randomness of the dice. In the real world, you might slip on ice, try to pick up an object that is too heavy, or go to a restaurant that is closed. These cases can be handled by trying things out and keeping track of what happened, and there is evidence that the forebrain does just that. The procedures and bookkeeping to make it all work are known formally as reinforcement learning, which is the main focus of this chapter.

Another aspect that has many loose ends is the way actions in a plan are coded. In this chapter, the specification of action will be very abstract. In tea making, "put the water on to boil" is actually a complex action involving many subactions. But for the purpose of planning, we will assume it is

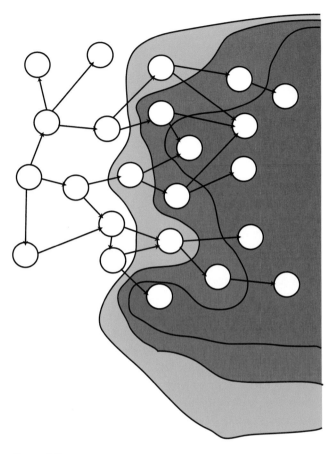

Figure 5.3
Finding a program by searching backwards from the goal. In the case of a reliable model, there is an efficient way to enumerate the possibilities. Start from the possible end states from a given state and back up to the state that you came from. Record the best action from the penultimate state. Next, repeat this process to the prior states. When the process reaches the initial state, the best program is known.

a primitive choice that the body knows how to execute. In fact, the body has enormous amounts of detail encoded in the spinal cord that can direct the execution of actions including complex reflexive contingencies. These details are important and crucial to understanding behavior, so much so that they will be described separately in part III.

Scoring a Program

Whereas something like an apple has primary reward—it has nutritional value for keeping an animal alive—when an animal has an elaborate plan to get the reward, the reward has not happened yet and may be far in the future. Yet to choose the plan, the animal has to have some representation of the expectation of getting reward. In other words, there needs to be an internal scoring mechanism that evaluates possible consequences of behaviors. In chapter 2, we pointed out that there is good agreement on the brain's scoring mechanism. Dopaminergic neurons are in the brainstem and innervate vast amounts of cortex, selectively emphasizing the most abstract areas. This vast internal dopamine-dispensing network has access to the brain's representations of potential plans of action and can evaluate them in terms of how much dopamine they are worth. Of course, many technical problems have to be solved in order for this to happen, but in broad outline there is an emerging consensus that the brain's way of solving this problem takes this route.

The common currency of dopamine is ingenious as it allows the relative worth of disparate programs to be compared. However, this solution is not without its downside. Most addictive drugs work by breaking into the brain's dopamine storehouse. Why work all week for a modest amount of dopamine when you can ingest some cocaine and effectively rob the dopamine bank? Drug addiction highlights the fact that the process of generating programs to control behaviors might be a mechanical process where programs are systematically evaluated. A program can be one that is socially destructive, but from the perspective of the person's internal bookkeeping system, it seems like the best thing to do.

With the concept of the internal reward, we can describe in précis a mechanism for evaluating programs. At the end of the program, it is easy to assign it a value based on its result. But with the states and action graph in hand, one can score the states of the program that are penultimate with respect to the end. It turns out the right thing to do is to assign these states a discounted value, typically the reward value multiplied by a scalar selected from a scale from 0 to 1, say 0.95. After doing this, all the red states in figure 5.3 will have a value. Continuing, one step further back,

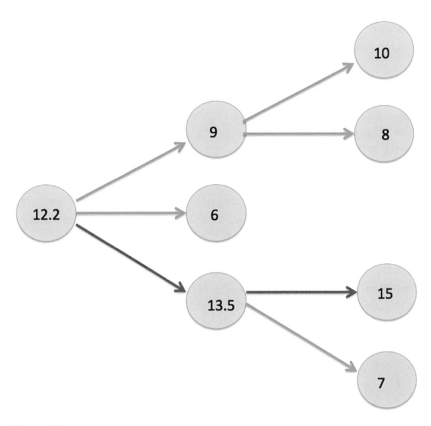

Figure 5.4
Working backwards: an example. At the beginning of the process, only the rewards associated with the goal states are known (10, 8, 6, 15, 7). Next, the discounted rewards for some of the penultimate states can be calculated (9, 13.5). Finally, the discounted value of the start state can be calculated (12.2). The actions associated with this value are indicated in red.

all the next level of connected states can be assigned a value. Repeating the process, eventually all the states will be assigned values, and when we start at the program's initial state, the discounted value of executing it is determined. The calculations for a small example are shown in figure 5.4.

Dealing with Uncertainty
In introducing the program generation and evaluation mechanisms, in the service of clarity of exposition, we have neglected perhaps the most important component, which is uncertainty. The reader will have realized by

now that we cannot assume the actions will always work as planned and that the brain's programs must have ways of dealing with this reality. There are many ways, but we will just present an outline of two levels of detail. The simplest model assumes that one knows the probabilities of different outcomes. Choosing a particular action does not mean that a given state is guaranteed to result, but that one can define a probability function over different possible states. The big assumption here is that one can enumerate the possible outcomes. It turns out that if one can do this, then the basic methodology described earlier still applies and programs can be evaluated.

Dealing with a repertoire of action outcomes adds complexity to program generation, but dealing with nondeterminism is even harder but absolutely essential. This situation occurs when the outcome of an action cannot even be predicted (figure 5.5). However, even this case can be managed, albeit with much more computational effort. The way to proceed is just to try out the actions and see what happens. This is certainly not the way to do math, but think instead of all the things you learn in development that are done

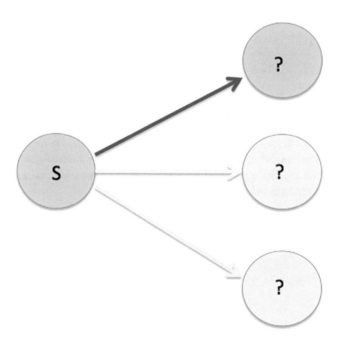

Figure 5.5
The hardest case for generating a program occurs when the outcomes of actions cannot be predicted.

with trial-and-error experimentation, such as learning to walk or eat with a spoon. These cases are where reinforcement learning algorithms shine. Formally we have entered the realm of *Markov decision processes* (MDPs), where the states and actions description is augmented by (1) a probabilistic transition function that determines action outcomes and (2) a reward function that scores the expected value of the action taken.

5.2 Reinforcement Learning Algorithms

At this point, we can talk about how the brain's programs might be learned. We will have to assume a lot, but we can still make some headway. The biggest assumption is that we can describe the venue of such programs in terms of particular states that the animal is in and particular possible actions that it can take from those states. In most of the models we come up with, we can be fairly sure that the states and actions are not exactly those coded in the brain, but the intent is that they mirror in some essential sense the actual states that an animal might use. Once we assume this much though, we can say rather precisely how to go about scoring those states and generating programs.

To see the computations involved, consider the very simple example of a rat following a maze shown in figure 5.6. The rat's formal problem is to learn what to do for this maze. Of course, what to do in this case is to follow the maze in a way to get cheese. The protocol in our case is that the rat is somehow returned to this maze again and again, so it had best save a map of what to do as a *policy* that will specify for each state the best action to move the rat toward the goal. The actions that the rat can use are to try to move in one of four compass directions, as shown in figure 5.6. If there is a wall in that direction, the action has no effect, leaving the rat in the state (maze square) from which the action was taken. You can form a cartoon image of the poor rat bumping into the wall and tottering backward if this helps keep the dry mathematics at bay. If there is no wall, then the rat progresses to the next square and consults the table again. Our cartoon rats are much smarter than their living counterparts, and it helps in this case because the rat knows where it is by reading numbers written on the floor. These are enough to define states.

These effects for the example maze are shown as a *transition function* $T(s, a)$ in figure 5.6. The final part of the problem specification is the reward function. The rat gets 100 "neuros" for getting the cheese, otherwise at any point along the way it gets nothing. So the reward schedule is

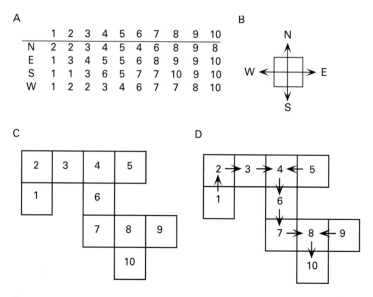

A

	1	2	3	4	5	6	7	8	9	10
N	2	2	3	4	5	4	6	8	9	8
E	1	3	4	5	5	6	8	9	9	10
S	1	1	3	6	5	7	7	10	9	10
W	1	2	2	3	4	6	7	7	8	10

Figure 5.6

A simple maze task illustrates the mechanics of reinforcement learning. (A) The transition function for the maze problem. The way to read this table is that, for example, if the rat is in state 5 and chooses the action "Go West," then it will end up in state 4. However, all other directions from state 5 will get nowhere as state 5 is a dead end. (C) The transition function. For each state listed in the top row, the effects of the four possible actions are given in terms of the resultant state. (D) The optimal policy for the maze problem. From every state, the arrow shows the direction in which to go to maximize reward.

$$R(s) = \begin{cases} 100 & s = 10 \\ 0 & \text{otherwise.} \end{cases}$$

Here, s is of course the state, and $R(s)$ is the reward given out in that state.

The result of reinforcement learning is an optimal policy, shown in pictorial form in figure 5.6 at the bottom. For every state, the *policy* stores what direction to head from that state. By repeatedly being placed somewhere in the maze and trying out actions, the rat eventually stumbles into the goal. The goal of a reinforcement learning algorithm is to calculate this policy. Real rats have exquisite senses of smell, so to make the problem make sense, we'll have to assume that our model rat has been genetically engineered to be without this capability. Thinking about smell is of great help though, because it allows you mentally to image what the solution found by reinforcement looks like. At every state, the value of following the policy is computed, and that value gets larger as the rat gets closer.

To be biologically plausible, it is best if the policy learning algorithm only has to make local improvements. Another reason is that all the rest of the program machinery also has this local property: To know what to do next, you only have to know the current state. What you did many time-steps ago is inconsequential. This *Markov* property allows actions to be chosen based only on knowledge about the current state and the states reachable by taking the actions available at that state.

Learning Values Directly

You might have already had the thought that a smart rat—remember these are cartoon rats—will solve this problem by making a map of the maze and going backwards from the goal. In this case, the algorithm would efficiently visit each mental state only once. You can do this as an exercise now: back up square by square, lowering the value of each square by the value of the square you came from times the discount factor. As we already discussed, if in fact you have a model, this is the right thing to do. But in more complicated real-life situations, we assume that you would not have a model either because the problem is too complicated to model or that your actions are uncertain in this case. Thus, in reinforcement learning, an assumption is that the rat is repeatedly plunked down in the maze and is forced to wander around. Such searches are called *forward iterations*. Again, forward iterations are necessary to learn the control policy because the transition function of the system cannot be known with certainty but must be learned by experiments. The only way to find out what happens in a state is to pick an action, execute it, and see where you end up. It turns out that this brute-force strategy works! In fact, the rationale for its success can be formalized in the *policy improvement theorem*.

The policy improvement theorem requires a specification of a policy and an additional bookkeeping mechanism to keep track of the value of states and actions. Figure 5.7 depicts the basic notation. Let $f(s)$ be the policy for every state s. The policy is simply the action a to take in state s. Denote $V_f(s)$ as the value of the policy. This is the expected reward for following policy f from state s. Now let $Q(s, a)$ be the *action-value* function, or Q-function, for policy f. This is the expected return of starting in state s, taking a particular action a, and then following policy f thereafter. The value function is related to the Q-function by

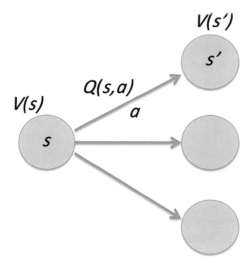

Figure 5.7
The basic vocabulary for reinforcement learning algorithms. The value function V rates each state. The action-value function $Q(s, a)$ rates the value of taking action a from state s.

$$V_f(s) = \max_a[Q(s,a)].$$

In other words, the value of a state is the value of the best action available from that state. Thus, it is easy to calculate V from Q.

Now we are ready for a crucial theorem. Suppose you have a policy, $f(s)$, which tells you which action to pick for every state, but it is not the best policy. Now consider a new policy, $g(s)$, which chooses the best action based on the Q-values, takes that action, and subsequently follows policy f. If that policy is better than f, then it is always better to choose g because after you arrive at the next state, you can apply the same reasoning to this new state and you will wind up choosing g again. The key point this result tells us is that we can repeatedly arbitrarily improve the policy locally and yet get increasingly close to the best possible overall policy.

The idea of choosing an action from a state, executing that action, and repeating the process from the state you wind up in has been formalized as *temporal difference learning*. Here, the progression of time is indicated by subscripts. The best way to understand temporal difference learning is to examine its learning rule. If s_t and s_{t+1} are two states where the latter is arrived at by taking an action from the former, we want to adjust the relative values of these two states so that they are related by the discount factor. That is, we want

$$V(s_t) = \gamma V(s_{t+1}),$$

so let's use

$$\Delta V(s_t) = \alpha[V(s_t) - \gamma V(s_{t+1})]. \tag{5.1}$$

After a lot of experimenting, the values will settle down to the point where they are all correctly related by the discount factor γ. Here, α is a scalar parameter that has to be chosen experimentally. What action should be chosen? When starting out and little is known about the consequences of the actions, the best thing to do is to pick actions more or less randomly, but as more and more information is acquired as to the better actions, they should be chosen more frequently. The exact way to do this is unsettled in general, but there are a lot of formulas that work reasonably well.

Another way of exploring the properties of reinforcement learning algorithms is to have a model system. Such a system can exploit advances in graphic environments to have humanoid avatars navigating a path in an artificial graphical world while avoiding obstacles as shown in figure 5.8. This model uses a representation of the state space that compresses far distances and expands nearby distances. After learning, each state has a value, and states right in front of the avatar representing possible collisions are of low value. The inset below shows the action policy, which is very straightforward. When obstacles are on the left, head right, and vice versa.

5.3 Learning in the Basal Ganglia

The complexity of the actual basal ganglia circuitry is extremely daunting, but nonetheless, the importance of reinforcement learning and the emergent evidence that the circuitry that carries it out is predominantly in the basal ganglia puts a premium on trying to constrain the physical locales of reinforcement learning pieces. Doya has made a number of important studies that attempt to situate features of reinforcement learning in the forebrain. In this process, he places the central update stage of reinforcement in the basal ganglia. To go one anatomic level deeper, we need to focus on two of the basal ganglia's principal parts: the striatum and the palladium.

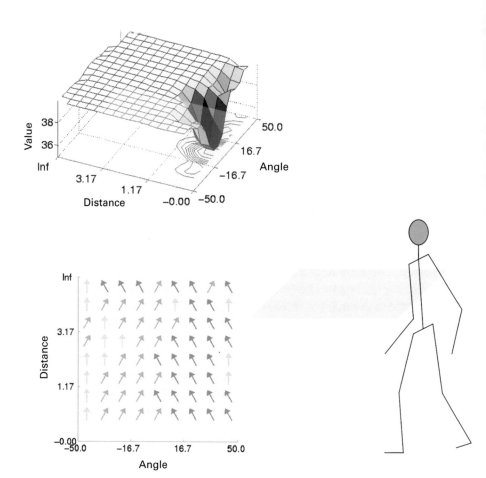

Figure 5.8

The results of reinforcement learning to avoid obstacles. (Right) A model humanoid pedestrian has an imaginary state space of headings coded with respect to the body. For any heading, RL computes two results: (1) the policy (i.e., which action to take) and (2) the value of that policy. (Upper left) The value of each point in the state space. Objects very near are dangerous and have greatly reduced relative reward. (Lower left) The value of the *policy* at each point in the state space is a direction to head given an object exists at that location.

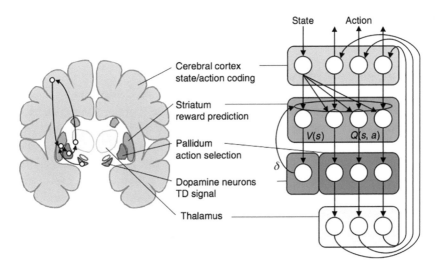

Figure 5.9
The Doya model of reinforcement evaluation in the cortico–basal ganglia–thalamic loop. The cortex is the site of remembered programs, which includes the evaluations of the worth of state-action combinations $Q(s, a)$. The striatum has the job of interpreting these programs. Successive steps provide the information to learn the values of these combinations by online adjustments. Reprinted from K. Doya and M. Kimura, *Neuroeconomics: Decision Making and the Brain*, copyright 2009, with permission from Elsevier.

Doya's model of the site of the all-important Q-tables, shown in figure 5.9, places them in the striatum and the update equation in the palladium. The figure also shows the cortico–basal ganglia–thalamus loop. Given the placement of the δ, the implication is that the reinforcement learning is run in simulation so that the current state information can be compared to the previous state information as needed by equation (5.1).

We will describe some evidence for this parcellation in a moment, but first it helps to revisit the big picture to motivate why one might want to have this part of the reinforcement learning algorithm just here. If we imagine that the cortex stores all of the programs in a coded form, one could imagine that it might be an issue to run reinforcement learning there. One huge technical problem would be to have a way of singling out a particular program in question. Given the interlocking web of cortical connections, there is the specter of cross-talk wherein the reward signals for the operant program leak over to another inactive program. From a computational perspective, the basal ganglia acts as a program interpreter, extracting the presentation from the cortex and "running" it in the basal ganglia, allowing it to be isolated so that its valuation can be simplified.

Action Codes in the Striatum

Now let's transistion to real animals. The Graybiel laboratory has conducted a tour de force set of experiments with mice that provide substantial evidence for the coding of actions in the striatum.[100] Mice are trained to run a maze while wearing a multi-electrode recording device. Figure 5.10A shows the maze structure. The mice waited at a gate and started when the gate was opened. Along the path, they received either a tactile cue or an auditory cue that indicated the direction to a food reward. The responses of individual neurons are shown in figure 5.10B sorted by locations along the maze path. These clearly indicate that neurons in the striatum respond to distinct locations. Additionally, they show stability to changes in the cue. The arrow indicates where the reward location cue was changed from auditory to tactile; the neurons show insensitivity to this task feature. The cue manipulation showed also that the overall responses were sensitive to the changes between auditory and tactile indicators, implying that the striatum was adjusting its codes to the differences in cue, but, concurrently, the striatal responses also reflected the overall structure of the task. While there have been a number of important studies implicating the role of the basal ganglia and the striatum in particular in the coding of actions (e.g., see refs. 101, 102), this is some of the cleanest evidence yet of the coding of actions in an extended task.

Coding Secondary Reward

The coding of actions in the basal ganglia is one part of the reinforcement learning story; the other component is the coding of reward itself. Some of the most compelling evidence that brains use dopamine to compute the score of a plan comes from Schultz's experiments (see ref. 23). He was able to record the responses from neurons in the substantia nigra—a part of the basal ganglia area rich in dopaminergic neurons—while monkeys were engaged in a task that required them to synthesize a short motor program to solve a reaching and grasping task. The purpose of the experiments was to use a task structure that could be modified so that by adding extra conditions, the time when a food reward obtained was less immediate.

Figure 5.11 summarizes the crucial findings of the experiment. Initially, monkeys are trained to reach into a box for a piece of apple. In these cases, immediately after the apple is obtained, the dopamine neurons fire, signaling that, as far as the brain's understanding of the plan is concerned,

A

B

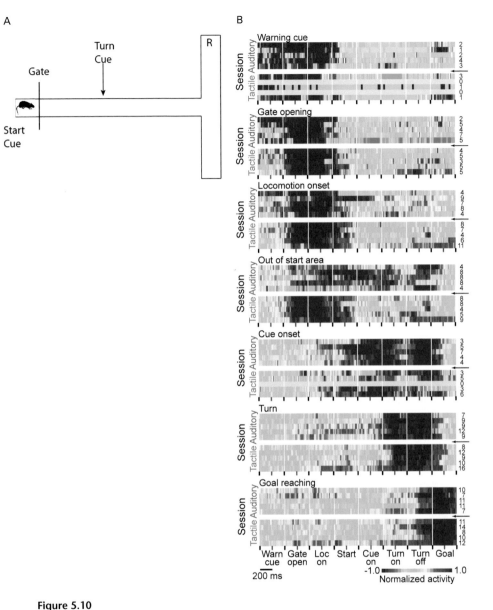

Figure 5.10

(A) The maze used to explore neural codings of a task reveals reinforcement-compatible sequences. A *start cue* lets the animal know that a trial is about to begin. Next, the *gate* is opened allowing the rat to go for a reward in one of the two arms (*R*). The *turn cue* is a tone that indicates which arm has the reward. (B) Neural recording summaries. Neurons subsequently grouped according to the stage of the task reveal that they are coding that stage by increased spike rates, indicated by dark red colors (see scale). From Y. Kubota et al., "Stable encoding of task structure coexists with flexible coding of task events in sensorimotor striatum," *Journal of Neurophysiology*, vol. 102, pp. 2142–2160, 2009.

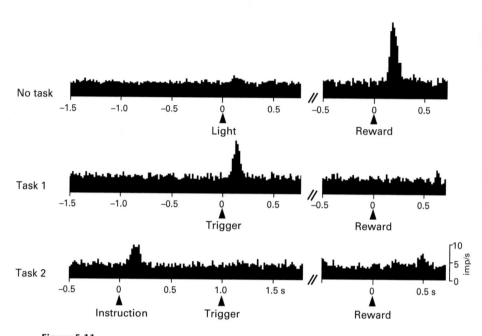

Figure 5.11

Histograms recorded from the substantia nigra of a monkey. The substantia nigra neurons deal out dopamine rewards for behavioral programs. Shown are recordings from the same cell under three different conditions. No task: The monkey reaches into a box for a piece of apple. Task 1: The monkey has to wait for a light trigger before reaching. Task 2: The monkey is given an instruction that indicates whether or not the box will contain an apple. The decreasing heights of the dopaminergic signal in these traces suggests that the reward for the more elaborate conditions is being discounted. From W. Schultz et al., in *Models of Information Processing in the Basal Ganglia*, ed. J. C. Houk et al., MIT Press, 1995.

reward was obtained. Then, the task was increased in difficulty by adding a trigger signal, so the monkey had to wait for a light signal to change before it could begin its reach. Surprisingly, you can see that the reward is now handed out at the time the trigger signal changes, in other words at the beginning of the program. The monkey has apparently encapsulated the new sequence of instructions and given it a score when it is initiated. Making the task still more difficult, the protocol was changed so that a light appears before the trigger that announces whether or not an apple piece is in the box. Now the reward signal shifts to the time of this signal, or the beginning of the more elaborate program. The brain doles out reward based on its expected consequences.

If you examined figure 5.11 closely, you might have already noticed that the dopamine signal is lessened depending on how far in the future the actual reward is. Researchers are excited about this feature as it may be evidence that the neurons are using *discounting*. The idea behind discounting is that future rewards are uncertain as the world has a way of intervening to thwart your goals: sometimes when we have had an important work assignment due date, we might be sidelined by a case of flu. Within the formal context of reinforcement learning, a way this can be handled mathematically is to pick a discount factor which factors in how much you value rewards as they occur more and more in the future. We already saw one example of how this is done in equation (5.1). The symbol γ for the discount factor ranges from 1(no discount) to 0 (complete discount). Low values of γ signify that to have any value, reward must be immediate.

The experiments summarized in figure 5.11 provide evidence that the monkey's program is paid in dopamine up front based on the expected result of executing it. However, the world is a complicated place, and as the estimate is expected but not certain, unexpected consequences may occur. Figure 5.12 shows the result of testing these cases with the beautifully concise interpretation that when the program completes, an adjustment in dopamine is made based on what actually happened. So when reward occurs that is unexpected, there is a large dopaminergic signal, but if that result was expected, no adjustment is necessary. But if the reward is not received, the adjustment is downward.

Up to this point, we have not discussed the value of the discount factor other than to note its relationship to the pitfalls of addictive behaviors. But low discounting can be valuable, particularly in time-limited situations. Inside a burning building is no place for long-term planning. Short-term planning in this situation is also related to risk. Exposing oneself to danger often enough makes a bad result more and more certain. Thus, it would be useful to have the possibility of recognizing such situations and having programs geared to short-term gains as well as long-term gains. Notably, some experiments suggest that discounting is explicitly signaled in the brain by different cells. In an experiment by Doya and Kimura,[103] human subjects performed tasks that had the same value, but this value was delivered in components at different times. Using an fMRI technique, they were able to show that different brain areas responded to different discounted values (see figure 5.13).

In concert with the representation of different discount factors in cortex and the association of low discount factors with increased risk, there also exist observations of serotonin concentrations. Emerging evidence implies

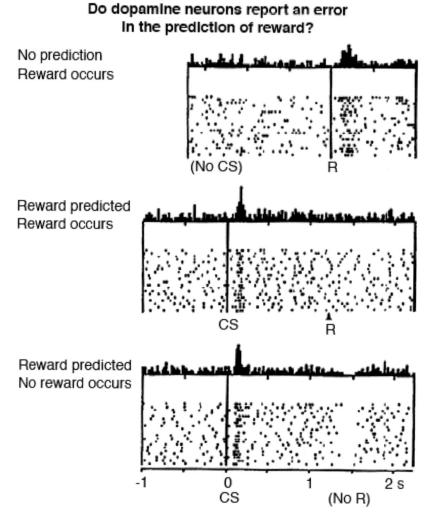

Figure 5.12
Evidence for the monkey's use of expected reward. When the reward is unexpected, it is recorded immediately afterward as shown in figure 5.11. Also, when the reward is predictable, dopamine is handed out when the prediction is made, and *not* when the reward actually occurs. However, in the absence of the predicted reward, a correction signal occurs at the time of expected reward. From Wolfram Schultz, Peter Dayan, P. Read Montague, "A neural substrate of prediction and reward," *Science*, vol. 275, p. 5306, 1997. Reprinted with permission from AAAS.

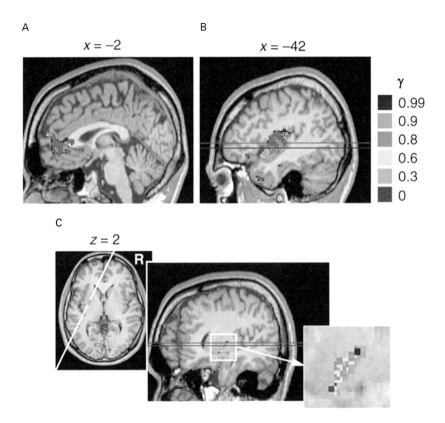

Figure 5.13
By manipulating a task with different temporal reward sequences, one can titrate the specific basal ganglia areas that are sensitive to different discount factors. To obtain these measures, the responses from the different situations are subtracted from each other. fMRI scans reveal the location of neurons that are differentially active in the different cases. Reprinted from K. Doya and M. Kimura, *Neuroeconomics: Decision Making and the Brain*, copyright 2009, with permission from Elsevier.

that low levels of serotonin are associated with increased risk taking. If dopamine is measuring expected discounted reward, serotonin might be implicated in its variance. Making a deposit at a bank would be a low-risk, high-discount-factor behavior. But robbing a bank is a high-risk, low-discount-factor behavior, just the opposite. And the spectrum of behavior offers examples from all the combinations in between. So if serotonin is associated with risk, and if subjects had their serotonin levels modulated, could these modulations show up as changes in the discount factor? One way to handle risk would be to make it less rewarding if not immediate. Evidence for this protocol comes from experiments by Doya and colleagues[104] where subjects ingested different levels of tryptophan, which has the effect of modulating serotonin values. As shown in figure 5.14, low and high levels of tryptophan produce different levels of activation in different parts of the basal ganglia.

5.4 Learning to Set Cortical Synapses

At this point, we understand how reinforcement learning works formally. Furthermore, the main parameter in the computational algorithm, the discount factor γ, has been related to the neurotransmitters dopamine and serotonin. Dopamine is associated with mean reward and serotonin more provisionally with the variance in mean reward. Furthermore, evidence is accumulating for the basal ganglia being the "processor" that executes actions, but additionally we can start to pin down the sites where the various suboperations needed in the reinforcement learning model might be.

However, you cannot have failed to notice the huge difference between the enormously complex state descriptions used by the cortex and their very simple abstract counterparts that appear in reinforcement learning formulations. It turns out this disparity potentially may be bridged by recognizing the hierarchical structure of the cortex. At the top of the cortex,

---

Figure 5.14
The effect of serotonin on discounting is seen in an experiment where serotonin levels are manipulated. Human subjects volunteered to ingest a substance that varied these levels and then performed a choice task in which different optimal choices were the result of different discount factors. Subjects showed different areas of the cortex differentially responsive to the different choices. Reprinted from S. C. Tanaka et al.,"Serotonin differentially regulates short- and long- term prediction of rewards in the ventral and dorsal striatum," PLoS ONE, vol. 12, p. e1333, 2007.

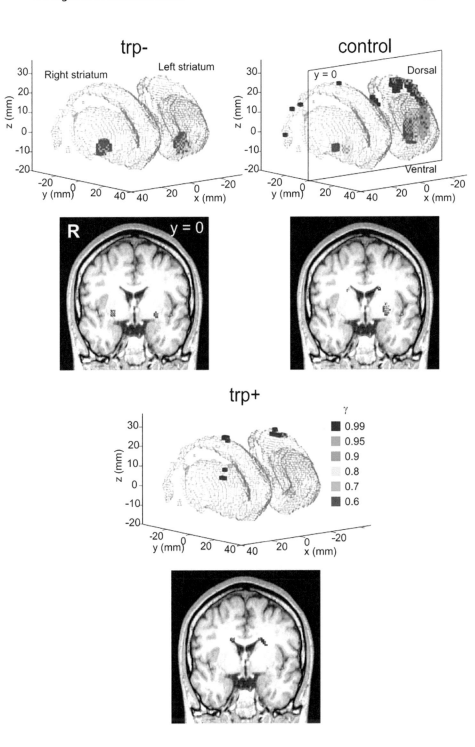

the neurons form the most abstract representations, so just focusing on this set of neurons provides a set of candidate state descriptors. Thus, one can put together the scoring ideas from this chapter with the representation ideas from the previous chapter.

Notably, these ideas come up in the context of a neural network invented by Tesauro that learns to play the game of backgammon.[105] Backgammon is a board game with two opponents that must advance their pieces using the roll of dice. It is basically a race between the two to see who can get their pieces around the board first and win. But, when we get into the rules in more detail, we will see the subtle interactions between the pieces that make the game interesting and difficult. Tesauro's network learned to play backgammon using the reinforcement learning algorithm just described. In fact, it learned to play so well that it plays on par with the world's best players.

To play backgammon, the value of different board configurations must be evaluated. If the board defines the state, we need to know how to estimate the value of that board state. For the moment, we will assume we know that value and want to train the synapses to reproduce it. The next section shows how to generate the value targets online.

In chapter 3, you saw that the synapses of a neuron could be set by requiring that they reconstruct the input as best they could when limited to using a small set of active neurons at a time. The setting there was neurons that were very close to the input stage, the lateral geniculate nucleus (LGN). But what about the other end of the abstraction hierarchy? Could we not have a circuit there that minimizes some abstract quantity such as reward expectation? We could, and to do so we will introduce an algorithm that is somewhat similar to the previous one but that has a different perspective.

Consider a single-layered network with two inputs and one output, as shown in figure 5.15. The network uses the same linear summation activation function that was used in chapter 4. Thus, the output is determined by

$$y = \boldsymbol{w} \cdot \boldsymbol{x}.$$

The problem is to "train" the network to produce an output y for an input pattern $\boldsymbol{x}$ such that the error between the desired output and the actual output y produced by the input pattern acting on the synapses is minimized. Of course, there would usually be several patterns, but this would complicate the notation a little bit, so we will stick with one pattern. If there are more than one pattern, the contributions from each pattern can be added together.

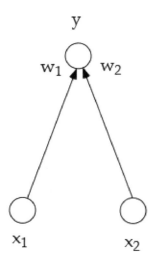

Figure 5.15
A simple case for optimization: a function of one variable minimizes error.

As before, the variables in the problem are the synaptic weights in the network. They are to be adjusted to minimize the error. Define an error function

$$E(\mathbf{w}) = \frac{1}{2}(y^d - y)^2.$$

An easy way to remember this metric is: "what you wanted minus what you got, squared." If the formula for y is included, it becomes

$$E(\mathbf{w}) = \frac{1}{2}(y^d - \mathbf{w} \cdot \mathbf{x})^2.$$

One way to minimize the cost function $E(\mathbf{w})$ is to use *gradient search*: Make a small change in a particular w_k and record the corresponding change in E. Next, use this information to adjust w_k in a way that would lower E. This can be done incrementally. Starting from an initial guess, use the derivative

$$\frac{\partial E(\mathbf{w})}{\partial w_k}$$

to improve the guess successively:

$$\Delta w_k = -\alpha \frac{\partial E(\mathbf{w})}{\partial w_k}.$$

The parameter α controls the size of the change at each step. The changes are discontinued when the improvements become very small. One way of measuring the improvement is to track the lsum of all the last round of changes and stop when this becomes less than some criterion (i.e., $\sum_k \|\Delta w_k\| < \varepsilon$). For an appropriately chosen α, this procedure will converge to a *local minimum* of $E(\mathbf{w})$.

Because E is a function of other variables besides $\mathbf{w}$, the *partial derivative* is used in the preceding equation to denote that we only want to consider the variations directly caused by changes in $\mathbf{w}$:

$$\frac{\partial E(\mathbf{w})}{\partial w_k} = -(y^d - y)\frac{\partial y}{\partial w_k}, \qquad (5.2)$$

which is simply

$$-(y^d - y)x_k.$$

This rule makes sense in view of the fact that it is similar to the earlier rule in chapter 3. For each input, x_k, each of its synaptic contacts can be grouped as a vector $\mathbf{w}_k$. So what the algorithm does is move this vector toward the desired pattern along the direction $y^d - y$, as depicted in figure 5.16.

To derive this, we had to use the chain rule from calculus. First, let $u = y^d - \mathbf{w} \cdot \mathbf{x}$. Then

$$\frac{\partial E(\mathbf{w})}{\partial w_k} = \frac{\partial E(\mathbf{w})}{\partial u}\frac{\partial u}{\partial y}\frac{\partial y}{\partial w_k}.$$

The first of these terms is just u, the second -1, and the third x_k. This particular formula for incrementally adjusting the weights is known as the *Widrow-Hoff learning rule*.

We described the rule for just one output, but if there were more, a vector of outputs

$$\mathbf{y} = (y_1, y_2, \ldots, y_n),$$

then the rule for changing the corresponding synapses $\mathbf{w}_k = (w_{1_k}, w_{2_k}, \ldots, w_{nk})$ would be

A

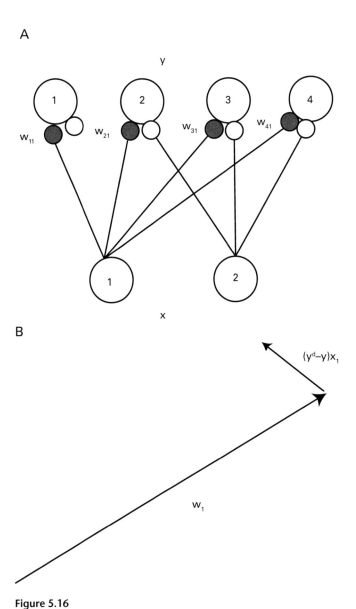

B

Figure 5.16

(A) The input to the neurons representing the output can be grouped such that they can be thought of as a vector. (B) The learning rule moves this vector in a direction that will lower $\boldsymbol{y}^d - \boldsymbol{y}$.

$\Delta \mathbf{w}_k = u(y^d - y)x_k$.

What has been done is to make the synapses at the final layer of neurons sensitive to error. The example network only has one layer of neurons with adjustable synapses. It turns out that networks that have multiple layers can do much better in reducing error, but they need to have a nonlinear activation function. A way to do this that is sympathetic to a neuron's actual input-output properties is to use the function

$$g(u) = \frac{1}{1 + e^{-u}}.$$

This function is zero when u is very negative and one when u is very positive. Thus, each output y is now given by

$y = g(\mathbf{w} \cdot \mathbf{x})$.

Of course, this change will affect the computation of the derivative, but the modification to the equation is straightforward. The main new problem to be dealt with is that now synapses that are not directly connected to the output layer need to be modified. They can be if the effect of the error is carefully tracked by the right derivatives. Figure 5.17 shows the neurons that contribute to the formula. The error at an internal neuron depends on the errors at the neurons it can effect. Furthermore, these errors can be "backpropagated" to the neuron in question. We won't produce the formulas here, but they can be found in Rummelhart et al.[106] The important point is that a multilayer network's synapses can be programmed by feeding an error signal into its final layer. This fact is going to be very important for the neural backgammon program, as the multilayer network is going to take as input the board position and produce as output the value of that position.

5.5 Learning to Play Backgammon

At this point, we have introduced the crucial tool we need to describe Tesauro's backgammon player, and that is a hierarchical neural network that can be trained by associating error with input patterns. The final versions of the player incorporated limited look-aheads to improve the evaluations, but in terms of the point that is important for understanding how brain programming might work, the first version of the network will have all the features we need.

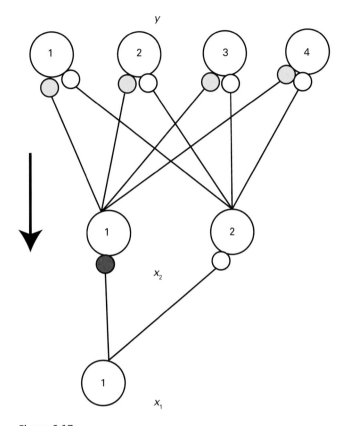

Figure 5.17

A schematic of the direction of error propagation adjustment. The output neurons are easy to adjust because the effect of each given input, given the current value of the synapse, is straightforward to calculate. By examination, one can see whether error is increased or decreased. However, at the penultimate layer, the relationship of the output to the error is not as straightforward, as the output interacts with the error through many synapses each of which are on different neurons. Nonetheless, a formula can be calculated that solves this problem. This calculation can be repeated for the next layer down, thus "backpropagating" the appropriate error adjustments.

In the backgammon board game, two opponents have 15 checkers each that move in opposite directions along a linear track. Figure 5.18 shows a version of the standard backgammon board. The player with the black checkers must move them clockwise from the position at the lower right toward the position at the upper right, and finally off the board. The player with the white checkers moves in the opposite direction. The movement of the checkers is governed by the roll of two dice.

Figure 5.19 shows a possible position in the middle of the game. Player 1, with the black checkers, has just rolled (6, 3). There are several options for playing this move, one of which is shown. Player 2 has a checker on the bar and must attempt to bring it onto the board on the next move. The game ends when all of one player's checkers have been "borne off," or removed from the board. At any time, a player may offer to double the stakes of the game. The opposing player must either agree to play for double stakes or forfeit the game. If the player accepts, then that player has the same option at a later time.

Reconsidering the board position in figure 5.19, the basic problem facing a player, given a throw of the dice, is to decide which of the several options is the best. Would it not be nice to have a program that would evaluate any board position and give it a numerical rating as to its winning chances? Then the problem is easily solved: just rate all the possibilities and pick the one with the highest rating. This motivates the design of the network like the one shown in figure 5.20, which has an input layer that codes the board, intermediate layers of neurons (only one is shown here) that describe important features of the board position, and a final layer to represent the value of the position.

Such a network could be trained if there was a way of representing the board for input and pairing it with a desired output. Representing the board is not too hard, and we'll get to that presently, but the real breakthrough is in representing the value at the output. Here is where reinforcement learning comes in. While there is not an absolute measure of value, there is a measure of relative value because, if the move chosen is to reflect that the best strategy is being pursued, the value of the resultant board should be related to that of the current board position value by the discount factor. That is simply stated as

$$\gamma V_{t+1} = V_t.$$

The extent to which this does not hold provides the measure of error that is needed. The remaining step is to make sure this error is attributed to the weight change in the correct way. For that, we can appeal to equation (5.2),

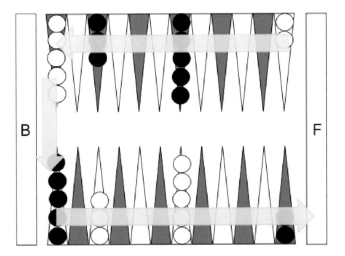

Figure 5.18
The backgammon board in its initial configuration. Opponents have checkers of a given color that can reside on the "points," shown as triangles. Players alternate turns. A turn consists of a roll of the dice and an advancement of checkers according to the dice roll.

replacing y by the pertinent variable in this case, V. The upshot is that the error adjustment becomes

$$\Delta w = \eta(\gamma V_{t+1} - V_t)\frac{\partial V_t}{\partial w}.$$

This is just a version of the backpropagation rule where the desired state is V_{t+1}. Thus, the network is doing a "smoothing" operation that tries to make successive states make consistent predictions. Where there is actual reward such as a win or a loss, then the actual value is used instead of the predicted reward. Thus, the network works just like Q-learning. Reward centers provide concrete centers that propagate back to more distal states.

The complete algorithm is given as follows:

Temporal Difference (TD) Algorithm

Initialize the weights in the network to random values.
Repeat the following until the policy converges:

1. Apply the current state to the network and record its output V_t.
2. Make the move suggested by the network, and roll the dice again to get a new state.

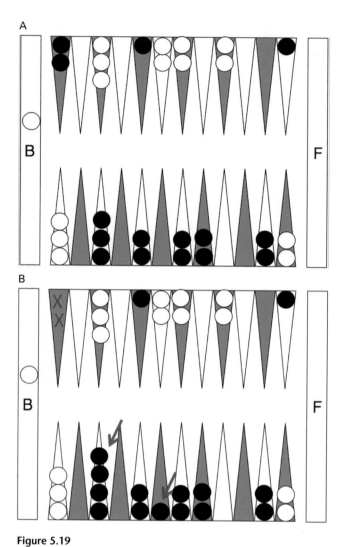

Figure 5.19
(A) Player 1, black, has just rolled (6, 3). (B) There are several options for playing this move, one of which is shown.

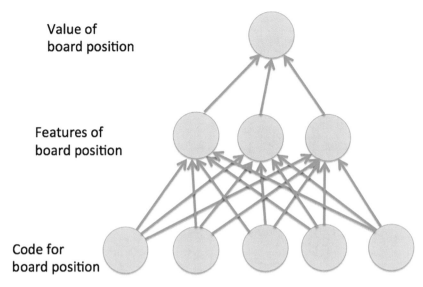

Value of
board position

Features of
board position

Code for
board position

Figure 5.20
The three-level network for evaluating backgammon moves.

3. Apply the new state to the network and record its output V_{t+1}. If the state is a win, substitute $V_{t+1} = 1$; if it's a loss, substitute $V_{t+1} = -1$.
4. Update the weights:

$$\Delta w = \eta(\lambda V_{t+1} - V_t)\frac{\partial V_k}{\partial w}.$$

5. If there are hidden units, then use the backpropagation algorithm to update their weights. The backgammon network (figure 5.21) consists of 96 units for each side to encode the information about the checkers on the board. For each position and color, the units encode whether there are one, two, three, or greater than three checkers present. Six more units encode the number of checkers on the bar and off the board, and the player to move, for a total of 198 units. In addition, the network uses hidden units to encode whole-board situations. The number of hidden units used ranges from none to 40.

It is interesting to inspect the features of the network after training, as they reveal the codes that have been selected to evaluate the board. For example, figure 5.21 (top) shows large positive weights for black's home

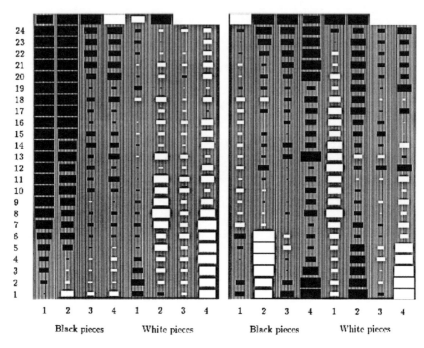

Figure 5.21

The board is encoded as a 24 × 8 set of "point" neurons denoting whether the occupants of a point are black or white as well as whether one, two, three, or more than three pieces are on the point. Additional neurons record whether pieces are on the "bar" (i.e., have been sent back to the starting line); white pieces borne off; black pieces borne off; white's turn; and black's turn. The weights from two hidden neurons after training. Black squares represent negative values; white squares positive values. The size of the square is proportional to the magnitude of the weight. Columns 1–24 represent the 24 positions on the board. The column at the end represents (from top to bottom): white pieces on the bar, black pieces on the bar, white pieces borne off, black pieces borne off, white's turn, and black's turn. These neurons represent crucial board features used in rating the position. Reprinted from G. Tesauro, "Practical issues in temporal difference learning," *Machine Learning*, vol. 8, pp. 257277, 1992, with permission from Springer Science and Business Media.

board points, large positive weights for white pieces on the bar, positive weights for white blots, and negative weights for white points on black's home board.

These features are all part of a successful black attacking strategy. Figure 5.21 (bottom) exhibits increasingly negative weights for black blots and black points, as well as a negative weighting for white pieces off and a positive weight for black pieces off. These features would naturally be part of an estimate of black's winning chances.

5.6 Backgammon as an Abstract Model

The backgammon example is worth spending time on as it provides a ready example of how the cortex might be organized to maintain and improve its programs. In the general case, sensory perception has to come up with an initial coding for the problem. We assume that is done here by choosing the inputs to the network to be just right for capturing the board. Next, we need a body to find and roll the dice. Again, that happens magically and on cue in our example. Given the dice roll, the different moves—the actions in reinforcement learning parlance—are used to adjust the board, and the resultant board configuration is evaluated by the network. The best move is taken, and the synapses to the neurons in the network are adjusted by the learning rule. But in the general case, the body commits to an action, and that action has consequences in the world that can be read as a new state. So rolling the dice and playing a specific move is analogous to taking an action the consequences of which may not be completely predictable. Backgammon is special, as with the dice the number of different possibilities increases so rapidly that the values of specific moves may be usefully summarized. By comparison, chess is different, as one particular sequence among many can turn out to be irrefutable. For that reason, computer programs routinely look more than 30 moves ahead. Humans can look far ahead when it is important, but not on every move. For that reason, it is thought that humans code similar promising board options as patterns that are somehow more compact. Support for the notion of coding comes from an interesting study by Dan Simons and Chris Chabris.[107] They asked a chess master to memorize board positions depicted on cards from short exposures to them. The master had little trouble with legal chess positions, but was totally flummoxed given random positions. In any case, pushing them forward means using cycles of the cortex–basal ganglia–thalamus loop.

Although the focus has been about games, the central idea is that any general problem can be solved the same way. The program that solves it has many possibly different steps, and these are represented in terms of a state-action table. This table has the current way of solving the problem represented as a policy. As we take actions recommended by the table, we are transported to new states and new parts of the table. New actions are chosen, and the process repeats until the goal state is reached meaning that the problem has been solved. The idea is that an everyday task like making tea can be broken up into discrete sensory/motor steps and that these steps can be translated into a state-action code and put in the table.

5.7 Summary

The past two chapters have sketched the core of the way the brain could use computation to manage behavior. The organization is fundamentally constrained by the slowness of the brain's neural circuitry. This slowness forces the brain to resort to a table lookup strategy in which tried-and-true behavioral sequences are stored so that they just have to be looked up much in the way we would find a telephone number in a telephone book. The set of such sequences that are relevant to the current situation can be scored with respect to reward or the expected gain from using them in this context.

So at this point, you have a pretty good idea, albeit in broad outline, of how programs in the brain might work:

1. You need to do thousands of different behaviors in order to survive.
2. Each of these behaviors can be broken down into ~300 millisecond steps.
3. The part of an individual step that controls the body is represented as a state-action pair.
4. The collection of state-action pairs for any given behavior can be represented as a table.
5. The cortex is the best candidate to be the neural substrate of that table.
6. Actions in the table are scored using dopamine.
7. The brain is designed to pick possible actions with the highest score.

In this broad outline, we have finessed the way neural codes actually get the body to do things, but we will have quite a bit to say about that in the next chapter. We also have glibly side-stepped issues that might arise in managing all the information in the tables. At any moment, should all the possible actions that could be taken compete to be chosen or should there be ways of triaging possibilities? This tricky issue will be taken up in chapter 8.

III Embodiment of Behavior

Although limited exposure to reduced stimuli in an isolation tank may be relaxing, prolonged complete sensory deprivation can cause the brain to hallucinate and/or trigger depression. The brain needs to be engaged in the world in order to function properly. But much more than that, the brain can take advantage of this coupling to use a huge variety of computational shortcuts that take advantage of the world's regular structure. The external world can serve as an indexable memory that triggers behaviors that are appropriate for the situation an animal finds itself in.[108] Primate brains with their huge storehouse of situations and responses are exquisitely designed to exploit this coupling.

Vision is especially organized to interrogate the world in this way. Remember the input to the thalamus constitutes only 6% of the thalamus's total neural connections, giving a compelling hint that the visual system is focused on questions raised by the ongoing agenda rather than on absorbing its raw image input. Chapter 6 develops this position, characterized as "visual routines,"[109] showing that it provides an elegant and succinct account of huge amounts of experimental observations.

The motor has the visual problem in reverse. It is extremely complex, being capable of synthesizing an enormous number of demonstrably different posture changes, and yet to achieve its goals, it needs some way of categorizing these so that it can activate the right movements at the right time. Hence, chapter 7 focuses on the organization of the motor system with the goal of establishing the concept of motor routines. These are encapsulated movements directed toward specific goals that can be used in synergistic combinations.

Having visual and motor routines in hand allows the exploration of the management of complex behaviors in an abstract form, which we do in chapter 8. This boon follows as the encapsulations of visual and motor routines allow them to be evaluated in summary form without examination of the details of their execution. In effect, the abstract computation is an "operating system" for the brain that allows its behaviors to be cast as concurrent and dynamic sets of indexable routines. This organization is particularly amenable to the reinforcement learning techniques introduced in chapter 5, as the encapsulations lead directly to huge savings in the evaluation computations needed for choosing behaviors.

6 Sensory-Motor Routines

The old theory had it that the topic of vision science is understanding how the brain builds up an internal picture. The subject of the new vision science is explaining why it seems to us as if the brain does this, when in fact it does not.
—Alva Noe, *Out of Our Heads* (Hill and Wang, 2009, p. 140)

When confronted with the myriad of complexity of the brain's structure, it is easy to lose track of its ultimate purpose of directing its body to interact successfully with the world. Up to this point, we have described the inside story about the brain's programs; that is, one that emphasizes the perspective of the internal forebrain operations. These programs sample the world in a cortical instant, somewhere between 100 and 300 milliseconds in duration, and use that information to compute the state that is used to decide what action to take next. The selection of the subsequent options is managed by the basal ganglia, and these alternatives are rated by the dopaminergic system. A lot of details have been suppressed, but in broad outline we have an initial picture of the brain's system for managing the programs that govern behavior. However, one of the most important issues that has been finessed is the computation at the *outside interface* of the all-important state needed to guide programs. Here, we will tackle the issue of how input is obtained and used, focusing on the visual system and to a much lesser extent the motor system. The latter will be tackled in much more detail in the following chapter.

Introducing sensory-motor computation requires a jump in abstraction level. In searching for an object to be grasped, it is important to refer to its identifying features such as color or gripping surfaces without referring to all the details of their cortical representation. The result is that the algorithm descriptions will look a lot like conventional program languages. Of course, there must be a way to translate these descriptions into the neural codes of the earlier chapters, but here this translation remains in the background. It is not quite taken for granted though, as we must always be mindful that at some point the translation has to be specified, and not being able to do it is grounds for rejecting the abstract proposal.

Why, in describing the mechanisms for computing state and the action dynamics sequencing through content addressable memory, was any discussion about how its contents were created postponed? The reason is that the body, which the brain finds itself in, plays an essential role in this context.[110] The body has been spectacularly designed by evolution to compute just those things that it needs to direct behavior. It can be thought of as a special-purpose computer, and the computation that it performs does not have to be repeated by the brain. Moreover, and this would be the more important point, given that the body has done a large part of the necessary computation, the computation that is left over for the brain to do is drastically simplified.

Of all the differences between classical silicon computing conventions and the brain's special tricks that allow it to function in real time with its slow circuitry, the body is perhaps the most stunning and unappreciated component. Computation need not be done only by a standard computer. In a wonderful example from Fischler and Firschein,[111] consider finding the shortest path between two selected points in a graph such as shown in figure 6.1. The standard expensive algorithm would start at one end and check all the path lengths to the nearby nodal points, and then repeat the process from each of these. But what about building a physical device that represents the particular graph veridically in terms of having strings of appropriate lengths connecting balls representing the nodes? Now to find the shortest path between any two nodes, you can just pick them up and pull them apart until the string is taut. The nodes on the taut path are the answer! The bodies of animals are like Fischler and Eschlager's special-purpose path finder in that they too are optimized for just the kinds of problems that come up in getting about in the world. The sensory-motor coordination that they do would be extremely expensive if done in a general-purpose way, but that is not what happens. Instead, evolution eschews general-purpose computations for blazing speed in a special-purpose computer simply known as the body.

6.1 Human Vision Is Specialized

In describing sensory input, we will concentrate exclusively on vision. It's not that the other senses of audition, touch, taste, and smell are not important; they are. Furthermore, each has its own set of commitments in terms

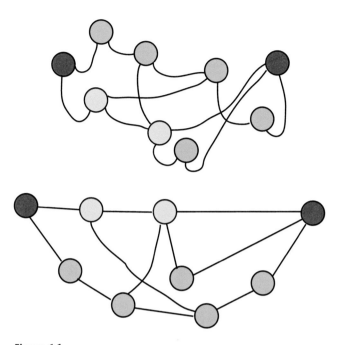

Figure 6.1
Finding the shortest path in a graph is easy if a physical embodiment of the graph is available. (Top) A graph with links between edges constructed from balls and string. The lengths of the string encode distances between the nodes. (Bottom) To find the shortest path between two nodes, grab the nodes and pull. Pulling the two ends apart results in the shortest path being horizontal.

of the transduction of transforming its raw sensed quantities into common neural representations that can be used by the cortex. However, vision has the unique combination of being able to detect stimuli that are very distant and to represent objects of great complexity (even though audition is arguably in a not-too-distant second place).

The hardest thing to appreciate about vision, and a characteristic shared to some extent by all the senses, is that in a deep sense, our phenomenal perception of it must be an illusion. Pause for a moment to take in your visual surround. The sensation is that you are immersed in the best theater possible with a greater than 180° field of view, a sensation of depth and color everywhere. Now consider some of the prominent features of the system that is creating this perception. The sensation of depth is created by the fact that you have two forward-facing eyes that see slightly different views, such that the brain can detect slight mismatches in the position of light and dark elements in each eye and interpret them as depth cues. We can be aware of this matching process by holding up two pencils one behind the other and spaced about a foot apart. We can fuse (match) the image of one of the pencils at a time but not both, because the central matching circuitry can only handle small image differences. In primates and especially humans, the fovea is very small indeed. A good way to remember its approximate size is the width of a thumb held at arm's length. Inside this area, the resolution is approximately 100 times that of the visual periphery. Figure 6.3 shows the fall-off of spatial resolution with eccentricity. The sensation of color is created by color-sensitive cells that are concentrated in a small visual area termed the *fovea* (figure 6.2). There are color cells in the retinas' periphery, but greatly reduced in number.

Given the visual system's special sampling strategy, you can understand immediately that it is important to be able to move the eyes quickly so as to foveate important objects in a scene. Human eyes are relatively lightweight with strong muscles capable of moving them at speeds of more than 700° per second and holding gaze stationary over interesting targets. Consequently, on average a human fixates more than 150,000 gaze points per day.

One human function the high-resolution fovea enables is reading, and it turns out that studying it can reveal a lot about vision in general. In reading, the eyes change their gaze points quickly to foveate almost all the successive words in a line of text. Pioneers in reading research used eye-tracking equipment to discover that only about 16 characters under the momentary gaze point have to be the correct ones.[112] The rest can be changed arbitrarily, as long as letter shape is approximately preserved, without slowing reading speed! Of course, as the eyes select a new gaze point, the computer monitoring the gaze has to change the fixated characters to the correct ones quickly.

It turns out that the reading result is indicative of how vision works in general. Figure 2.12 of chapter 2 introduced Yarbus's seminal work demonstrating that the overall gaze points showed patterns from a subject viewing a famous painting that varied markedly depending on the question asked. It turns out, as we will soon show, each gaze point is chosen for very specific information. You might be already thinking that if this is the case, and the gaze points are chosen at about 3 per second, then there must be a lot of information left out. This is true and has been revealed in ingenious experiments that test "change blindness," which show that we have insensitivity to large unexpected changes in a scene.[113] Subjects viewing two almost identical alternating frames of soldiers lined up to board a

A

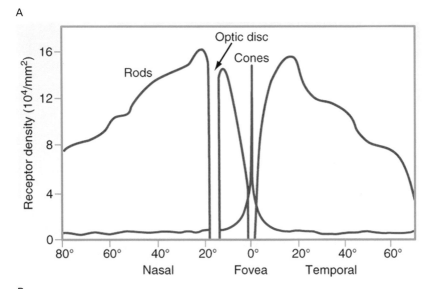

B

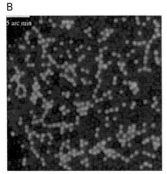

Figure 6.2
(A) The distribution of light-sensitive rods and color-sensitive cones on the human retina. (B) In a very small area in the fovea, the distribution of red-, green-, and blue-sensitive cells can be revealed. (The coloring of the cells is for illustration only after they are identified by other methods.) (B) courtesy of David Williams.

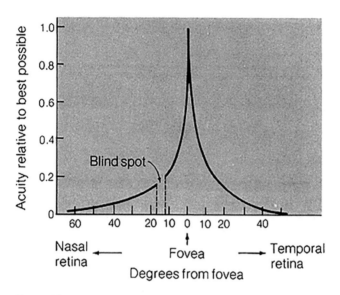

Figure 6.3
The falloff of spatial resolution with eccentricity from the gaze point. The blind spot occurs as the neural axons that carry the signal to the thalamus are grouped into a neural "bus" termed the *optic nerve*, and there are no photosensors over the bus area.

troop plane do not notice that in one of the frames, the plane's jet engines are missing.

This result has been replicated in many different venues. From the vantage point of our prior beliefs about what we can notice, the change blindness result seems astonishing. It is only when we take a different vantage point and posit vision as only an information-gathering device that one can appreciate these observations. So, given that all this dynamic behavior is going on, how can we have the sensation of a seamless visual world? The answer many scientists are gravitating to is that this sensation is a by-product of being able to obtain information on demand. In this explanation, it helps to appreciate the relative timescales involved. For a helpful analogy, recall early television set technology. To make the image on a cathode ray tube, an electron gun sprayed electrons in a line-by-line pattern on the screen, the instantaneous strength of the beam being proportional to the desired lightness of a point. It was only because the eyes integrate that signal that a whole image is seen instead of a rapidly moving miniature electron fire hose. In the same way, it may be that the brain's internal "cognitive" circuitry is significantly slower than the circuitry getting

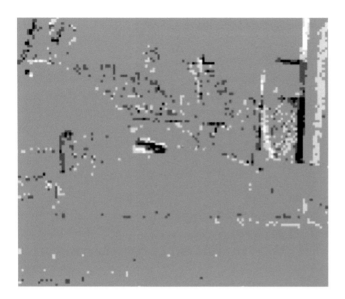

Figure 6.4
An image made during driving from the analog VLSI camera of Dulbreck's laboratory at ETH, Zurich. The camera circuitry is especially designed to turn gray-level
differences into a spike code. A single frame proves difficult to interpret, but temporal sequences of spikes readily convey spatiotemporal structure. Courtesy of Tobi
Delbruck.

information at the sensory periphery: Before the answer to any visual question is required, the visual system has it ready.

One final point to appreciate is that, even if you hold fast to the
notion of seeing the world in all fidelity, just interpreting the retinal spike
code is not straightforward. Figure 6.4 shows the image of a spike code
made with an integrated circuit camera that turns analog measures of
temporal changes into a spike code. Looking back at figure 3.11 from the
salamander retina, you might be disappointed by the sparsity of the spikes,
but remember that the image in that figure was flashed so that there were
temporal changes throughout the array. In figure 6.4, the continuous stimulus is the image seen while driving a car, so temporal changes are fewer
and farther between. Naturally, if you have the full temporal record, it is
much easier to get a sense of the spatial structure, but even as it is, we can
see that the brain has a lot of inferencing to do to make sense of the spike
data.

6.2 Routines

With this initial sketch of the primate visual system under our belt, let's revisit the central point. Because the ultimate behaviors are motor acts and these are carried out by a particular body, the codes that the brain uses, even the ones that support phenomenal perceptions, only have to have the fidelity to direct that body. This point is so important that we will try and emphasize it with what first might seem like a digression: the use of bodies by insects, in particular ants.

Routines in Weaver Ants

Ants are among the most remarkable of creatures (and almost the most numerous: two thirds of the biomass is composed of ants and termites). A tremendous amount of what we know about them comes from the work of Hölldobler and Wilson.[114] Let's just focus on one genus of ant, the weaver ants. These ants build nests out of large tropical leaves by bending them over and gluing the result. The problem of bending the leaves is complicated by the fact that they are much larger than a single ant. What the ants do is form an interlocked multi-ant chain to grab the end of the leaf and slowly crank it over. Then there remains the problem of gluing. The ants' own larvae make glue. So the ants have to pick the larvae up and maneuver them back and forth, squeezing them to get the glue in the right places. Figure 6.5 shows this process under way. The amazing gluing behavior of ants is easy to describe as an abstract program. The gluing program is the equivalent of:

Find larva.
Pick up larva.
Move to the right spot on leaf.
Squeeze larva.

Ants use vision, but most of them also have an arsenal of chemicals that allow them to signal other ants and mark locations. These instructions work because of the ant's body. They are designed for it and depend on its construction and attributes. They also are functional. Consider the use of larvae in our tentative model program. What is a larva to an ant? We don't know, but we can be sure that it is a very minimal functional test that works because on a leaf there aren't any competing alternatives. To drive this point home, consider that horseshoe crabs—an evolutionary relic that is shaped rather dome-like—will mate with an overturned bowl. It's not that they are not fussy; it's rather that their test for a mate is rudimentary,

Figure 6.5
A wonderful example of a computational routine from the insect world. A weaver ant using a larva as a glue source. For the insect, most of the routine is implemented in "hardware," but for mammals, routine-like processing is handled by the forebrain. Photograph © Mark Moffett/Minden Pictures.

and the bowl passes muster. In the same way, the larva test can be very spartan. In fact, the tests can be different for each of the steps. For example "squeeze larva" can be actually just "squeeze," which works fine at the point where a larva is being held over the right spot on the leaf. We also know that the sequence of instructions is very rigidly linked.

The main point is that it's the ant's body that makes this minimalism work. Its biomechanics has been refined by evolution to work in its ecological niche, and the constraints that are needed to get the job of existing and procreating done are hard coded—literally—in the degrees of freedom of the ant's exoskeleton.

Human Visuomotor Routines Exploit Context

Researchers have posited for some time that vision was not solely for the phenomenological experience of seeing but also essential for functional problem solving in space and time. Carpenter and Just showed that, when comparing two patterned blocks of different orientation, the time needed to make a decision as to a match was proportional to the number of block transformations.[115] Kosslyn used mental imagery experiments to show that, for example, when estimating the transit of an imaginary moving clock hand, the time to answer depended on the angle that the hand rotated through. Furthermore, fMRI experiments implied that cortical area V1 was heavily involved in these calculations.[116] Ullman argued that the Marr proposal for cortical representations did not have a functional component and proposed the use of a library of visual routines (see ref. 109).

The implication of all these experiments is that ecological visuomotor computations are functional, and their result often has to be computable in a few hundreds of milliseconds. Thus, for the most part, the computations are task-dependent tests, as those can use prior information to simplify the computations. Suppose you are looking for your favorite cup that happens to have an unusual color. Then rather than any elaborate parsing of the visual surround, you can simply look for a place that has that color. Motor routines need to be efficient for the same reason that visual routines are. Because they are also goal directed, they can make extensive use of expectations that can be quickly tested. If you are searching for the same cup on a nearby table in the dark, you can use the remembered shape of the cup quickly and efficiently to interpret haptic data from your fingers and palm.

Consider another example of walking down a sidewalk with instructions to stay on the sidewalk, avoid objects of a blue color, and pick up objects of a purple color. As table 6.1 shows, each of these tasks can be handled separately, and each of the corresponding vision and motor routines is very simple.

Figure 6.6 shows three different visual tests used in three separate tasks on the same image. The bottom line is that human visuomotor behaviors are directed in a compact, purposeful way. Like the ants, they can also be

Table 6.1

Different tasks have specialized uses of vision and specialized motor responses that avoid the need for complete models of the environment

Task	Visual Routine	Motor Routine
Stay on sidewalk	See if sidewalk edge crosses midline	If so turn away
Pick up purple objects	Find closest purple thing	Head toward it
Avoid blue objects	Find closest blue object	If its heading is unchanging, veer

A

B

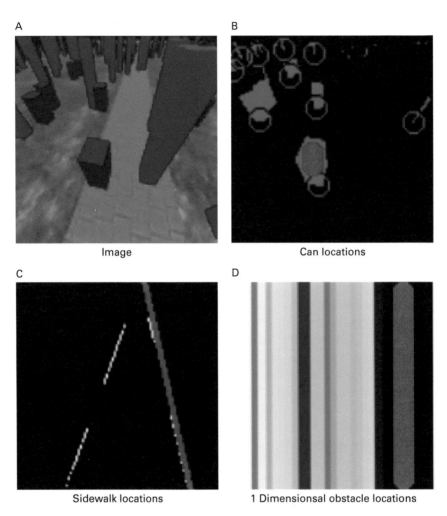

Image   Can locations

C   D

Sidewalk locations   1 Dimensionsal obstacle locations

Figure 6.6
Walking down a sidewalk in a virtual world. Three different tasks can be solved differently by special-purpose visual processing. Finding a litter box that is known to be purple can be handled by using a purple color template. Staying on the sidewalk can be handled by finding the edge of the sidewalk, which is of a known range of orientations with respect to the viewer. Avoiding obstacles can be handled by using a depth map to extract nearby obstacles that are impeding the walker's trajectory.

solved by extracting just the information needed to orchestrate the immediate subtask. The huge difference is that the human brain is programmable and can learn an enormous library of routines from experience.

6.3   Human Embodiment Overview

Thus far, we have covered two major points. The first is that the human
body has been especially designed by evolution to solve data acquisition
and manipulation problems. The second is that the solutions to these problems are very compact owing to the fact that they are purposive. But a
further important point is that these do not have to be re-solved by the
forebrain. Instead, the forebrain can access and direct the form of these
solutions using its own very compact, abstract representations that are tailored to the body's design. Furthermore, as we will elaborate, the vision and
motor system solutions have essential cortical features in common.

Dynamic Reference Frames

Any representation that involves geometric data has to use some kind of
reference frame to describe the data. Huge debates have ranged over the
coordinate system used in vision: Is it head based, eye based, or otherwise?
The fixation system shows that it must be dynamic. Depending on the task
at hand, it can be any of these. Imagine using a screwdriver to drive a screw
into hardwood. The natural reference frame for the task is the screw head,
where a pure torque is required. Thus, the gaze is needed there, *and* all the
muscles in the body are constrained by this purpose: to provide a torque at
a site remote to the body. To develop further the idea of a dynamic frame
of reference with another example, recall that the visual system codes for
depth binocularly by matching the left and right images. In the early stages
of the cortex, the neurons are sensitive to small differences or *disparities*
between the images in the left and right eyes. Roughly speaking, neurons
that have disparity-sensitive receptive fields code for zero, negative, and
positive disparities. But to where in external space do zero disparities refer?
This measurement is produced when the two images match, and this is on
a surface that goes through the fixation point, a locus not in the body at
all, as depicted in figure 6.7.

Like visual routines, motor routines also use dynamic frames of reference. The screw-driving example used for vision also applies for motor
control. The multitude of muscles work together to apply an appropriately
oriented pure torque at the screwdriver end. Another example is balance.
For an upright stable stance, the motor system must make sure the center

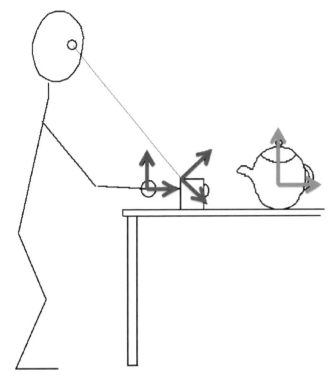

Figure 6.7
Most of human hand-eye coordination can be directed by discrete gaze and hand targets. The hands and eyes use egocentric frames of reference (red frames), whereas objects in the world have their own natural allocentric coordinate systems (green).

of gravity is over the base of the two feet. But the center of gravity is a dynamic point that moves with postural changes and on these excursions can easily be outside of the body. Nonetheless, inside or out, the control system must refer its computations to this point.

Consider the act of grasping a drinking tumbler. The best way to do this would be to orient the hand's grip so that the axis of the grasp aligns with the axis of the tumbler. But of course, the hand's orientation typically does not start out that way at the initiation of the grasping action. The problem is to align the two frames of reference. The tumbler frame is termed an *allocentric frame* as it is outside of the body. Conversely, the hand frame is termed an *egocentric frame*. The visual system also uses an egocentric frame, but as just discussed, it is not attached to the body either but floats in space at the point of fixation.

Cortical Organization Anticipates Embodiment

Visually guided grasping is a very appropriate example because our hand-eye coordination ability sets primates apart from the rest of the animals. Unsurprisingly, in our cortices, both the central area of the visual field and one the central areas of the body, the hands, have expanded representations. The small 1° of central vision that has more than 100 times the resolution of the periphery is known as the fovea. Unsurprisingly, the hand area also is hugely exaggerated with respect to the rest of the body (excluding the face area) in motor cortex area M1. There is not really a name for this, but it's a kind of "movement fovea." At this point, we can compare the visual and motor cortices and see that they have a parallel organization. Vision has two phases: (1) orienting the gaze and (2) extracting information from the gaze. It uses peripheral egocentric information to orient the gaze located in dorsal cortex, and then it uses the temporal cortex to extract allocentric information from the fixated area. Think of this as a "point and read" strategy. The readout of the fixation system locates the tumbler with respect to current position of the body, and the readout of the foveal routines supplies necessary allocentric information. This information is connected to the motor side via the abstract areas of the cortices. In this connection pattern, the motor cortices are similarly organized. The egocentric information computes how to change the body's posture to grasp the tumbler in the large, and the allocentric information helps determine the fine points of the grasp. This organization is summarized in figure 6.8.

Gaze Stabilization

The visual system has six separate systems to stabilize gaze, and their functions are summarized in table 6.2. Having so many systems for one function is an indication of just how important it is to achieve gaze stabilization, but the consequence is that the visual computations can be simplified as they do not have the burden of dealing with gaze instability. Despite the fact that the human may be moving and the object of interest may be independently moving as well, the algorithms used to analyze that object can assume that it remains in a fixed position near the fovea.

Human vision uses fixations that have an average duration of 200 to 300 milliseconds. The fixational system brings home the key role of embodiment in behavior. Although the phenomenological experience of vision may be of a seamless three-dimensional surround, the system that creates that experience is discrete. Furthermore, as humans are binocular and make heavy use of manipulation in their behaviors, they spend

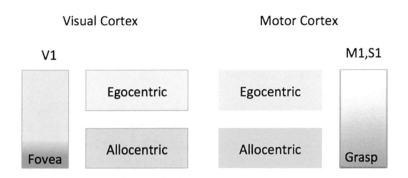

Figure 6.8
The organization of the cortex in human hand-eye coordination shows similarities that reflect the overlying organization of the body. Both the visual and motor areas have exaggerated central representations that are used to extract and coordinate allocentric information.

Table 6.2
Six different methods of stabilizing gaze

Stabilization	Description
Fixation	Hold gaze steady on a target
Saccade	Move gaze quickly to a target
Pursuit	Track a moving target
Vergence	Verge the two eyes on a target
VOR (vestibular-ocular reflex)	Compensate for head movements
OKN (Oculo-kinetic nystagmus)	Compensate for whole-field motion

most of their time fixating objects in the near distance. That is, the centers of gaze of each of the eyes meet at a point in three-dimensional space and, to a first approximation, rest on that point for an average of 300 milliseconds.

Compliance in the Motor System
The adult human motor system is composed of an extensive musculoskeletal system that consists of more than 200 bones and approximately 650 muscles. One of the most important properties of this system is the passive energy that can be stored in muscles. This spring-like system has at least two important properties: (1) it can lead to very economical operation in locomotion, and (2) it can be used in passive compliant collision

strategies that save the system from damage. Moreover, it can be driven by a discrete strategy whereby setpoints for the spring-muscle system are communicated at a low bandwidth. Simulations show that reasonable motions can be obtained with sampling intervals approximating the fixation interval.

There are many examples that could be mentioned to illustrate how the computation done by the motor system makes the job of motor routines easier, but one of the most obvious is the extensive use of passive compliance in grasping. If the motor system was forced to rely on feedback in the way that standard robot systems do, then the grasping strategies would have to be far more delicate. Instead, grasp planning can be far easier as the passive conformation of the multifingered hand provides great tolerances in successful grasps. Figure 6.9 shows a virtual figure demonstrating that grasping can be straightforward once the hand is oriented to cope with slipping and slight alignment mismatches.

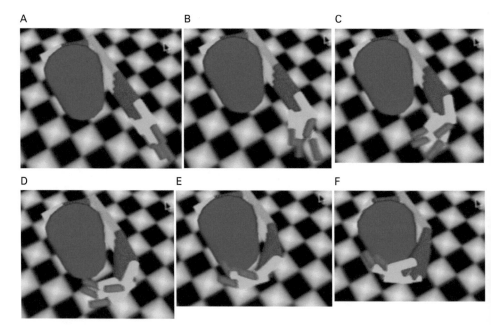

Figure 6.9
The natural springiness of the musculoskeletal system coupled with the use of discrete targets can lead to very simple grasping strategies even when dealing with slippery objects. Here, a model human hand simply has to aim for the center of a cylinder and open and close the fingers. The natural compliance properties of the hand do the rest.

Visual Routines for Identification and Localization

Computation can be greatly simplified by recognizing that the properties of an object can be divided into those that are associated with the object itself and those that are associated with the object's relation to other objects. The former are related to the object's identity, and the latter are related to the object's relative location. Tackling these computations separately greatly simplifies the costs involved. Figure 6.10 shows an example using color features.

You will have recognized that the identification task corresponds to the use of allocentric features gained at the point of fixation, and the location task corresponds to egocentric features measured using the full field of view. Arguments break out among scientists as to whether these distinctions can be so orthogonal, but, appealing to the central theme of computational abstraction, the answer depends on the particular level. One level down leads to the content addressable memory (CAM) models of cortex, and there the different maps should interact. However, the algorithmic behavior can suggest ways in which the distinctions might be made. To locate a pattern, one can fix the pattern's cortical code and let the dynamics settle accordingly. Neurons will be firing in locations that match as suggested by figure 6.10A. Similarly, to identify a pattern, fixating it has the effect of fixing the input code at the fovea, which is selectively exaggerated in the ventral maps, so that the dynamics could settle to a labeled internal model when abstract maps have pronounced firing patterns as in figure 4.10 of chapter 4. A final point to make is that the examples in figure 6.10 are illustrative of the point that very different visual routines would be used for different purposes, and as a consequence there must be a mechanism that uses the cognitive agenda to choose between them, but they are far from the state of the art. The Poggio laboratory, using a form of receptive field model of 1,000 combined V1- and V2-like responses and a high-dimensional state-space classifier, has shown that these features perform competitively to other competitors in benchmark classification tasks in multiclass problems.[117] Thus, it is easy to imagine that, given a restricted context such as provided by a focused agenda, the capability of extracted cortical features is more than adequate to classify foveal image samples. The tricky part left over at this point is to specify the circuitry that does this. However, at this level of abstraction, the job is to suggest constraints on its structure.

6.4   Evidence for Visual Routines

When we think of movements, it seems obvious that the motor program makes choices and communicates those choices to the spinal cord. But it

Figure 6.10

An illustration of the segregation of routines into different tasks using the feature of color. (A) In a location task, the target object's colors are known, but the position needs to be determined. This can be done by computing where the target colors are clustered in the scene. In the upper left image, a dense cluster is a probable location of the colored object. (B) Two examples of an identification task being solved. Here, the location is known to be the foveal area, and the colors from that location—in these examples the entire image—are compared to colors associated with internal models of objects. The insets show the top three models matched.

would be a little strange to think of the sensory systems, in particular the visual system, as working in the same way. Or would it? In fact, evidence is mounting that the visual system may work in just that way. The brain runs a program to do a certain task. That program needs information about the visual world in order to proceed. So visual tests are commissioned to get the requisite information.

The view that vision works this way is tough to swallow because our conscious experience suggests that it works in a different way. The visual world seems constantly present in a rich tableau that we can inspect at our leisure. That this is not the case was given dramatic voice in experiments done by Simons and colleagues that we mentioned earlier under the heading "Change Blindness." These experiments were specifically directed at trying to pin down just what we notice about our surroundings in everyday situations. A student would approach a professor on the Cornell campus and ask for directions. Cornell has an extensive campus, so it took time to get the directions right. While the professor was answering, two rude students carrying a door would barge between the professor and his interlocutor. Unbeknownst to the professor, the three students were in cahoots: As the two passed through, the rear door carrier and the interlocutor would change places. They had practiced this maneuver and carried it off smoothly. The surprising result was that 50% of the professors did not notice anything unusual and carried on talking to the new arrival as if nothing had happened. Since that time, Simons has explored many different variants of this situation. In one setting, a customer checking in at a hotel sees the clerk duck below the counter for a moment only to be talking to a new person who pops up in his place. In another, subjects watch movies where the contents of the scene are changed between cuts. But the results are always the same. Large numbers of people are not aware of any changes.

We need only be surprised at these results if we are lured into adopting the expectations of conscious experience: We think we see everything, so if something changes, gosh darn it, we should notice. However, if we adopt the program-centered view, then the visual processing done is at the behest of the program for very specific purposes. The hotel guest only has to be sure that she is having a conversation with a clerk that will lead to being checked in. The expectations of the interaction are that the clerk is swiping the credit card and coming up with the room key. A moving thing that passes the person test as well as the card and key particulars will be good enough to do the job. Basically, the strategy is the same as that used by the insects. The difference is that the tests can be composed as they are not as directly tied to the motor system.

If routines are composable, what is the brain's programming language? Ullman was one of the first to articulate the logic of visual primitives (see ref. 109), but we are still looking for the right set. Nonetheless, there is dramatic evidence that primates can decompose visual tasks into separate parts. Let's review some of the evidence.

Cortical Evidence for Routines

If the brain is working this way, then given what we know about the cortical moment, the time for these alleged tests is pinned down: They should be carried out within the periods where gaze is fixed. Some beautiful experiments by Roelfsema's group in Amsterdam suggest just that. They trained monkeys to do a visual line-tracing task depicted in figure 6.11. The monkeys had to hold their gaze fixed and determine which of two lines was connected to a central point. They were rewarded for making a saccade to the end of the connected line. Both lines terminated near the point, so it was not too easy. Next, the monkeys had to look at the end of the line that passed the test. The rationale behind the experiment is that the monkey in computing the end point would trace the locus of the correct line. Because the neurons in the cortex form a visual map, a neuron on the line's locus should be activated by the tracing process. The trick in the experimental manipulation is that it takes the monkey a few hundred milliseconds to program the gaze change so that during that time, gaze is fixed and the visual world projection onto the cortex is stable. Thus, they could record the output of a neuron along the tracing path with the hope that they would see a tracing-related signal.

To the experimenter's delight, at 90 milliseconds after the trigger signal, the neuron about halfway along the path fired, suggesting that the line was indeed being mentally traced. Next, the experimenter introduced a more complicated variant of this protocol. The ends of the lines and a nearby central dot had colors. The monkey now had to look at the end of the line that terminated with the same color as the central dot. The hypothesis of course was that the internal program now has three steps; namely,

1. determine the line with the matching color;
2. trace that line to the other end;
3. look at the other end.

If this hypothesis was correct, then the line-tracing operation would be done slightly later as it would have to await the result of the color-matching step. Sure enough, in this case the neurons along the path fired about 60 milliseconds later. This work is some of the most convincing evidence

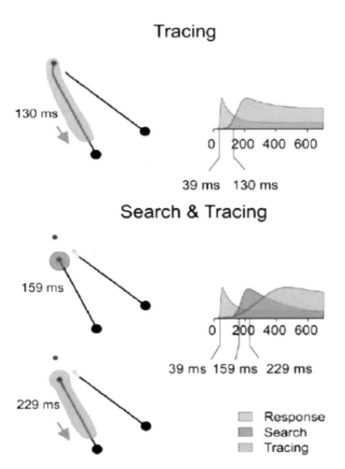

Figure 6.11
An experiment with primates provides evidence for visual routines. The top panel shows an experiment where a monkey has to stare at a point and then, when a pair of lines appears, look at the end of the line that is attached to the point of fixation. The fact that programming the eye movement takes 250 to 300 milliseconds means that the eyes are still for the interim. Recordings from edge cells along the path show elevated firing beginning at 130 milliseconds suggesting that the monkey's brain is using a visual routine to trace the path. In the bottom two panels, the task is made harder, and the result is that tracing evidence starts later (see text). Reprinted with permission from P.R. Roelfsema et al., "Subtask sequencing in the primary visual cortex," *Proceedings of the National Academy of Sciences*, vol. 100, pp. 5467–5472. Copyright 2003 National Academy of Sciences.

yet on the use of specific tests in the course of visual problem solving. In this case, the task has been arranged so that the best way to solve it is to do specific tests in a specific order. But the thought is that in general, it may be that visual processing is broken down into specific tests like these.

Routines and Visual Search

An enormous amount of effort has been expended toward studying routines in visual search tasks, following Treisman's groundbreaking experiments.[118] These showed that, in searching for a visual target in a display with distractors, some configurations were easy to search and others difficult. One class of models of the search process represents the items individually (e.g., Wolfe et al.[119]), whereas subsequent experiments argue for a signal-to-noise effect governing difficulty.[120–123]

Figure 6.12 shows a subject's data for a task where the subject had to remember a small icon and then report present or absent in a larger display via key-presses. In one presentation, the subject saw a very short preview of the image to be searched, and in another no preview occurred. With the preview, subjects could remember the locations of all the items and fixate the right one directly, whereas without it, they had to resort to a different kind of search. Zelinsky et al. showed that a template match model could account for these observations (see ref. 120). Subsequent research has explored fixations during search for categorically specified targets.[124]

Visual Routines in Robotics

Of course, when robots use vision, it is exclusively purposive in that the measurements are subserving some very distinct goal. Nonetheless, some of the techniques used by the roboticists are very suggestive of methods that could be used by humans as well. A computational experiment by Salgin tested the use of a looming routine to stay a constant distance behind a car in a virtual environment. The key trick was to transform the image from Cartesian x and y coordinates into an equivalent polar system that uses an angle θ and the logarithm of the radius (i.e., log r). The latter system is close to that used by the cortex and has some nice properties, an important one being that an expansion or contraction of the image becomes a shift that is very easy to detect algorithmically. Figure 6.13 shows how it works. The outline of a braking truck is just shifted to the right because as it comes closer, its radius gets larger. A visual routine can easily measure the shift. The point is that the problem can be solved without taking the expensive path of recognizing and organizing the objects in the scene. One just has to track the followed stuff and check for expansion (danger of collision,

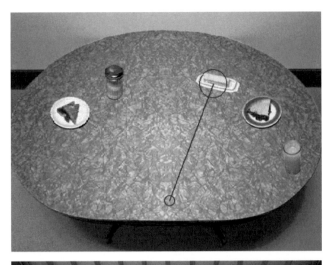

Figure 6.12

(Top) When subjects have previewed the display, they can memorize the locations of individual items and pick out search targets with a single saccade. (Bottom) Without the preview, subjects are forced to use a multiple-saccade strategy that can be explained via a template-matching model. Reprinted from R. Rao et al., Eye movements in iconic visual search," *Vision Research*, vol. 42, pp. 1447–1463, 2002.

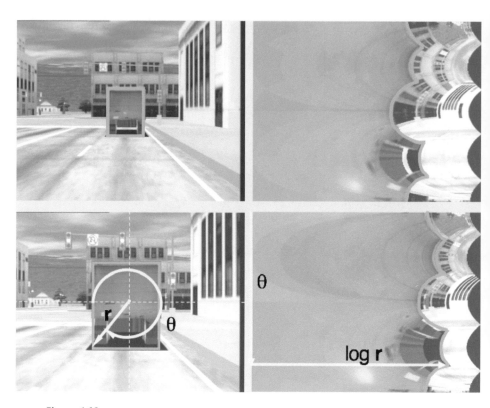

Figure 6.13

A model of the human ability to follow a car tested in a virtual environment. The image is transformed into a log-polar representation whereby the coordinates x, y become θ, log r using the relationships θ = arctan (y/x) and $r = \sqrt{x^2 + y^2}$. In the transformed coordinates, the contour of the followed car simply shifts to the left or right depending on whether it is expanding or contracting, respectively. This relationship can easily be detected by cortical circuitry in principle.

so brake) or contraction (falling behind, so accelerate). A key point for the translation to humans is that the looming computation only works when the target is being fixated, so that if, say, the cortex is doing this, it has to somehow "know" when the observer's fixation is locked onto the target. However, studies of human drivers driving in a simulator show that, when following a lead car, gaze is invariably locked on that car.

Psychophysical Evidence for Routines
The monkey experiments are a tour de force, but still, it would be good to have corroborative evidence that humans are doing the same thing. It turns out that there are a number of highly suggestive psychophysical experiments all pointing in the same direction.

In introducing model visual routines, we used the example of multitasking in human walking. Thus, figure 6.6 showed the computations used by a model of the humans' behavior. But by using immersive virtual environments, actual humans can carry out identical scenarios. It turns out that they have different routines for picking up objects and avoiding them and reveal this by using different fixation patterns in each of the two cases. Figure 6.14 shows the fixation points as dots in each case.

Even when integrating vision with manipulation, the results are the same, as evidenced by a set of experiments conducted by Triesch et al.[125] using the capability in virtual reality at the Rochester Computer Science laboratory, which includes the ability to create artificial sensations of force when picking up a virtual object using a dual Phantom force-producing system. In Triesch's experiment, subjects picked up virtual blocks and put them on one of two conveyor belts whereupon they were whisked away. The subjects sorted the blocks by a size feature. The subjects were told that the software running the experiment was buggy and to report any changes. Three separate sorting conditions were used that varied according to different initial instructions. The differences were chosen with the view to changing the subject's internal program. Here they are:

1. "Pick up the blocks in front to back order and put them on the closest conveyor belt."
2. "Pick up the tall bricks and put them on the conveyor belt. Then pick up the small bricks and put them on the closer conveyor belt."
3. "Pick up the tall bricks first and put them on the closer conveyor belt. Then, pick up the small bricks and put them on the distant conveyor belt."

On 10% of the trials, the size of the block was changed en route. Because the trials occurred in a virtual reality environment, this can be done

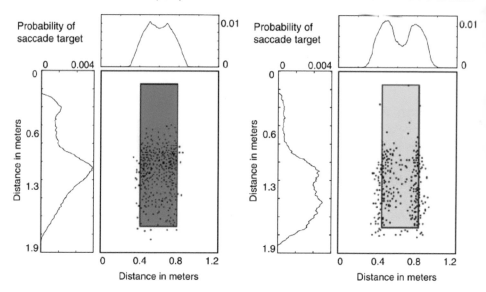

Figure 6.14

An experiment where subjects walk through a three-dimensional virtual environment and have to approach purple targets and avoid blue targets. Even though the targets have identical shapes, a subject's gaze is used differently in the two cases. When approaching a target, the gaze is directed toward the target's center. When avoiding a target, the gaze is directed toward one of the edges, the subject keeping his distance and by implication using a very different algorithm than when approaching.

seamlessly. If you think about these instructions for a moment, you will no doubt guess the hypothesis. The importance of size is varying in the different instructions. In the first case, size is irrelevant; in the second, size is important at the time of pickup; and in the third case, it is important both at the time of pickup and release. So if subjects are using an internal program that tests for size, they should notice more of the changes when size was relevant. This is exactly what happened. The average of two independent measures showed that subjects noticed about 10% of the changes in case 1; 25% of the changes in case 2; and 50% of the changes in case 3.

This paradigm was expanded by Droll and Hayhoe,[126] who used separate cues to signal the pickup and release action. For example, subjects might be cued to pick up a red block from a choice of red and blue, and then at the conveyor belts they would be cued to put red on the left belt. Subjects are instructed to throw any changed blocks that they notice in a virtual trash bin, depicted as a square dark hole in the middle of the display. One of the virtues of this design is that it provides information on the subject's internal state for missed changes. Suppose a red block turns to blue and the change is unnoticed. Will subjects put the block on the conveyor belt indicated by the red cue or will they put it on the conveyor belt indicated by the blue cue? In the first case, they registered the color as red and did not repeat the check. In the second case, they apparently would have read the release cue but forgot the color at pickup. It turns out that subjects overwhelmingly sort missed changes by the first cue color. Figure 6.15 shows stages in the experiment with the subject's gaze superimposed on the display.

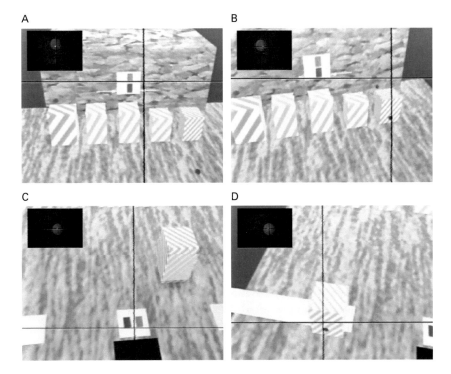

Figure 6.15
(A–D) Four frames from the moment-by-moment eye fixations recorded as a subject places a block on a conveyor belt in a virtual environment. The red circles show the position of the fingers. Even though fixating the block that has changed color, the subject fails to notice the color change. The color is registered by the retina but apparently not monitored by the observer.

These results fit in with the general theme that vision is the result of explicit tests in the agenda of an ongoing program. From this perspective it all makes sense; any program should not keep testing for color because the way the world is organized is that objects do change color. To keep checking would be a waste of effort. But you can see that we are back to the ants. We share the enormous efficiency of just probing the parts of the world that we need to drive behaviors. This view might seem counterintuitive at first, but as evidence for it accrues, we will have to get used to it, just as we accept that the stuff of the world is made up of atoms or that atoms are mostly space. But we still need a computational account for the phenomenology of conscious experience. This will be developed extensively later, but for the moment we will have another taste of the Bayesian viewpoint, which has rapidly become mainstream. One example of this was the motion-perception illusion in chapter 4, which used the all-important notion of prior statistics. Think for a moment about the inside of your refrigerator. Is it light or dark in there? We have a prior that the light is switched off when the door closes, but we have only ever seen its inside state with the door open. In that case, the light is on. In fact, every time you check, the light is on! This is a good analogy to help with the mechanisms that drive stable visual perception. Anytime we think of a way to check our perceptions, the test always comes back in a way favorable to the way the world actually is. In the case of the block, the supposition is that surface reflectance is invariant.

6.5   Changing the Agenda

It is tempting to finish with this lofty analogy and move on. But there is one hugely important issue that has been glossed over, and that is the bugaboo of a purely agenda-driven system. How does one ever change one's agenda? Computer systems have this problem, too. If a peripheral device needs to be monitored, how should this be handled? It would be expensive to keep polling the device to check its status. It would be a little like repeatedly asking of the held block in Droll and Hayhoe's experiment, "Are you still blue?" The way this is handled is to have the device interrupt the computer's operating system with a request for service. So we know how to do this for silicon computers, but how is the function handled in the brain? We will have a lot to say about this two chapters hence, but for now let's introduce the main idea that has been thought of so far, which is *image saliency*.

A   B   C

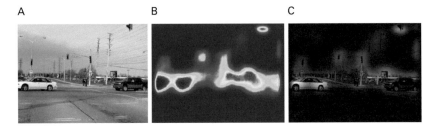

Figure 6.16
The Bruce and Tsotsos image salience model is representative of attempts to define potential image locations on which to focus attention. (A) Original image. (B) Saliency map. (C) Saliency map used to highlight image locations. From N. D. B. Bruce and J. K. Tsotsos, "Saliency, attention, and visual search: An information theoretic approach," *Journal of Vision*, vol. 9, pp. 1–24, 2009.

In 1985, Koch and Ullman[127] proposed the idea of visual saliency for grouping places in an image that exhibited an overlap of features such as texture, color, and motion as important. This idea proved enormously compelling, and a huge effort has been devoted to improving the metric used in deciding on interesting feature combinations. An example of recent work is that of Bruce and Tsotsos,[128] shown in figure 6.16.

The saliency model in figure 6.16, which weights feature clusters in terms of their probabilities, can account for a significant portion of human observations, but, as you might expect from all the evidence of vision subserving a visuomotor agenda-driven perspective, there are things left out. One is the use of all the prior information that the brain has about complex interrelation of visual objects. Another is the role of the task in modulating the purely feature-based approach. A final example will sensitize us to this point. Rothkopf et al.[129] had subjects walk down a virtual sidewalk and carry out the tasks illustrated in figure 6.17. Before the start signal was given, the subjects fixated objects that were irrelevant to the task, but while they were executing the task components—staying on the sidewalk, picking up litter objects, and avoiding obstacles—they ignored the distractors almost completely. Of course, the subjects may have an agenda when surveying the distractors, but it is hidden from us. But the conclusion that the suite of behaviors changes between the before-task and during-task intervals seems inescapable. The implication of the direction of visually guided behaviors using task-specific visual routines is that there must be some way of managing the particular sets of routines that are chosen at any instant.

A

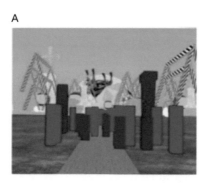

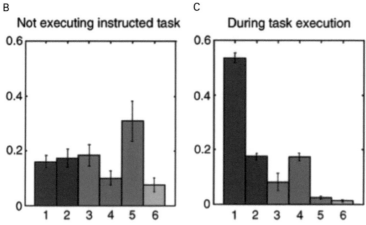

Figure 6.17
A walking experiment in a virtual environment. Code: 1 = litter, 2 = obstacle, 3 = sidewalk, 4 = grass, 5 and 6 = distractor objects. While subjects are waiting to start, they fixate distractor objects, but when their transit begins, they focus almost exclusively on task objects.

And, importantly, there is the problem of choosing this suite in a changing world with varying demands. It would take us too far afield to summarize all the additional work addressing these issues, but for a system that addresses the issues of prior knowledge, see Oliva and Torralba,[130] and for another that introduces task constraints, see the work from the Itti lab.[131]

6.6 Discussion and Summary

The concept of routines is a very useful abstraction that is wedged between two very different levels above and below it. In chapter 8, where whole behaviors are summarized, the results of such routines will be taken for granted. Here the cortical mechanisms for obtaining the information are glossed over, but nonetheless the interface between the two levels is important, as the usefulness of the routines abstraction depends on having a way of translating between them. The discussion of cortical representations focused on the feedforward pathways that connect low-level representations to more abstract representations. Such direction follows the transformation of image data into objects and situations. The reason for revisiting this discussion here is that the routines logically send information in the opposite direction. To conduct a test, for example the search for the specific location of a colored shirt depicted in figure 6.10, the abstract description of the shirt in terms of its set of colors needs to be communicated to neurons in the front of the visual system, such as V1 and probably the LGN, where the test is carried out. This idea was initially speculative, but the evidence for it is increasing.[132–135] Note that the result of the test is logically used at the most abstract level in the cortical hierarchy, wherein it is used to direct a behavior. Thus the logical transit of activation in the cortex for a visual routine is first of all top-down and secondly bottom-up.

Given that the resolution of a visual routine suggests a two-way cortical transit, one can wonder whether the feedforward initiated use of cortex also has a role for two-way communication. As noted in chapter 4, a focus in the feedforward pathway has been in fast recognition. However some caveats are necessary in this venue. First off, the transit times in the cortical hierarchy are in the worst case on the order of a hundred milliseconds, leaving ample time for a two-way transit within the cortex's putative 300-millisecond processing window. Moreover, the times between areas can be on the order of a few milliseconds, so that the number of possible transits may be considerably greater. A similar issue arises in repeated trials wherein feedback effects from prior trials can condition the network to be adapted to anticipate the current trial's input. This routing of information has a ready interpretation in a Bayesian framework. Feedforward neural activation can trigger abstract priors high in the hierarchy that subsequently modify the lower level signals. In this analysis, the power of abstractions comes to the fore. The transit order for visual routine signals, being feedback first, leads directly to the consideration of the feedforward case, and suggestions for enhanced signal processing.

The use of visual routines advances a utilitarian view of visual processing that is very different from the traditional Cartesian view of recognizing objects and relations between them.

In particular:

1. Information is sampled directly from external three-dimensional space by utilizing the eye's ability to focus on distant points.
2. The information acquired is specific to the brain's internal task agenda
3. In order to navigate around three-dimensional space the brain must keep some kind of representation of it, but the extent of this representation is an open question.
4. In order to be sensitive to new situations and changes in agenda, the visual system must have some way of sensing novelty. This has been extensively studied and there are some promising directions, but to date a satisfactory solution is still unresolved.

7 Motor Routines

What could possibly motivate precise and highly coordinated manipulations with a needle, a carpenter's plane, a chisel, a microscope, a micromanipulator/or calligraphic pen had these movements become at some point self-contained not only with respect to their motor composition but also with respect to their purposeful structure and meaning?
—N. A. Bernstein, *On the Construction of Movements* (Medgiz, 1947; p. 43 of David Adams translation)

The previous chapter argued that the most important aspect of vision is its deployment of visual routines that acquire just the right information needed to direct the brain's ongoing agenda. Here, the case is made that the motor system is organized in a similar way, composing "motor routines" to achieve action goals. The pioneering Russian neurophysiologist Nikolai Bernstein makes this point obliquely in the above quote, stressing that it would be inconceivable to understand the specializations in observed human movements if they did not have a useful generality. However, deconstructing the human movement system into primitive routines is a daunting task owing to its sheer complexity. There are about 650 muscles, some of the principal ones being shown in figure 7.1. The muscles span 206 bones that interconnect via various kinds of joints. Such joints typically provide several possibilities of motion characterized formally as degrees of freedom. For example, to a first approximation, the shoulder is a ball-and-socket joint that has 3 degrees of freedom: rotations in a two-dimensional space of pivots and a limited rotation about the pivot axis. The elbow has 2 main degrees of freedom: It is primarily a hinge joint, although the bones of the forearm allow the wrist to rotate about the forearm axis as well.

A particular setting for all the joints results in an overall posture, and movements can be understood as changes in posture, which require changing the settings of all the degrees of freedom. The degrees of freedom

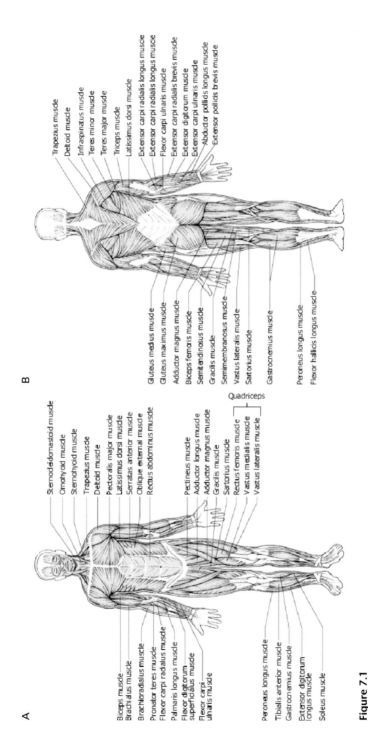

Figure 7.1

(A, B) The human motor system has a huge array of 650 muscles and approximately 300 degrees of freedom of motion that are coordinated in posture changes. From *Current Procedural Terminology*, Revised 1998 Edition, AMA, 1998. Copyright 1995–2014 American Medical Association. All Rights Reserved.

themselves are controlled by changing the length of muscles, and this process in itself is very complex. At the very lowest level of molecules, this control is implemented with a beautiful design. Neural signals in the spinal cord activate tiny gear-like molecular structures that rapidly ratchet over each other to change muscle length. At a much more macroscopic level, the muscle tissue is also important for its passive properties, forming a complex damped-spring system that can store energy and facilitate complex interactions with surfaces.

With respect to the vision system, we can see right away that the general system is far beyond the eyes' movement system in complexity. The eye-gaze system allows a model of a low-inertia pair of spheres driven ballistically by relatively very powerful muscles. In contrast, the rest of the movement system is far more complicated and importantly has to handle significant mass and inertias, with the consequence that large-scale motions that are governed by dynamics need to be dealt with. If you are running at full tilt, it will take you more than one stride to stop.

As a result of all this complexity, there are many different computational approaches cast at many different levels of abstraction. However, these can be divided into two main foci: robotics and humans. The one that has received the most computational attention is the focus on robotic systems and tends to sidestep biological details to expose the fundamental constraints from an information-theoretic perspective. The second broad approach makes the actual biological features the centerpiece and tries to deconstruct the complexities of the animal system into manageable components. We will briefly touch on the basic computational issues, but our primary emphasis will be on descriptions of the movement system's musculoskeletal components. Computational approaches within this focus take biological constraints into account. For example, a standard technique uses memory to speed up computations that must be done with slow neural circuitry. Hence, the main exposition will describe how the movement system can be thought of as a library of remembered movements, implemented by the spinal cord as posture changes. Analogous to the visual routines of the previous chapter, these can be thought of as "motor routines" that can be activated in service of goal-directed behavior. Bernstein's seminal observations can be interpreted as an appeal to motor routines, but the first specific model, which we will discuss in due course, was developed by Feldman.[136,137] Routines have since been given an elegant abstract interpretation by Bizzi and Mussa-Ivaldi[138,139] and most recently by Ting.[140]

7.1   Motor Computation Basics

Whether human or robot, all physical systems have to obey Newton's laws of motion, and for massive systems with multiple degrees of freedom, the associated equations can be very complex and not easily solved. The equations involved in movement generation can be factored into kinematics, the geometric degrees of freedom of the system, and dynamics, the movement of the system accounting for mass and inertia. Let's briefly introduce the main issues that make this problem difficult.

Kinematics

Human kinematics studies position, velocity, and acceleration of joint angles.[141] The forward kinematics problem consists of finding the position and orientation of some configuration of end effectors in a three-dimensional world. Thus, a relationship is sought between the laboratory coordinates, which is modeled as a Cartesian space, and the joint configurations of the skeletal system. In the case that the skeletal configuration is known, the Cartesian configuration of the end effectors can be obtained by a well-defined transformation. However, in movement control, the goal of the movement is usually described in Cartesian space, and a coordinate transformation is required to find a solution in joint space. This problem, known as the *inverse kinematics* problem, is much more difficult than forward kinematics because the task space is usually the Cartesian world with its six coordinates, while the number of joints is very much greater. It is not easy to solve, and all the methods for doing so have to introduce extra constraints.

Dynamics

Given a kinematics solution, the next step is to calculate the forces and torques required to actually drive the skeletal system to move along the kinematics trajectory. Similar to kinematics, the dynamics step can also be framed in two ways: the forward dynamics and the inverse dynamics. The forward problem is to determine the trajectories expressed in joint values, velocities, and accelerations, given a series of forces and torques applied to the skeletal system. The inverse problem is to determine the required time sequence of joint forces and torques to produce the desired time sequence of positions, velocities, and accelerations. As with kinematics, it is the inverse dynamics problem that is much harder than the forward dynamics problem, but this is the one that needs to be solved to be able to make movements. It is hard because the equations are nonlinear and typically

Table 7.1
Difficulty of solution of Newton's equations by robots and humans

Equations	Robot	Human
Kinematics	Easy	Easy
Inverse kinematics	Possible	Hard
Dynamics	Easy	Easy
Inverse dynamics	Possible	Hard

Note: Directly solving Newton's equations poses problems for humans and robots. Robots can solve them in principle, but typically lots of computation is required. Without additional constraints, inverse calculations are impractically expensive for humans.

have to be solved using many iterations that take considerable time. This makes them an option for very fast silicon computers used by robots but is a basic nonstarter for humans unless, as with the inverse kinematics problem, extra constraints are added. Table 7.1 summarizes the situation.

Surface Contact
A special problem arises with *manipulative* motor control, and that is the interaction problem. One needs to control how the end effectors react when their surfaces make contact with the designated objects. To motivate the discussion, let's consider collisions. Assume that the sensory feedback on the end effector takes about 150 to 200 milliseconds to return to the brain and another 150 to 200 milliseconds for the motor commands to be sent back to the muscles. During this 300 to 400 milliseconds, a rigid arm moving at a speed of 10 cm/s would have penetrated into the object for 3 to 4 cm. Naturally, the fingers would be damaged for most objects. Fortunately, the human skeletal system is relatively light and has forms of compliance in the form of the passive spring-like properties of muscles and the cushioning of skin (table 7.2).

If the damage is to be avoided in conventional robot feedforward control models, such models need to be extremely accurate and work at very high bandwidth to send correct motor commands. But as noise and other disturbances are always present in the real world, motor commands are rarely able to bring the end effector to the accurate position. Another potential way of avoiding damage is to use feedback control. But the time element means that the servo rate on any controller has to be very high. For example, the servo rate of the hydraulically controlled arm of the SARCOS robotic system is 40 kHz, a rate that is unrealistic for biological systems.

Table 7.2
Comparison of basic features between robots and humans

Feature	Robot	Human
Strength-to-weight ratio	100:1	1:1
Bandwidth per degree of freedom	10,000 Hz	10 Hz
Compliance	Reactive	Passive, programmable

Note: A comparison of basic features shows extensive differences between humans and robots. Reactive compliance uses high-speed servoing. Passive compliance uses material properties that can be adjusted (i.e., are programmable).

In contrast, biological systems use layers of different approaches. One of the most basic is the reflex system mentioned at the outset. This is a bit similar to the robotics strategy. For certain emergencies, such as touching a hot plate, the spinal cord circuitry bypasses the forebrain and directly accesses a much faster connection to a retractive motor response. Higher levels have more considered approaches. When planning the movement, a person's motor system can anticipate the kind of interaction with an object that is coming up and preset the musculoskeletal system to be either stiff or compliant as appropriate. For example, when splitting wood trunks for firewood with an ax, one wants to have the grip stiff at the start of the stroke, so the rapidly accelerating ax is stable, but one wants the grip to be relatively pliant at the wood-contact moment to avoid chafing and blisters.

7.2   Biological Movement Organization

The foregoing discourse made headway by focusing on the computations that had to be done without worrying about the details of exactly how the biological system performed these calculations. But a very different and complementary perspective emerges when we consider the biological substrate. Because computational demands are severe in relatively low degree-of-freedom mechanical systems, the prospects for scaling classical optimization techniques up to the human musculoskeletal system, which has approximately 650 muscles and much more than 300 degrees of freedom, are remote. Conventional robotic systems like the SARCOS need to control each degree of freedom at rates exceeding 40 kHz. Given that the neurons in motor cortex that are directing the posture changes produce spikes at rates that are maximally in the 50- to 100-Hz range, it is very unlikely that the cortex can produce these rates. The neural "bus" to the spinal cord, at ~5 × 10^6 axons, is slightly larger than the 10^6 "bus" of the optic nerve, but keep in mind an additional overhead. The silicon computers are using high-precision floating-point numbers, whereas the neurons are sending noisy single spikes.

Humans are an existence proof that somehow they must solve this problem, as they have exquisite motor coordination. At the same time, because their physical properties are so different from those used by standard robotics platforms, one must be prepared that the humans' solution might look very different than those used by their robotic counterparts. The biological solution is conveniently thought of as being composed of two main parts, as shown in figure 7.2, which is slightly modified from Squire et al.[142] One component addresses the key way to go faster, which is to precompute movements that are needed and store them in the spinal cord; the problem of computing the movements cannot be circumvented, but it can be amortized by stringing it out over the timescales of evolution and development. This is denoted in the bottom half of the figure in green. The second part of the movement story consists basically of the varied roles of forebrain and cerebellum, which manage important issues related to the selection of movements and the management of their execution, as summarized by the section of the figure colored in red.

Let's elaborate on this dichotomy in reverse order, starting with the top. We will elaborate on all these issues subsequently, but discussing the provisional functions of the subsystems in overview provides a framework for ancillary dependent issues. First the cortex. In choosing a movement, there must be some way to translate the goal of the movement, which is typically specified in an allocentric coordinate frame, into an index that allows the body to access appropriate sequences of posture changes and dynamic parameters such as stiffness to achieve it. This is a thorny problem but easier than the original unstructured problem, as we now have a prescription for how the basic elements of the movement are going to be represented: they are to be stored and looked-up using some index when needed. Because it will be impossible to store a movement that would accomplish a goal exactly, there must be some method of adjusting stored movements for the particular circumstances. One way this could happen would be if it were possible to interpolate the movement parameters linearly. That is, if the stored movement is parameterized somehow, the difference between the stored movement outcome and the desired outcome could be small enough so that by adjusting the parameters, a revised movement would achieve the goal. The bottom line is that the cortex specifies the movement plan in body coordinates.

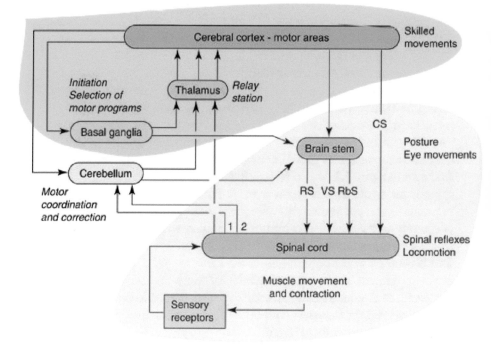

Figure 7.2
Squire and colleagues' concise diagram showing the organization of the movement
control system. At the top level, the cortex plans movements in segments, carrying
out the translation from world to joint coordinates, and different possibilities are
evaluated by the basal ganglia. Movement segments come with feedback expectations, which are checked by the cerebellum. The bottom level consists of a very
elaborate spinal cord reflex system that (1) is fed by the brainstem and (2) monitors
sensory receptors. Commands from the forebrain are organized into a functional
hierarchy: VS, vestibulospinal inputs instigate tonic activity in antigravity muscles;
RS, reticulospinal inputs activate stereotypical movements and control upright body
posture; CS and RbS, corticospinal and rubrospinal inputs transmit commands for
skilled movements by modulating the spinal cord. Adapted from L. Squire et al.,
Fundamental Neuroscience, 2nd ed. Academic Press, 2003.

The basal ganglia is very important for the valuation of planned movements, but as it was discussed extensively in chapter 5, it will be given short shrift here. In recapitulation, the basal ganglia is important for learning programs and their value and can learn by acting them out or by simulating them. So if there are different movements to choose between, the basal ganglia should be playing a pivotal role.

The basal ganglia is one set of circuit loops that compute important parts of the movement execution demands in terms of value, but another part is to predict in detail what the consequences of a movement should be. To save time, rather than make a movement and then evaluate its consequences in a *tabula rasa* state, one can predict the consequences ahead of time and just verify whether or not the right thing happened. In the case of a string of successes, the execution of the movements can be quickly checked and the whole process sped up. Furthermore, if the model parameters need adjusting to improve the prediction, the information to make this adjustment is ready at hand. The loop that does this goes through the cerebellum. Thus, the essential components of movement planning, valuation, and evaluation can be provisionally mapped onto the cortex, basal ganglia, and cerebellum, respectively.

You will have noted that the cerebellum was uncolored in figure 7.2. The main reason for this is that its functionality is not completely settled. One part of the controversy turns on whether it functions as a slow unclocked circuit that checks the end point of movement segments or whether it is a faster time-locked control loop, micromanaging a movement as it happens. The former view has long held sway but recent data invite the latter.[143]

Now let's turn to the issue of movement precomputation and storage. The way the basic human musculoskeletal design can improve things is to use muscle synergies to augment the basic mechanical degrees of freedom. By using set contractions, the muscles can guide the complete motor system through a series of posture changes. These are designed to both respect the basic point limitations of the mechanical system and at the same time be efficient. Efficiency can be enhanced by exploiting the passive properties of the muscle system. A central feature of the muscle systems of mammals in general and humans in particular is that they have a large number of physical structures that can store energy passively. Passive storage can also make the movement dynamics very energy efficient. Thus, during the execution of the movement, the extra momentum when the end effector contacts the object is passively absorbed by the elastic muscles without hurting the limbs. For the cat, it has been estimated that 70% of the energy used in walking comes from the natural elasticity in its muscles, as the passive energy stored in the muscles can be used as a springboard for the next action step. A certain percentage of momentum energy is transformed into potential energy stored in muscles when the feet make contact with the ground. This potential energy is reused in the next cycle to push the feet forward.

The strategy of choosing just the movements that are needed for behavior is refined in mammals, but a lot can be learned about basic movement coding by studying simpler animals. Giszter, Mussa-Ivaldi, and Bizzi have studied frog reflexes extensively, and what they found coded in the frog's spinal cord are movement primitives. In the frog's spinal cord, the neurons are encoding whole time-dependent movement patterns, which can be used in combination to produce variations in a basic movement.[144] If a frog wants to scratch itself, perhaps to get rid of a parasite on its back, the location of the scratch site can be adjusted by changing the combination of spinal neurons that activate the movement. Figure 7.3 shows their original result.

The researchers found this out by varying the point of electrical stimulation in the spinal cord and observing that the scratch site changed systematically. Also, co-stimulating two points resulted in the frog interpolating its scratch site between the two. If it helps, think of (musical) chords in the spinal cord to remember this idea. A candidate for the world's most brilliant and prolific composer is Wolfgang Mozart. If you attempted to capture all of his music by digitizing the concert sounds, it would take an enormous amount of space, but if you instead saved the sheet music, all of his compositions would fit on one compact disc. Crudely put, the forebrain has the sheet music but the spinal cord, combined with the body, makes the music. For reflexes such as scratching, the spinal cord can play its own tune.

In mammals like us, movement reflexes are even more complex. Early experiments on cats showed that a cat has highly sophisticated control systems in its spinal cord. If most of the brain of a cat is removed, sparing the spinal cord and brainstem, that cat can walk, with a little support help on a treadmill. Thus, the basic coordination of the gait and support of the weight of the body is all coded in the spinal cord. Even more amazing is that when a small obstacle is placed in the path of the cat's foot, the moment that foot detects the obstacle, it is retracted and lifted in one smooth motion so as to clear the obstacle and return to the treadmill base. Much subsequent research has confirmed that the spinal cord has an extraordinarily complex system of central pattern generators (CPGs) that can be modulated to produce different movements. Thus, as the review by

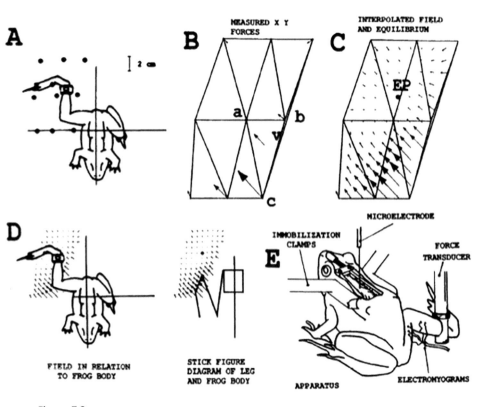

Figure 7.3
The frog's hindlimb, when stimulated in the spinal cord, will swipe its back at a
certain central point. Restraining the limb away from this point induces a restorative
force that increases with displaced distance. Thus, wiping reflex can be seen as having an equilibrium point (EP on the figure). It's as if the activated muscles can be
thought of as being rubber bands with a natural stretching point, such that when
deviated from the EP, they are configured to return to it. From S. F. Giszter et al.,
"Convergent force fields organized in the frog's spinal cord," *Journal of Neuroscience*,
vol. 13, pp. 461–491, 1993.

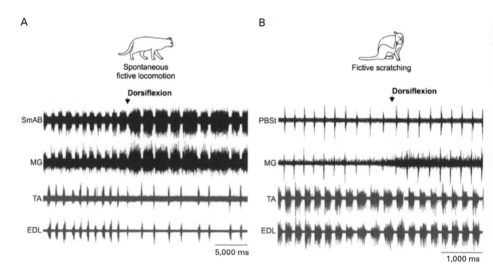

Figure 7.4
Stored primitive movements in central pattern generator circuits in the spinal cord
can be combined to generate different kinds of complex movements. Here, the CPGs
are modulated differently to produce the muscle activations for locomotion and
scratching. From A. Frigon, "Central pattern generators of the mammalian spinal
cord," *The Neuroscientist*, vol. 18, p. 1. © 2012 SAGE Publications. Reprinted by permission of SAGE Publications.

Frigon[145] shows, the same muscles may be used in locomotion and scratching, as long as they are modulated differently (figure 7.4). Runners who
have inadvertently tripped can be grateful for these reflexes, which make
balance-restoring movements that they may not have time to think about
and consciously correct.

A further constraint to introduce is that the movements need to have
natural breakpoints signifying the beginning and end, such as changes in
contact with surfaces. The reason for this is that, if the movements are
going to be learned posture changes, then the combinatorics of variations
in long complex movements would overwhelm any system for remembering them. Thus, there has to be a natural length scale for movements that
could allow them to be economically stored and composed into more complex forms. To summarize so far, movements are coded as a series of posture
changes, and the dynamics between the changes is controlled by a combination of active control and passive stored energy.

A last helpful constraint to consider is the mechanical design of the
skeleton itself. The value of having just the right mechanical design is

Figure 7.5
The mechanical walker from the Ruina laboratory can "walk" unpowered down a gentle slope using only the counterbalanced moving parts, a demonstration that sensitizes us to the fact that a lot of the solutions to difficult problems of human movement are built into the skeletal design itself. Courtesy of Hank Morgan.

illustrated by work at the Ruina laboratory at Cornell,[146] which has created anthropomorphic bipedal designs that can successfully walk downhill in a completely passive mode. Figure 7.5 shows one of their early designs. This device speaks volumes about the human design in that much of the machinery needed for the movements necessary to survive has been incorporated in the mechanical design itself. This point needs some elaboration to be appreciated. We know that the human system is extremely complex and is capable of many different kinds of movements, but when cataloging its movements as a fraction of all possible movements, one can see that this fraction is very small indeed. There are various ways to do this, but let's consider just two contractile states for each muscle. Mathematically, there should be at least of the order 2^{300} possibilities, but the actual number of movements humans make is just a minuscule fraction of this estimate, as the degrees of freedom are used in concert and combined in small repertoires (e.g., in walking). Given its rarity in the space of possibilities, we can appreciate what a design achievement the musculoskeletal system is. Innovative simulations on small scale walking systems demonstrate that it is not easy to get all the muscle attachments right via computer experimentation.[147] Moreover the actual attachments in the human system serve to enhance its postural stability.[148]

With the spotlighting of the vital role of the skeleton in movement, our introductory tour of the movement system is complete. However, to develop the big picture, many essential details were glossed over. Now that we have a context, we can develop a computational perspective via a final top-down traverse through this complicated system. Let's get started with the cortex.

7.3 Cortex: Movement Plans

Although a skilled movement executed by a human is extremely complex, coordinating the 650 muscles of the body in a sequence of routines, such as done by an ice-skater carrying out her program, one has to expect that the cortical representation of movement would somehow have the ability to refer abstractly to these complexities. The organization of the visual system is a useful comparison in this respect. Primate vision has very precise methods for quickly stabilizing gaze on targets in the visual field. Most of the connections use that stability to analyze the foveated area with elaborate precision. This organization is highlighted by Lennie,[149] who makes the point that the connections in the neural hierarchy are heavily weighted toward the analysis of the pattern with respect to the fixated coordinate system. In an analogous way, the movement patterns of the body have to be coded with respect to set frames of reference. In playing the piano, the finger movements heavily depend on the dimensions of the keys and the position of the seat to be in a very precise arrangement. Similarly in most sports, players use the position and orientation of the head as a coordinate system with which to access standard body movements. Golfers have to keep their heads still if they are going to compete on the PGA tour. Thus, the motor system turns out to be unexpectedly similar to the visual system in that the emphasis is on elaborately coded patterns with respect to standard frames of reference. With this in mind, let us turn to the organization of the cortices that encode movements.

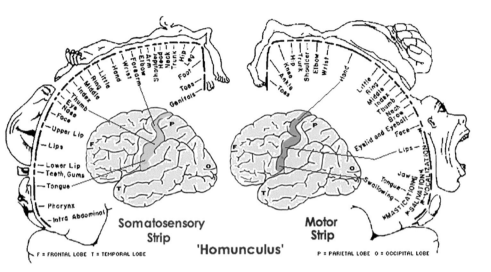

Figure 7.6
The standard characterization of sensory motor representation in the cortex. Stimulation of single neurons on either side of the central cortical sulcus reveals a body map in terms of observed responses. The left side of the figure shows area M1, which handles motor output. The right side of the figure shows the map for area S1, which handles body feedback. By permission from Penfield/Rasmussen. The Cerebral Cortex of Man. © 1950 Gale, a part of Cengage Learning, Inc.

Like all cortex, the cortices representing movement are organized into hierarchies, with progressively more abstract areas dedicated to representing successively more abstract motor planning issues. However, to make the main points about motor coding, we can focus on the "front end," or most concrete of the areas. The bulk of motor commands are sent from primary motor area M1. As shown in figure 7.6, M1 is located in a strip anterior to the central sulcus, a deep fold in the cortex. Like all of the cortical areas, it represents a map of something, in this case the set of neurons that can drive the body's musculature. If stimulated independently, muscles in different parts of the body can be selectively activated. Thus, one can draw a caricature of the anatomy on the motor strip representing the parts that responded. As shown by this depiction, some parts of the body are represented with increased resolution over others. The diagram is termed a *homunculus*, or "little man," for its human-like organization.

On the posterior side of the central sulcus is the all-important somatosensory area, which also has a body map. Its job is to represent the sensed information as to the state of the joints and haptic inputs. Area S1 is the first cortical stage that receives feedback as to the change in posture accomplished by a movement. When this area is damaged, a person loses the sensation of the position of the damaged area and has to rely on other senses, primarily vision, to maneuver, but the resultant disability can be quite devastating.

As one might suspect after the example of the visual cortical hierarchy, the motor side is similarly organized into hierarchies representing abstractions needed to stitch together purposeful movement sequences. This organization has been studied by a number of researchers but notably Rizzolatti and colleagues.[150] However, our main focus is that of motor routines, and its exposition centers around points that can be made at area M1, the lowest cortical level. The homunculus organization has stood the test of time, as the basic finding that stimulation along the motor strip can produce movements predictably mapped to the body has been replicated many times. From the standpoint of what one might expect of movements though, it raises a number of issues, the chief one being that purposive movements usually coordinate many different muscles in the body, and it is difficult to see how this is done if M1 is so distributed. For this reason, some researchers have recently questioned the homunculus organization, notably the Graziano laboratory.[151] By using larger than usual electrical stimulation in the monkey, they were able to produce multijoint movements that could be classified in purposeful ways. Figure 7.7 shows Graziano's classification of the end points of stimulated movements. You can see immediately that

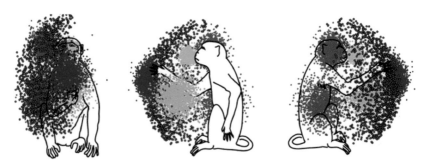

Figure 7.7
Graziano's characterization of human hand-eye coordination movements. By stimulating motor cortex with higher alternating current levels, he was able to excite whole-body movements. Light blue, hand-to-mouth; dark blue, reaching; red, defense; green, manipulation; purple, climbing. Reprinted from M. S.A. Graziano and T. N. Aflalo, "Mapping behavioral repertoire onto the cortex," *Neuron*, vol. 56, p. 2, copyright 2007, with permission from Elsevier.

these are larger, multijoint movements that cluster in different regions. From a computational standpoint, this change in movement coding is profound; by virtue of the larger, goal-directed movements, the cortex is implicated in solving the allocentric-to-task reference frame problem rather than leaving it for some downstream process.

If the cortex may be specially organized for purposive movements, the relevant neurons must pick a way to describe them to the spinal cord. The main issue to be tackled is the low cortical spike rates. These rates preclude use of high-bandwidth strategies employed by robots; there must be another way. One likely possibility is that the movement primitives that are needed are already in the spinal cord, and there just needs to be a way of indexing them. These observations motivate the construction of a computational model of movement primitives.

Motor Cortex Measurements Support Table Lookup Hypotheses

Multicell recording electrodes now allow the simultaneous recording of large numbers of neurons, with the result that one can now study how they interact in real time. As a pertinent example, the Shenoy lab at Stanford[152] has been able to record the activity of 80 to 90 cells in and around motor cortex during a monkey's preparation and execution of a reaching task. The monkey holds fixation on a central cue while a reach target appears on a screen. The dimming of the fixation cue is a "GO" signal for the monkey to reach and touch the target.

Figure 7.8 shows some of the results. The simultaneous firing of all the neurons constitutes a high-dimensional, time-varying vector, but its basic elements can be observed by projecting onto the right two-dimensional space by the principal components method. Three separate trials sampled every 20 milliseconds are shown on the display. Every trajectory is divided into a fixation period, then a target onset period, a GO cue period, and finally a movement onset period. Each trajectory all the while is in motion. However, you notice that most of the transits revealed in the trajectories, which represent changing firing rates in the neural code, occur *before* the physical movement onset. Given the low bandwidth of the neurons, it is compelling to think that what is occurring is the table lookup process. Thus, it could easily be the case that the cortex is computing parameters representing the index of the movement to be made; perhaps the kinematic parameters after the target onset and the dynamic parameters after the GO cue. A further argument in support of this view would be that, to a major extent, the cortex does not have the bandwidth to steer a fast movement appreciably. Any racquet sports player can attest that the ball strike has an

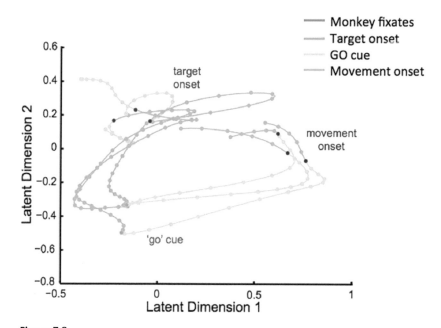

Figure 7.8
Experimenters have used a 100-element recording array over motor cortex to extract
the signals from approximately 80 to 90 neurons. The data are then presented in
a rotated coordinate system to highlight the two components that show the most
variation. What the data show is that for a substantial time before the onset of the
movement, the neural coordinates are changing considerably. One possibility is that
they are translating the movement from task coordinates into body coordinates. Reprinted from A. Afshar et al., "Single-trial neural correlates of arm movement preparation," *Neuron*, vol. 71, p. 3, copyright 2011, with permission from Elsevier.

open-loop character. The racquet swing is so rapid that it is very unlikely
that it can be served by the cortex en route. Consequently the movement
has to be coded before the swing occurs.

After all this development, you might be wondering how this
formulation helps get around the basic nonlinearity of the dynamics
equations. The reason is that the cost of the iterations can be amortized
over the body's lifetime. During the extended development period, one
can explore possible movements and save the ones that work. Thus, any
particular movement can be found using the cortex's vast table lookup
capacity. Small adjustments may need to be made, but these are easier to
handle.

7.4   Cerebellum: Checking Expectations

The problem of computing a desired posture change is formidable indeed, but the assumption at this stage is that this can be done in some way. The next issue is that, although the brain might have access to the solution of the general equations of motion, there are still issues in monitoring the body's ability to execute the control policy. Having a model of this process gives the brain enormous leverage. It can generate expectations of the consequences of actions that can be checked by comparing them to sensory motor feedback. This is the primary job of the cerebellum. We saw this process in chapter 1, with the adjustments to prism goggles, but this example just highlighted what the cerebellum does all the time. If expectations can be explained by an adjustment, then that ongoing calibration process does its job. This process has an elegant computational formalism in *optimal feedback control theory* (OFCT), which addresses the issues facing the brain in implementing the control policy that does these adjustments.[153-155]

Figure 7.9 shows the overall model of the monitoring system. Formally, the problem can be described as producing a general trajectory $x(t)$ of all the degrees of freedom, and also to compute the desired control policy $u(t)$ that ideally would achieve this change. The state estimate x^e allows the control policy to access the desired control signal u_t, which is an abstract command that is to direct a posture change. Focusing on the bottommost pathway, that command interacts with the body in the world to produce a new state estimate x^e. You might think that this step would complete the process and that the new state could be used for the next control policy lookup, however there are two extremely important additional steps. One arises because state is not available directly but must be measured via the sensory systems. The symbol y emphasizes that the measurements may be partial. The second arises because the brain can predict its own internal estimate x^p of what should have happened in the world via an internal model of the effects of the policy. These two estimates can be combined to produce the new state estimate. You can see this process in action by having a willing subject balance a weighty book on an upturned hand. If you ask the subject to remove the book, the hand will remain in place. Since the subject is directing the action, their system knows the weight change is coming and adjusts the support force accordingly. However, if you distract the subject and remove the book yourself unexpectedly, the hand will fly upward as the subject's motor system could not predict the action and responds reactively to the unweighting. This example highlights an issue

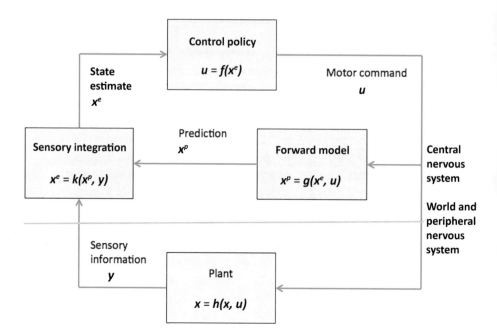

Figure 7.9

Optimal feedback control models are supported by extensive evidence that the control signal uses the difference between sensory feedback and internal state estimates. A classical feedback loop would eschew the middle row of computations, send the control signal to the "plant" (read muscles), and then estimate the consequences via sensors to obtain a new state estimate. However, such simple architectures contain delays, which can lead to instabilities. Instead, evidence suggests that the brain shortcuts these delays by using an internal "forward" model to predict what the sensory consequences should be. This allows a sensory integration stage to check the feedback immediately upon arrival. Redrawn from R. Shadmehr and S. Mussa-Ivaldi, *Biological Learning and Control: How the Brain Builds Representations, Predicts Events, and Makes Decisions*. MIT Press, 2011.

with the application of the theory, and that is the timescale of the control. A standard view is that movements are organized into large-scale episodes, such as a stride, and the expectations are built around the termination of the episode. You might have had the experience of misjudging the number of stairs by one and stepping into empty space, thinking you were at terra firma. Although recently, there has been some evidence that the timescale might be a lot finer (see Scott[156]), this idea is still in the testing stage.

The OFCT model has proved enormously insightful as to the brain's control system that extends far beyond the movement venue. Rather than accept the feedback from its action as is, the brain has its own internal model that predicts what the feedback will be and then reconciles that estimate with what is actually received. One important consequence is that this approach makes it easy to signal unexpected situations when the feedback diverges from the internal estimate. Also, this arrangement is congruent with the reinforcement learning measurements by Schultz that we saw earlier where the brain's dopaminergic system codes an internal reward signal in terms of deviations from expectations.

7.5 Spinal Cord: Coding the Movement Library

How does the brain solve these problems and generate motor commands to drive the limbs? The initial approaches used feedback. Adams[157] first proposed the closed-loop control system, which used sensory-motor feedback signals to minimize the errors between the desired output and actual output. However, the delays in pure feedback systems make it difficult for them to account for human performance. Thus subsequent proposals, derived from robotics research, calculated feedforward solutions based on inverse dynamics. As we saw though, more recent proposals emphasize internal models that include both feedback and feedforward control.[158] The feedforward control directly obtains motor commands from an inverse model, given a motor task. In the feedback control, the actual sensory data are compared to the desired sensory data, and errors are used to improve the inverse model. However, these solutions are in the form of algorithms. To get further along, we have to introduce the special structure of the body. The dynamics equations are universal, but different physical instantiations mean that there are different solutions possible. Very small insects can exploit the surface tension of water to navigate on its surface. Humans cannot do this, but they have options that follow from the high strength-to-weight ratios of their limbs as well as their passive compliance and energy storage properties. However, the sine qua non of animal motion is the central pattern generators in the spinal cord. These pattern generators compose different movement patterns in terms of oscillations that are the basis of all vertebrate movement. The characterization of the neural circuitry that generates these patterns has been the subject of many decades of research,[159] but given that our focus is in abstractions, we will sidestep any elaboration of this circuitry and instead make the key points with a set of oscillatory functions that can serve as a guide for the more detailed neural understructure. Such functions may in fact turn out to be muscle length and stiffness settings. The primary reasons for this are two-fold. One has already been mentioned; the cortex does not have the temporal bandwidth for high-speed servoing. The other is related. The world interface is unpredictable, as any runner knows. Unexpected footfalls thus have to be handled by the spinal cord's reflex circuitry. Given this setup, let's describe a way that the cortex could communicate with the spinal cord at low bandwidth.

Gabor Functions as Oscillatory Primitives
Given that the spinal cord has oscillatory primitives, the computational problem becomes finding a way that they can be characterized. Motivated by studies of more basic animal systems, a natural way of characterizing movements is in terms of temporal oscillations that are time limited. This constraint in turn motivates the choice of a multiscale Gabor dictionary as a means to decompose signals in both time and frequency space. The advantages are threefold:

1. Human movements can be decomposed into time-limited temporal segments owing to changing phasing and contact relationships.
2. Human movements owing to inertia tend to have a temporal smoothness that can be approximated nicely with sinusoids of varying frequencies.
3. The reduced bandwidth required to drive the movement allows for abstract control strategies.

Another drastic simplification that we will make is to choose the length change of a muscle as the stand-in for the actual control parameter for controlling muscle length. We make this approximation to sidestep the details of how motor neurons activate muscles to create specific length changes in natural movements which would add unneeded complexity. The goal is to demonstrate that tremendous bandwidth compression can be achieved if the forebrain can *modulate* the CPGs.

Following Chen et al.,[160] let's define a dictionary $D = \phi(t)_\gamma$, $\gamma \in \Gamma$, where Γ denotes the set of Gabor basis functions, or *wavelets*, that we will use, and

γ is an index that ranges over the set. Then, the signal $x(t)$ can be encoded with a linear combination of the basis functions as

$$x(t) \approx \sum_{\gamma \in \Gamma} \alpha_\gamma \phi(t)_\gamma \, . \tag{7.1}$$

The Gabor functions are attractive as they can simultaneously index time and frequency and are optimal if one wants to minimize noise in both. Finding the optimal decomposition can be very computationally expensive, but fortunately for biological systems, the problem is made simpler by solving for an approximate decomposition. Generally, *overcomplete* dictionaries, which have many more entries than the minimum that would be technically needed, are used as they provide greater flexibility in capturing the data. A larger set of more specialized basis functions means that relatively few are required to represent any particular signal. Here, as in the spatial coding in chapter 3, a *sparse representation* is a result of the presence of specialized basis functions that can closely represent the original signal.

For the Gabor dictionary, the index $\gamma = (\omega, \tau, \theta, \delta t)$, where $\omega \in [0, \pi)$, represents the frequency, τ represents the location in time, θ is a phase, and δt is the duration. The atoms of the dictionary are defined as

$$\phi_\gamma(t) = e^{\frac{-(t-\tau)^2}{\delta t^2}} \cos[\omega(t-\tau)+\theta].$$

For fixed δt and $\theta \in \{0, \pi/2\}$, a discrete Gabor dictionary would be complete for $\omega_k = k\Delta\omega$ and $\tau_l = l\Delta\tau$, for sufficiently fine $\Delta\omega$ and $\Delta\tau$,[161] where k and l are integers (see also ref. 159). For coding purposes, we would like the dictionary to be adapted by the learning algorithm so that it spans the space of movement signals in the data. To initialize the dictionary, we set $\Delta\tau$ as a small multiple (l) of the sampling rate of the signal, and $\Delta\omega = 2$, resulting in a sufficient resolution in the frequency space. Figure 7.10 illustrates example dictionary atoms of the Gabor dictionary. The wavelets were initialized with dyadic scales, starting from a scale that does not eliminate the smoothness of the Gabor (e.g., first column of figure 7.10), and ending in a scale that enables the Gabor to extend to the whole movement span (e.g., second column of figure 7.10). The exact structure of the dictionary at any time, however, would be determined by the learning algorithm, which is just a variant of the one used to find receptive fields for image data in chapter 3, with the exception that the modification is carried out with respect to the parameters defining the curves rather than the curves themselves,

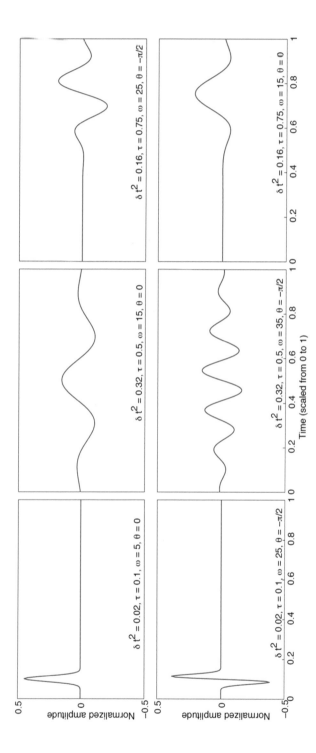

Figure 7.10

Gabor dictionary atoms across time. Scale, frequency, time shift, and phase are the four variable parameters. The learning algorithm controls all parameters except the phase, which is set to 0 and $-\pi/2$. Shown here are example atoms of the initial dictionary with time scaled from 0 to 1, where 0 is start of movement, and 1 indicates end of movement.

as this strategy is much more efficient. Although each dictionary atom is active only at a specific time, it is defined for the entire movement time (it is zero for the nonactive part), easily to enable a scaled superposition of different atoms when reproducing a movement signal.

With this development, we can now turn to the main focus, which is using superpositions of scaled basis functions to code muscle lengths. Figure 7.11 illustrates a representative result, which is that the length change of the gracilis muscle, one of the leg muscles used in walking, can be fit with extraordinary precision by using just 25 coefficients. With respect to the model, these coefficients would be the α_γ in equation (7.1). The functions denoted by ϕ_γ are assumed to be represented in the spinal cord. The 25 coefficients sent in a little over 2 seconds represent a rate of about 40 Hz, a 4,000-fold savings over the greater than 10 kHz used in conventional robotics. Using such a code, the cortex has a lot less to do.

Per the earlier discussion, you will notice that this model has taken liberties with respect to the problem in that the length of the muscle was fit. What biology uses is not length but force. However, because the muscles work by contracting and thus creating forces, there will be a relationship between force and length. Because it will be systematic, fitting length is a demonstration that biology should be able to solve the slightly harder problem of determining forces by the same means. Figure 7.11 represents the traverse of just one muscle. Combining all the muscles' length changes allows the reconstruction of the movements of the many joints. In a perfect world, these desired length changes would bring about the posture changes that the animal needs. However, because the world that we move in is not perfect, there is a lot for the spinal cord to do in implementing muscle length changes so that movement goals are achieved. At the very least, the spinal cord has to have a way of modulating the speed of contraction and interpreting the cortex's muscle stiffness recommendations. A third and important feature of spinal cord control is that of organizing the motor control signals so that they can make the best use of the body's natural dynamics. To these issues we now turn.

Equilibrium Point Control

The advantages of using stored energy was recognized early by Bernstein[162] and led to the Feldman equilibrium point (EP) theoretical framework.[163,164] This states that limb movement can be controlled by shifting limb posture, represented as equilibrium, from one position to another. Conceptually, it's as if the desired posture had damped springs attached that were in equilibrium. They then are stretched to get to the current posture so

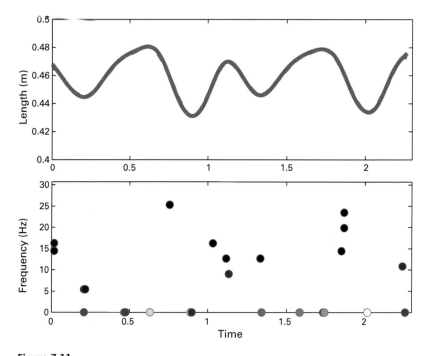

Figure 7.11
Reconstruction of the length of gracilis muscle using basis pursuit for different number of coefficients. The bottom plot marks the particular dictionary atoms used in
reconstructing the data. Each circle marks the time at which the atom was chosen,
and the ordinate value indicates the frequency of the atom. The colors of the circles
indicate the relative magnitude of each coefficient, white being the highest. The time
is in seconds.

that once released, the system returns to the goal posture. Subsequently,
researchers in areas such as physics, mechanical engineering, and brain and
cognitive science have successfully built passive walkers or walking models
using damped springs as muscles. The first was Raibert's robotic systems
models[165,166]consisting of a number of movement systems using damped
springs. This approach has also been studied in simulation by Gunther
and Ruder,[167] who programmed a bipedal model consisting of 14 muscle-
tendon complexes per leg and demonstrated stability margins in different
planetary gravitational fields.

Although the above work has successfully modeled walking using the
concept of passive energy, the methods are restricted to walking only and
are not general enough to be applied to other human movements or to

more complex situations. Gu[168] proposed that humans can generate complex movements by first dividing them up into segments and then using EP models to calculate equilibrium points for each segment by simulating the movement. That is, the brain first plans the end point in musculoskeletal space before the initiation of movements. Gu shows that this can be used not only to control simple movements such as reaching and walking, but also to use motor synergies to generate complex movements with composite goals, such as walking together with reaching, sitting with balancing, and other common abilities humans have.

The model has two main advantages. First, it separates motor planning from execution. During the planning stage, movements are divided into one or more segments. For each segment, the end point in joint space is calculated, given a task description in Cartesian space using the gradient descent method of Torres and Zipser.[169] Based on the simulations, the model is sufficiently general and robust to be applied to both simple and complicated movements. The second advantage is that EPs can be used in synergies to satisfy *multiple* goals. If a motor control model represents a genuine human movement model, it should be easily extended to satisfy multiple goals, such as walking to pick up something or sitting down and reaching for something, as these are common abilities humans possess. She shows how our simulation method can be extended to multiple simultaneous movements.

The planning method also extends to complex movements such as walking. To do this, one cycle of walking is divided into four segments, and each segment has an end-point configuration. Natural and realistic walking is generated by shifting from one end point to the next. Figure 7.12A gives the four equilibrium points that are defined for one cycle of walking. For each leg, three damped springs are used to control the contraction and extension of three joints: hip, knee, and ankle. Figure 7.12B presents four snapshots from the walking simulation. A comparison of the joint profiles obtained from simulation uses the human data from Boston Dynamics' graphic human, shown in figure 7.12C, which plays back the walking data captured from human subjects. Figure 7.12D shows the data-fitting results for hip, knee, and ankle during one cycle of walking. Given that only four EPs are applied for the walking, the data fit is quite close.

You must have noticed that the length versus time profiles are a little like the muscle profiles in terms of smoothness and frequency, and they are. So an obvious next step is to fit them with basis functions, too. A crucial thing to notice though is that the control points for the movements are exterior to the body, in the human's goal space. So now the cortex would be faced

A

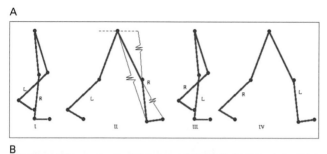

B

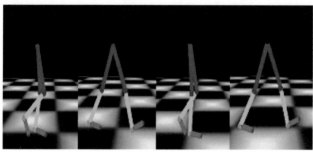

C

D

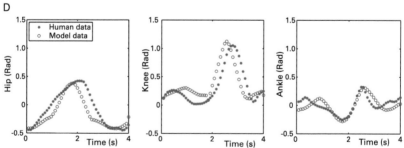

Figure 7.12

Walking simulation and comparison to human walking data. (A) Three abstract muscles are used to control the hip, knee, and ankle. One cycle of walking is divided into four segments, and each segment has one equilibrium point. Walking is generated by shifting from one end point to the next. (B) The snapshots from the walking simulation. (C) Boston Dynamics Inc. graphic human. The virtual agent walks via playing the walking data captured from human subjects. (D) To quantitatively evaluate our model, we compared the joint profiles in simulation to the human data from the company.

with relating the coefficients in the exterior coordinate frame with those in the body frame of reference. But the nice thing about this problem is that transformations between coordinate frames are relatively easy to compute.

Despite the elegance of equilibrium point control (EPC), it has been controversial. Gomi and Kawato showed that a model of EPC did not track human performance in a reaching task; however their result may have been sensitive to their stiffness model.[170] An earlier result by Polit and Bizzi had monkeys rotate a manipulandum toward a goal point. Midflight on different trials, they perturbed the manipulandum both toward and away from the target, only to see their trajectory recover and reach the goal.[171] The expectation under the EPC model was that the helpful perturbation would be absorbed into the ongoing trajectory. However, both of these experiments have the same caveat. As Feldman initially stated, one can expect the parameters of the movement to be slowly changeable during the motion. Thus, the dynamic stiffness during reaching can be expected to vary, allowing alternate explanations of the data in both experiments. In the reaching task, it is possible that the dynamic stiffness is quite different from the static stiffness measurement used in the model. In the perturbation experiment, per Feldman, the EP can transit from the start to the goal, and the perturbation is not helpful for its midflight position. This comes up in the informal example of spitting wood with an ax: The stiffness at the beginning of the trajectory needs to be quite different from the stiffness at its end. A nice mathematical approach that uses basis functions that addresses this capability is the recent work of Ijspeert et al.[172]

7.6 Reading Human Movement Data

The upshot of our brief introduction to the complexities of modeling dynamics is that, like the rest of its issues, the brain has to somehow solve these problems with slow circuitry. If we are to sort things out, we at least need to measure the essential generators of human motion to reveal how the dynamics operates in the prosecution of movements. One of the long-standing ways of doing this is by muscle electromyography (EMG). During muscle activation, the electrical neural activity can be directly measured. This technology has become very sophisticated with 48-channel commercial systems available with wireless data transfer. However, EMG systems, for the most part, use surface electrodes that pick up the summed activity of the muscles within their sensor range. Also, they do not readily differentiate between the net torque at a joint and the stiffness setting, the latter brought about by co-contracting opposing muscle groups.

Partly because of these issues, breakthroughs in the direct computational modeling of complex muscle systems have been enthusiastically welcomed. One of the foremost of these is the OpenSim system, originally developed by Loan and greatly refined by Delp at Stanford.[173,174] This software allows the simulation of muscle activations and their dynamic consequences in models of up to 300 muscles and 150 degrees of freedom. Both the kinematics and dynamics can be solved through optimization of their equations treated as constraints, but so far not in real time.

One can get partway toward a solution by estimating muscle lengths by measuring human motion directly. Such measurements can be routinely made by use of a motion-capture technology that incorporates marked points on the body that a surrounding array of cameras can identify and triangulate to recover the three-dimensional coordinates of the points. Motion capture is a standard technique and has been used extensively in human action modeling (e.g., the Carnegie Mellon University extensive motion-capture database at http://mocap.cs.cmu.edu/). Once three-dimensional coordinates of points on the body are available, they can be used to recover the movement of a model of the underlying skeleton, which has model muscles attached. These two stages are depicted in figure 7.13A. Once the muscle temporal excursions have been fit, the result can be played back as shown in figure 7.13B. The fit is sufficiently exact that the replay cannot be easily distinguished from the original.

The OpenSim modeling system is a major advance and has allowed the exploration of human dynamics in many different situations, most importantly comparing normal and abnormal conditions for the purpose of diagnosis and treatment. However, the system has limitations in that it must specify muscle activation profiles, and the exact form of these is still an open question. Keep in mind that the force-length curves in standard textbooks are typically obtained under conditions of isometric contraction and are usually not representative of the curves used during natural movements. The bottom line at present is that the solution of the inverse dynamics equations that model muscle forces can be done via a number of approximation techniques, but in each case it is computationally expensive.[175]

To that end, effort is being expended on systems that eschew muscle models and compute net joint torques directly, glossing over the stiffness issue. For general constraints, such systems currently can be near real-time. They are faster because the inverse dynamics for high degree of freedom systems can be readily computed if the problem can be specified in body kinematics differentially. This solution was developed for graphic objects[176] but has been extended to robotic models.[177] The solution depends on

A

B

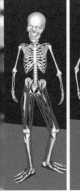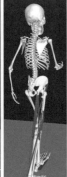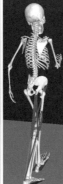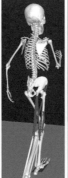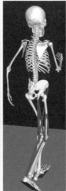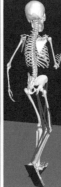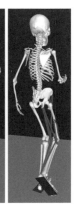

Figure 7.13
(A) Light-emitting diodes (LEDs) attached to a body suit are recognized by cameras in the PhaseSpace system. These are then turned into muscle lengths using Stanford's OpenSim software system. (B) Human walking using muscle reconstruction with Gabor dictionary. The color represents the muscle length compared to its resting length, where red indicates that the length is greater, blue indicates that the length is less, and white indicates that the length is close to the resting length.

having a way of solving the kinematic problem, but that can be easier than solving the entire problem.

Computing Joint Torques

One particular development of this methodology uses motion capture to have human subjects generate the body kinematics in the course of solving natural tasks that involve whole-body movement.[178] Thus, human subjects communicate the motor plan by acting it out. The overall idea behind the method is straightforward. For each human subject, construct a dynamic model within an ordinary differential equations (ODEs) physics simulator (see http://www.ode.org/), and force that model to follow human motion-capture data. The act of following the data leads directly to the recovery of joint angles. The only way that the dynamic model can track these angles is to generate the correct joint torques, so they too can be recovered. To see how this works, consider the very simple example of a single link that is rotated by a torque at its joint. Thus, where τ is torque and I inertia, we can write its Newton equation as

$$\tau = I\ddot{\theta}. \tag{7.2}$$

Now suppose that we know that the observed motion can be described by an oscillation $\theta(t) = A \cos(Bt)$. Differentiating this twice results in

$$\ddot{\theta} = -AB^2 \cos(Bt),$$

which allows the solution of equation (7.2) for $\tau(t)$. This is a tremendously simplified system, and it is possible that the additional complexities of a humanoid system could introduce new problems, but this does not turn out to be the case in numerical simulations. This concept was originally demonstrated in two dimensions for human walking by Faure et al.[179] Cooper extended the method to the significantly more demanding case of 48 degrees of freedom in three dimensions and arbitrary posture changes. The principal difficulty is that the constraints in the high degree of freedom three-dimensional model present many delicate numerical issues for the ODE solver that need to be addressed.[180] Let's consider Cooper's larger human model system.

Kinematics

First build a simple humanoid model with dimensions close to that of the subject. Then, map each marker to a point on the model. The markers are then introduced into a physics simulation and represented as kinematic bodies without collision geometry. In the physics simulator, each marker kinematic body is effectively treated as having infinite mass, so that when another dynamic body is attached through a joint constraint to a marker, only the dynamic body's trajectory can be changed by the constraint.

The humanoid model used has 48 internal degrees of freedom as shown in figure 7.14A. The shape and dimensions of the body segments shown are the same as those used by the simulation for both collision detection and the calculation of mass properties. Mass and inertial properties are computed from the volume of the body parts using a constant density of 1,000

A

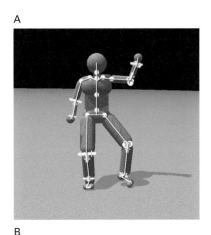

B

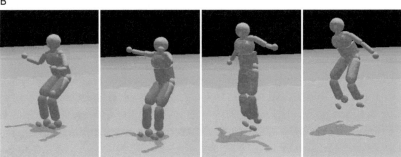

Figure 7.14
(A) The humanoid model used in this work has 48 internal degrees of freedom: Four ball-and-socket joints connect five body segments along the spine from the head to the waist. Ball-and-socket joints are also used at the collarbone, shoulder, and hip. Universal joints are used at the elbows, wrists, knees, and ankles. Hinge joints connect the toes to the heels. All joints limit the range of motion to angles plausible for human movement. (B) Four frames from a jump sequence captured and replayed.

kg/m³, the density of water. This value is a good first approximation given that many humans can either float or sink in water simply by inhaling or exhaling. The dimensions and articulation of the model loosely correspond to that of the human whose motion-capture data are used. The joints themselves are designed to allow the model to reproduce most movements the human can make. For example, joints at the elbows have 2 degrees of freedom to reproduce the hinge movement of the elbow as well as the twisting movement of the radius and ulna bones in the arm. Joints are also limited to prevent some impossible movements such as reverse bending of the elbows or knees.

Using the PhaseSpace motion-capture system,[a] one records the three-dimensional positions of specific human body locations marked by active PhaseSpace LEDs over time. Thirty-six of these markers are mapped to appropriate locations on the simulated character. When simulation is stepped forward, a constraint-solving algorithm attempts to find a body state that satisfies the internal joint constraints, the external marker constraints, and other constraints such as ground forces, joint stiffnesses, and conservation of momentum. As long as the markers are mapped to body segments reasonably well and the segment dimensions are somewhat close to those of the human actor, this produces high-quality body pose information at video-frame rates, even with unfiltered motion-capture data, making it suitable for real-time interaction scenarios as in virtual reality environments.

This construction allows us to use the physics engine itself to solve inverse kinematics problems in a forward manner. An initialization phase creates kinematic bodies to represent the motion-capture markers and then creates a constraint between the markers and the dynamic bodies of the articulated skeleton we are animating. Next, the physics engine constrains the corresponding point on the dynamic body to have the same effective velocity as the kinematic marker body. The physical simulation steps forward at the same rate as the motion-capture data (30 frames per second for the results shown here), assigning the kinematic marker body the linear velocity necessary to take it from its current position to its future position. The articulated character maintains its momentum and thus naturally extrapolates through brief gaps in the data. Constraining the articulated character to conform to motion-capture marker data produces a sequence of kinematic body-pose data. The degrees of freedom that are not fully determined by the marker trajectories can be manipulated by adjusting joint stiffnesses or applying external forces, but, in general, the character can only imitate the exact action that the human trainer demonstrated.

Dynamics

A second step produces dynamic behavior in the character, again using physical simulation and constraining joint velocities in such a way as to achieve the next frame's target joint angle. Joints in ODEs have a simple motor mechanism, which allows one to set a target velocity for a given joint. This velocity is accounted for in the constraint satisfaction step that moves the simulated system forward in time. After the system is stepped forward, the amount of force actually used to achieve the constraint can be retrieved from the system. The forces used account for ground and other collision forces as well as momentum and other perturbations. These retrieved forces can be used to generate feedforward force profiles for actuating the character.

As an illustration of its generality, the Carnegie Mellon University (CMU) computer science graphics laboratory motion-capture data was analyzed using this method. To make the correspondence between the data and the torque recovery software, (1) the CMU laboratory marker conventions were used, and (2) the inertial properties of a standard model were assumed to be appropriate for the kinematic data. The result for one subject in the act of playing the drums is shown in figure 7.15. The arithmetic signs of the data are shown in pseudocolor with red being positive. Additionally, the data are normalized to visualize more readily the smaller torques of the outboard joints in the human kinematic chain.

Stabilization Torques

If the model were sufficiently accurate, then replaying the torque commands would reproduce the motion. However, owing to various small differences between the model and its corresponding human, the torques can be different enough to destabilize the motion. This issue is addressed by adding small corrective torques as follows. In addition to the joint angles, the position and orientation of the waist body segment is recorded. An angular motor to this segment applies torques and forces needed to reproduce the recorded movement fully. Without them, the humanoid eventually falls down, but using them, a wide variety of different movements are easily stabilized. Figure 7.16 shows a comparison between the magnitudes of the external stabilizing torques.

Calibration

The computed torques from the method together with the stabilizing torques are adequate to reproduce motions, but do they correspond with the actual values used by the human? One way to check this is to compare

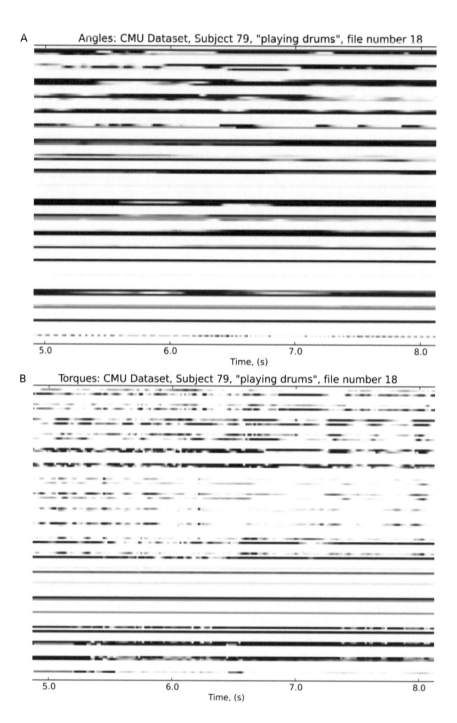

A

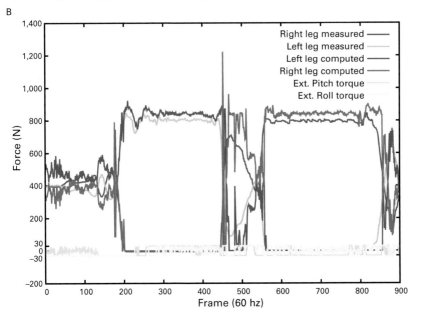

B

Figure 7.16
Torque comparison during a shifting stance. (Top) During a simple movement requiring dynamic balance, a human standing on two force plates goes from double
stance, to left foot, to double stance, to right foot, and then ends in a double stance.
(Bottom) The stabilizing torques are only a small fraction of the ground forces used
to support the body. The vertical ground forces measured by the force plates closely
match those computed through the physical simulation during single stance and
even remain relatively close during double stance. Stabilizing torques less than 30 Nm
were sufficient to maintain the stability of the model during this type of movement.

Figure 7.15
A 3-second time segment of torque data obtained by our torque recovery method applied to human drumming from the CMU motion-capture database. (A) Joint angles
read from the CMU data file. (B) Joint torques computed using the fast estimation
method. Once the CMU marker placement convention has been interfaced to our
system, the torque computation is straightforward. The torque data clearly show
the human subject's drumming pattern. Multiple lines per joint indicate different
degrees of freedom for that joint.

an independent estimate of a crucial value with that computed by the model. One such value is the ground force produced by the human subject. This can be measured independently by interfacing two standard Nintendo Wii force plates to a computer. The human subject varies the relative forces needed to balance by shifting the center of gravity back and forth between the plates. During this time, the ground forces as determined by the force plates were compared to the ground force as measured by the algorithm. As figure 7.16 shows, the two forces are in good agreement during the stance and in qualitatively good agreement during the stance shifts.

The OpenSim and specialized joint torque models illustrate a developing approach for understanding movement in that these approaches can serve as quantitative testbeds for testing proposals for the organization of the movement system that are difficult to get at with purely analytical approaches. The dynamics of human bipedal posture changes is extremely complicated and is very difficult to appreciate intuitively. Consequently, many questions as to the organization of human movement are still very much open. The simulation systems provide a potential avenue for closing some of these gaps.

7.7 Summary

Despite all the leverage gained by its parallelism, the brain's neural circuitry is fundamentally slow, with the result that every possibility must be exploited to create a system that can perform in real time. This chapter covered many important elements that allow this to happen. Chief among these is the hierarchical organization of movements, which can be characterized by different timescales. The most time consuming is the planning of movements done by the cortex and basal ganglia. This takes time because the goal of the movement has to be translated into body posture changes, which is computationally difficult. Once a posture change has been chosen, then that movement has to be monitored to see that it has been carried out successfully. This is the role of the cerebellum. In between these movement landmarks is the fastest process, and that is the execution of the posture change itself, the job of neurally modulated pattern generators in the spinal cord.

Besides this overall organization, there are many ancillary features of movements:

1. The musculoskeletal structure of the body can be regarded as a special-purpose computer that has been refined by evolution to handle many very difficult aspects of the Newtonian dynamics.

2. The orientations of the body introduce critical frames of reference that make the subsequent computations easier. In particular, the head provides a ready coordinate system for hand-eye coordination.

3. Movement computations are further simplified by being purposive, allowing them to ignore aspects of the environment that are not relevant to immediate goals.

4. The spinal cord's production of complex movements can be abstracted using a library of motor functions to make it accessible to the cortex. These functions can be coded so as to be indexable at relatively low temporal bandwidth commands.

5. Simulation tools such as OpenSim, are becoming increasingly important for modeling the human movement system. Newer systems take shortcuts to estimate the body's joint torques in real time.

6. At a foundational level, the body is the engine that allows its owner to achieve its goals. Thus the cost and constraints on movement are of basic importance.

8 Operating System

[T]he cognitive task we perform at each moment, and the efficacy with which we perform it, results from a complex interplay of deliberate intentions that are governed by goals (endogenous control) and the availability, frequency and recency of the alternative tasks afforded by the stimulus and its context (exogenous influences). Effective cognition requires a delicate, just-enough calibration of endogenous control that is sufficient to protect an ongoing task from disruption, ... but does not compromise the flexibility that allows the rapid execution of other tasks when appropriate.
—Stephen Monsell, "Task switching," *Trends in Cognitive Science*, vol. 7, p. 134, 2003

Visual and motor routines provide us with a basic understanding of how the brain interacts with the external world, but somehow the brain has to organize all this activity in ways that subserve its basic behavioral needs of its human phenotype. What makes the organizational job challenging is that it has many competing components that have to be handled in a timely manner. Monsell's quote captures the olio of competing temporal demands nicely: multiple ongoing tasks are trying to finish, while at the same time new potentially more important tasks may appear on the cognitive horizon. All of these exigencies evoke the structure of a silicon computer's operating system, which is exactly tailored to meet the time-varying tasks of multiple programs as well as interrupts from new programs and signals that must be dealt with. Thus, the silicon architecture, in an abstract form, can provide a useful framework for thinking about the issues that the brain must handle.

Of course, brain operating system issues have not emerged ex nihilo. Very early on, it was recognized that to realize the sophisticated decisions that humans routinely make, their brains must utilize some kind of internal model.[181] Internal models allow the brain to make sense of interactions with the world by having prior expectations about what should happen. Such models are useful caricatures of the world that capture its essence. To come to more concrete terms with the basic idea, think of the mental book-keeping you would need to have to keep track of preparing a meal when you are in the middle of executing a recipe: How much information at any moment do you need to take the next step in the process?

One of the pioneering figures in defining this effort was Neisser, who formalized the idea of an internal cognitive architecture.[182] Subsequent architectures specialized in codifying prior knowledge that came from experts.[183–186] Such "expert" systems codify mental processes in terms of fine-grained rules with variables that are bound to external data by pattern matching. Their broad intent is to cast a behavior as a problem to be solved and find a sequence of rules that will do the job. However, the pattern-matching approach has proved formidable in that the rules are normally hand crafted and the matching process can easily become very expensive.

Despite the enormous difficulties involved, expert systems have achieved notable successes, particularly in intellectual problems where requisite symbol bindings can be readily modeled, such as in algebraic problem solving.[187] However, from the broader perspective of visual and motor routines, that is of perceiving and acting in natural real-world environments, rule-based descriptions have proved challenging. The current best-known rule-based system, Anderson's ACT-R, perhaps being sparked by early ideas from Triesman (see ref. 117) and Pylyshyn,[188] presupposes that the image can be segmented prior to the introduction of the cognitive agenda. This choice also may have been influenced by Marr (see ref. 2), who specifically eschewed the idea of task influence on image segmentation with his "principle of least commitment." Exactly how much task-independent organization the brain does is still disputed, but as argued in the previous two chapters, it is probably very much less than initially supposed.

The regulation of vision and action to secondary levels of importance is consistent with the view of brain function as being composed of separate sequential stages of perception, cognition, and action.[189] This characterization has the role of vision to extract information from a scene that is then analyzed by a decision-making process that finally orders an appropriate motor response (figure 8.1A). The central assumption is that a central "cognition" component can carry out its machinations undisturbed. Evidence for this view comes from Pashler et al.,[190] who point to a cognitive bottleneck in a dual-task paradigm. Cognitive limitations block the flow of information from perception to action. However, the sequentiality as defined by this stance may require hierarchies to interpret. When following a car in a

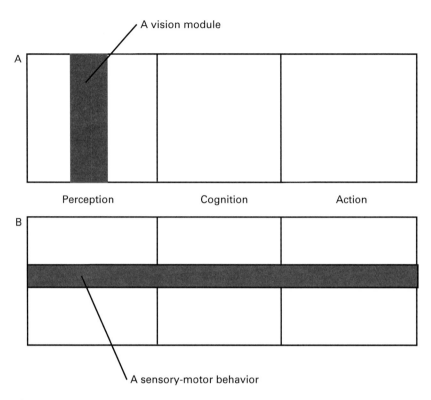

Figure 8.1
Two approaches to behavioral research contrasted. (A) In the Marr paradigm, individual components of vision are understood as units, compatible with the separation of vision, cognition, and action. (B) In the Brooks paradigm, the primitive unit is an entire behavior that ties specialized visual routines with specific actions and can operate with varying degrees of autonomy.

driving environment or navigating an obstacle field, the task may be ongoing, with successive actions that have no discrete beginning or end. When responding to cues on a display, decisions are typically done in blocks, so that the expectations as to the response to any particular trial may have been encoded long before its perceptual cues appear. Even in single-trial situations, responses may be driven predominantly by previously learned priors. Thus, the pure perception-cognition-action, while attractive from the point of view of descriptive compactness, may be relatively uncommon in natural behavior.

An alternative to the perception-cognition-action formalism is that of a behavioral primitive, wherein the three components are integrated into a complete behavior, as depicted in figure 8.1B. Think of a trophic behavior such as following a light, where the motor system servos doggedly toward the source. In this example, the light source coordinates are piped directly to the motor system without any additional computational overhead. The integrated-behavior view is especially associated with Brooks.[191] His original formulation stressed the idea of *subsumption* wherein primitive behaviors could be a component of more complicated behaviors by adding additional components that could achieve economies of expression by exploiting, or subsuming, the functions of existing behaviors. This approach has had a huge impact on robotics and cognitive architectures, and for no small reason in that it explicitly recognized the need for abstraction in behaviors, whereby complex behaviors could be constructed as arrangements of simpler behaviors. However, the choice of representation placed a premium in coding behaviors directly in circuits, and complexities in the composed circuits may have contributed to the difficulties encountered when scaling to more and more abstract behaviors.

To circumvent the scaling dilemma, diverse communities in robotics and psychology have been working on cognitive architectures that take an alternate embodied approach to vision and action. Both groups have recognized that the ultimate architecture will have a heterogeneous hierarchical structure where different levels in the hierarchy are qualitatively different and have very different functions. Robotics researchers in particular[192–194] have gravitated to a three-tiered structure that models strategic, tactical, and detail levels in complex behavior based on Bonasso's architectural design,[195] summarized in table 8.1.

The focus on heterogeneous architectures, together with the appreciation of the advantages of situating them in a particular body, has become crystallized under the rubric of *embodied cognition*.[196–198] Embodied cognition integrates elements from all the robotic advances but in addition places a special stress on the body's role in computation. It emphasizes that the brain cannot be understood in isolation, as so much of its structure

Table 8.1
Bonasso's three-level architecture, which has been widely adopted in various elaborations for robot control

Level	Description
Planner	A deliberative planner that coordinates the selection of skill sets based on goals and time constraints
Sequencer	Activate and deactivate sets of skills
Reactive skills	Use set sensory-motor responses

is dictated by the body it finds itself in and the world that the body has to survive in.[199-204] This has important implications for cognitive architectures, because the brain can be dramatically simpler than it could ever be without its encasing milieu. The reason is that the brain does not have to replicate the natural structure of the world or the special ways of interacting with it taken by the body but instead can have an internal structure that implicitly anticipates these commitments.

Embracing the embodied cognition paradigm is not without challenges. The very abstract levels of description[205,206] risk losing a direct connection to the human body's particulars. In contrast, starting with very low level models (e.g., at the neural level[207]) faces difficulties in managing the brain's enormous numbers of degrees of freedom. This chapter chooses a middle ground of abstract exposition that explicitly addresses issues of scheduling information acquisition using a body model that has important features, such as a foveal retina. We will use a specific hierarchical cognitive architecture that uses small collections of special-purpose behaviors to achieve short-term cognitive goals. These behaviors are defined at the intermediate tactical level of cognitive representation in terms of MDP *modules*. Modules are self-contained and can operate independently, but most importantly they can be composed in small groups. For example, in driving, active modules might consist of lane following, speed monitoring, and collision avoidance.

While the specification of modules could potentially take several forms, we use reinforcement learning (RL) as a basis for defining them. The RL formalism has several advantages, which will be highlighted here.

1. The formalism provides an understanding for task-directed fixations as a method of increasing reward by reducing task uncertainty, so we can have a rationale for gaze changes.
2. The formalism provides a way of computing the value of new tasks by executing them within the context of subsets of other tasks whose reward values have been already estimated. This is a tremendous boon as it allows the incremental evaluation of new modules.
3. The RL formalism provides a way of estimating rewards from observations of subject data. This allows us to probe the behavioral weightings used by human secondary reward systems.

8.1   A Hierarchical Cognitive Architecture

Hierarchies are the essence of any internal cognitive model. Newell pointed out that any complex system that we know about is organized hierarchically. Furthermore, as one proceeds up the hierarchy toward more abstract representations, its combined components take up more space and run slower, simply because the signals have farther to travel. In silicon, we have the ready hierarchy of: gates, VLSI circuitry, microcode, assembly language, programming language, operating system. For each level, a crucial understanding is that one must respect the abstractions at that level. For example (one we use a lot), in microcode, multiplication is deconstructed into sifts and adds, and those parts can be inspected. At the more abstract level of assembly language this deconstruction can no longer be done, as multiplication is a primitive.

Newell's dictum about working within a level applies to the cognitive hierarchies as well, and the fact that the different levels deal with different kinds of issues is one reason why the hierarchy ends up being heterogeneous. Although it is easy to define several more abstract levels,[208] our operating system hierarchy consists of four levels—*debug, operating system, task, and routines*—but we focus on issues related to the last three levels. When something goes very wrong, the job of the debug level attempts to find out what is wrong by simulating what should have happened and reprogramming the responsible modules. This would be loosely coextensive with high-level attention. How it happens is very much an open problem and will not be pursued here. The operating system level has the job of selecting an appropriate suite of modules, including an appropriate agenda-changing module, for the current set of behavioral goals. How this can happen is also an open problem but is more accessible when cast in terms of RL modules, as these have an associated rating as to how effective they should be in their applicable context. The task level describes a module dedicated to achieving one specific goal and is programmed by reinforcement. Per the MDP formalism described in chapter 5, a task's module will have a collection of states $s \in S$ and actions $a \in A$, each of which can direct progress toward a rewarding goal. Each action earns a scalar reward $r \in R$ that may be positive or negative. At the bottommost level in the hierarchy (for the embodied level) are routines, which do the job of linking the abstract states and actions to the real world. Figure 8.2 summarizes this organization.

Correspondences with the Posner Model

To appreciate the value of our particular hierarchical organization, it helps to review some key concepts related to attention as conceptualized in psychology. A huge amount of work has been done in the psychological field directed at characterizing how the brain manages resources under the general rubric of *attention*, for which there are basic resources.[209,210] The primary measures have tended to be behavioral features such as reaction

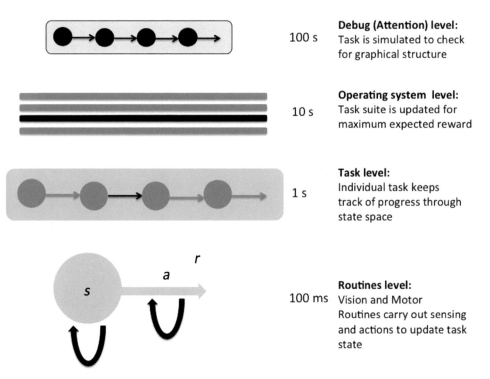

Debug (Attention) level:
100 s   Task is simulated to check for graphical structure

Operating system level:
10 s   Task suite is updated for maximum expected reward

Task level:
1 s   Individual task keeps track of progress through state space

Routines level:
100 ms   Vision and Motor Routines carry out sensing and actions to update task state

Figure 8.2
Four levels of a hierarchical cognitive architecture that operate at different timescales. The central element is the task level, wherein a given task may be described in terms of a module of states and actions. A *thread* keeps track of the process of execution though the module. The next level down consists of visual and motor routines (arrows) that monitor the status of state and action space fidelities, respectively. Above the task level is the operating system level, whereby priorities for modules are used to select an appropriate small suite of modules to manage a given real-world situation. The topmost level is characterized as an attentional level. If a given module is off the page of its expectations, it may be reprogrammed via simulation and modification.

times. However increasingly there has been increased appreciation that "under the hood," the brain's system is considerably more complex than can be characterized with reduced descriptions and thus the openness for considering more elaborate synthesis models, such as ACT-R and the hierarchical model described here. The latter, in particular, resonates with the triage of attention used by Posner et al.,[211,212] which is particularly interesting as its elements line up with the features of the modules-based computational hierarchy. They characterize attention as having three separate neural networks that handle *alerting*, *orienting*, and *executive* functions. These functions have also been associated with specific neurotransmitters norepinephrine, acetylcholine, and dopamine, respectively. To summarize this psychological view:

1. *Alerting* is associated with sensitivity to incoming stimuli.
2. *Orienting* is the selection of information from sensory input.
3. *Executive* attention involves mechanisms for resolving conflict.

It turns out that each of these components of the broad characterization of attention can be seen as addressing an implementation issue in the modules' cognitive architecture. However, at the specific tactical abstraction level of modules, the specific enfleshment of these concepts results in interestingly different interpretations from those used in the psychological domain. Let's elaborate.

The central module hypothesis is that any ongoing behavior can be seen as a possibly dynamic composition of a small number of active modules, each of which is associated with a separate task. For example, in driving a car in a city, different modules could be active for staying in a lane, monitoring for pedestrians, and dealing with other traffic. A central constraint is that of capacity. From neural principles the number of instantaneously co-active unmonitored modules can be around 30. Thus, we would reinterpret Baddeley's classical result on the limitation of working memory to a small number of monitored states.[213] Rather than being about "items," the modules perspective has the bottleneck being about *the state information necessary to keep track of different co-active modules*. Importing Luck and Vogel's result for items,[214] we take the limit on the number of co-active modules whose state can be monitored to be about four. Computer science uses the term *thread* to denote the unique sequence of states needed to describe the time course of a process, and in that sense working memory may be usefully characterized as being about threads.

The exacting constraint of a small number of modules places a premium on a method for choosing between them and thus provides the primary raison d'être for an *executive*. The executive's function is to evaluate the effectiveness of the existing modules with respect to other modules waiting in the cognitive wings to see if there are grounds for a productive exchange. The executive also adjudicates conflicts. If several active modules require the use of the gaze vector, the executive mechanism has to resolve this competition.

In order for the executive to evaluate the prospects for an inactive module, there must be a calculation of these prospects that falls short of activating the module. There are many possible ways to do this. Our module design posits a twofold index that (1) expresses that relevance of a module and (2) estimates the worth of its potential contribution. These calculations would be putatively handled by a special alerting module whose function would be to trap any of a set of unexpected conditions. In this description, we are close to a saliency map,[215, 216] but our saliency map needs to have the combination of features, priors,[217] and task-relevance measures. The focus of psychology has been on exogenous visual features on the optic array, but natural behaviors are mediated by many factors. The signaling feature may be in the form of an auditory, tactile, or olfactory cue, and its value may be signaled by a variety of factors that may have their roots in endogenous as well as exogenous sources. In our modular scheme, a special *alerting* module provides a short list of contenders, and the executive decides which ones are in or out.

The final issue is associated with the set of active modules. The virtue of a module is that it is a program fragment that encapsulates detailed knowledge about dealing with the world to achieve a specific goal. In this effort, it usually deploys very specialized resources for extracting just the knowledge it needs from the surround.[218, 219] To take a visual example, a visual routine, deployed by a module, may need the color of a region of space and in that case may need to deploy gaze to obtain an *orienting* measurement with enhanced foveal resolution.

8.2 Program Execution

All the ideas about hierarchies and links to psychology are fine in the abstract, but how are we going to make them concrete and test them? Fortunately, new tools have become available. One is instrumentation that allows the monitoring of human head, hand, and eye movements. This capability allows us to track the actions of a person engaged in an everyday cognitive activity. We can then test cognitive architectures by having them predict the observed human behavior. If the model is a good one, it should

Table 8.2
The organization of human visual computation from the perspective of the behavior model

Task Code	Description
BT	Put the bread onto the plate
PLF	Take the lid off the peanut butter jar
JLF	Take the lid off the jelly jar
KH	Pick up the knife
POB	Spread the peanut butter on the bread
JOB	Spread the jelly on the bread
PLO	Put the peanut butter lid on
JLO	Put the jelly lid on
KT	Put the knife on the table
FB	Flip the bread to complete the sandwich

be able to predict what is going on, given sensory data and its internal program.

To see a specific example, let us consider the down-to-earth example of making a peanut butter sandwich.[220] To model this process, Yi[221] chose a level of description that had the major steps denoted as in table 8.2. You might quibble with this level as being too abstract, as each of the steps could be broken down into finer-grained steps. For example, spreading the peanut butter onto the bread may involve more than one trip to the jar depending on the size of the bread and the size of the glob of peanut butter on each transit of the knife. The exact way to handle such hierarchies is not completely settled, and of course the steps combine both visual and motor routines, but whatever the final result, it should be cast in terms of state-action table lookups. Keep in mind also that as complicated as this task is, conceptually it can be tracked with a single thread that accesses the necessary context to pinpoint the stage in the recipe.

Table 8.3 summarizes the scheduling of 10 tasks in making a peanut butter sandwich by three human subjects. There are some coding assumptions such as that the knife is picked up only once and is not put down until spreading finishes. Despite that some chronological constraints (e.g., BT, PLF, and KH must precede POB and JOB) have ruled out most of the 10! task orders, the number of possible orders remaining is still a large number. If we divide the 10 tasks into three stages, {BT, PLF, JLF, KH}, {POB, JOB}, and {PLO, JLO, KT, FB}, we have at least $4! \times 2! \times 4! = 1,152$ different orders; how-

Table 8.3
Scheduling of tasks for subjects A, B, and C

Task list	Sequential Time Intervals									
	1	2	3	4	5	6	7	8	9	10
BT	ABC									
PLF		A	C		B					
JLF		BC				A				
KH			AB	C						
POB				A	C	B				
JOB				B		C	A			
PLO					A				B	C
JLO									C	AB
KT							C	AB		
FB							B	C	A	

Note: Sandwich making is decomposed into 10 tasks including BT (putting bread on table), PLF (taking peanut butter lid off), JLF (taking jelly lid off), KH (grabbing knife in hand), POB (spreading peanut butter on bread), JOB (spreading jelly on bread), PLO (putting peanut butter lid on), JLO (putting jelly lid on), KT (putting knife on table), and FB (flipping bread to make a sandwich). Placement of letters A, B, and C identifies the orders of tasks taken by the three subjects A, B, and C; for example, in the first two steps, subject C put bread on the table and took the jelly lid off.

ever, as shown in table 8.3, the orders picked by human sandwich makers display common features.

If the sandwich construction is proceeding smoothly, then the maker's standard program can be followed, but in the event of a mishap, some other program has to be invoked to correct the error. How that happens is unsettled, but a way station that is needed for its solution is some way of detecting that sandwich construction is going according to plan. This component of the problem has been modeled (see ref. 221), and we can gain a considerable amount of understanding by following its structure. The task is to recognize each stage of the sandwich construction process and, in particular, to identify the task under way. Figure 8.3 shows two frames from the online recognition algorithm. In line with the routines assumption, the algorithm tests for color information in the foveal region and also classifies the momentary trajectories of the two wrists. It also keeps track of the temporal stage of the construction. All of this information provides evidence, and the computational task is to turn this evidence into a task estimate. It turns out that the Bayesian framework for managing evidence introduced in chapter 4 for estimating velocities works especially well for this problem also. It just has to be scaled up to the more elaborate venue. In the velocity estimation algorithm, three sources of information, two velocity estimates and a prior, were combined in another node. Here, the problem is similar in that dependent sets of evidence must be managed. The main addition is that the task is ongoing, so that evidence accumulates over time. The sandwich-making program can be expressed graphically as shown in figure 8.4. The temporal dependence makes it a *dynamic Bayes net*.

Armed with the Bayesian framework notation, the larger graph for sandwich-making of figure 8.4A can be easily interpreted. The shaded nodes represent data that are directly measured by a routine. Once these measurements are taken, their evidence can be quickly propagated throughout the graph. The graphical nodes are shorthand for the different numbers of state values as shown by the inset table. Figure 8.4B shows how the overall construction is handled. Suppose the first task done is t_1. Then there is a probability p of doing t_2 next and a probability q of doing t_4 after that. Each time a task is posited, the evidence for it is evaluated.

Using this model, the sensory feedback data can be classified at successive points in time. For the result shown, the chosen sampling interval is 300 milliseconds. Figure 8.5 compares the actual stage (dotted red) with the classification determined by the model (blue), showing that almost perfect classification is achieved. At this point, the whole problem of making a

A

B

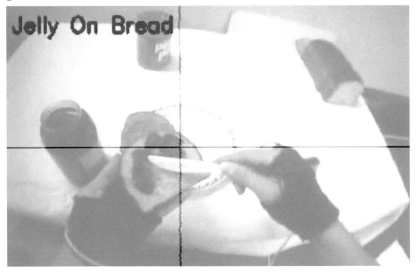

Figure 8.3
Steps in sandwich making recognized by a computational model that uses Bayesian evidence pooling to pinpoint steps in the recipe by observing the sandwich constructor's actions. The algorithm has access to the central 1° of visual input centered at the gaze point, which is delimited by the crosshairs. Also, the position and orientation of each wrist is measured. The label in the upper left is the algorithm's estimate of the stage in the recipe.

A

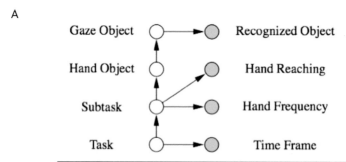

Node Name	# of States	Node Name	# of States
task	80	time frame	20
subtask	10	hand frequency	2
hand object	4	hand reaching	2
gaze object	5	recognized object	5

B

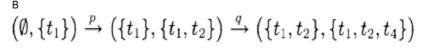

$$\left(\emptyset, \{t_1\}\right) \xrightarrow{p} \left(\{t_1\}, \{t_1, t_2\}\right) \xrightarrow{q} \left(\{t_1, t_2\}, \{t_1, t_2, t_4\}\right)$$

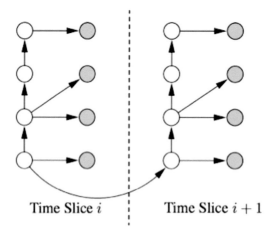

Figure 8.4
The computational model behind the sandwich recipe recognizer. (A) A graph similar to the Bayes network described in chapter 2 but more elaborate is used to weight
measurements from sensors (gray nodes) and propagate it to a "Task" node. The
graph shown is a convention for one that has many nodes. The exact numbers used
are shown in the inset table. (B) The sandwich making is coded as a sequence of
large-scale actions, such as spreading peanut butter. Such tasks are denoted by a specific index. The kth task would be t_k. The task graph by itself codes all the different
sequences of tasks that can be used in sandwich making.

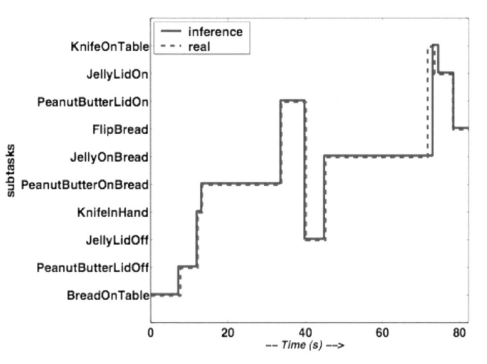

Figure 8.5
The graphical model of the task in the form of the Bayes net shown in figure 8.4. The net classifies that task every 300 milliseconds given the peripheral sensory evidence from vision and motor feedback and can categorize the steps almost exactly.

sandwich is far from being solved, however now there is some insight as to the steps in the process. As the actions are generated, an ancillary algorithm can monitor their outcomes. If the desired next stage, where a task has been achieved, is communicated with a high probability, all is well; otherwise some fault handling must be invoked. The dynamic Bayes net simulation is encouraging in that just a small amount of sensory information in the form of visual and proprioceptive routines proves sufficient to track the progress of a sandwich-making "thread," but what about the case of multiple threads? To address that venue, we turn to avatar models.

8.3 Humanoid Avatar Models

The use of a human to generate data was extremely useful in shaping an internal model, but another very different modeling path exists. It is now possible to create an artificial human that can perform behaviors, and the capabilities of such avatars are becoming increasingly sophisticated. As we saw in the previous chapter, research programs that focus on embodiment have been facilitated by the development of virtual reality (VR) graphics environments. These VR environments can now run in real time on standard computing platforms. Their value is that they allow the creation of virtual humanoid agents that implement complete visuomotor control loops. Visual input can be captured from the rendered virtual scene, and motor commands can be used to direct the graphical representation of the virtual agent's body. Terzopoulos and Rabie[222] pioneered the use of virtual reality as a platform for the study of visually guided control. Embodied control has been studied for many years in the robotics domain, importantly in RoboCup soccer,[223] but virtual agents have enormous advantages over physical robots in the areas of experimental reproducibility, hardware requirements, flexibility, and ease of programming (so much so that RoboCup soccer now has a simulation league).

Like the robots in soccer, the humanoid avatar is on its own. Its internal model must direct a wide variety of tasks, each of which requires its own perceptual and motor resources. Thus, the avatar brain's "operating system" must have mechanisms that allocate resources to tasks. Understanding this resource allocation requires an understanding of the ongoing demands of behavior, as well as the nature of the resources available to the human sensorimotor system. The interaction of these factors is complex, and that is where the virtual human platform can be of value. It allows us to imbue our artificial human with an internal model that has a particular set of resource constraints. One may then design a control architecture that allocates those resources in response to task demands. The result is a model of human behavior in temporally extended tasks that may be tested against human performance.

For a demonstration, we will use Sprague's virtual human model "Walter"[224] created using Boston Dynamics software. Walter has physical extent and programmable kinematic degrees of freedom that closely mimic those of real humans. The crux of the model is a control architecture for managing the extraction of information from visual input that is in turn mapped onto a library of motor commands. The model is illustrated on a simple sidewalk navigation task that requires the virtual human to walk down a sidewalk and cross a street while avoiding obstacles and collecting litter. The movie frame in figure 8.6 shows Walter in the act of negotiating the sidewalk, which is strewn with obstacles (blue objects) and litter (purple objects) on the way to crossing a street.

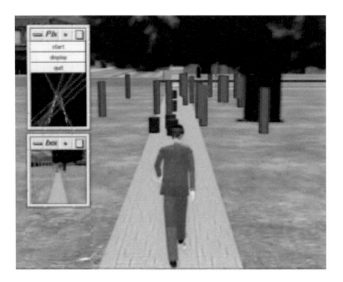

Figure 8.6
The Walter simulation. A humanoid graphical model allows the testing of the inter-action of vision and motor commands in simulated real time. The insets show the use of vision to guide the humanoid through a complex environment. The upper inset shows the sidewalk-finding visual routine that is running at this instant, and its read lines delimit possible sidewalk borders. The lower inset shows the visual field in a head-centered frame.

State Estimation Using Visual Routines

The first question that must be addressed is how individual programs map sensory information to internal state descriptions. This information is gathered by deploying visual routines. For example, litter collection is based on color matching. Litter is signaled in the simulation by purple objects, so that potential litter must be isolated as being of the right color and also nearby. This requires combining and processing the hue image with depth information. Regardless of the specific methods of individual routines, each one outputs information in the same abstract form: the state needed to guide its encompassing behavior (figure 8.7).

Programs as Reinforcement Learning Modules

Once state information has been computed, the next step is to find an appropriate action. Given the modularization of behavior, it's possible to learn the appropriate actions via reinforcement learning (RL). As intro-duced earlier, one way to make RL practical is to divide a complex behavior

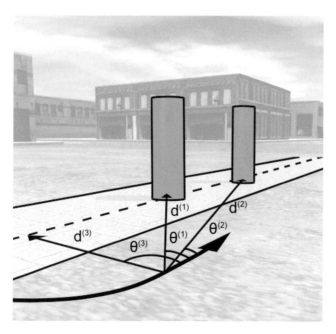

Figure 8.7
The conventions used for the three different state spaces are similar. For example, the litter task uses the heading to the nearest litter object, the obstacle task uses the heading to the nearest obstacle, and the sidewalk task uses a heading to the closest point of the sidewalk's central axis. Because Walter walks at a constant speed, the crucial action parameter to determine is his heading.

into smaller specialized modules. Each module works at a level of abstraction wherein its program can be stored in a state-action table. Such tables can be learned by reward-maximization algorithms: Walter tries out different actions in the course of behaving and remembers the ones that worked best in the table. It might seem counterintuitive to posit reward as the basis of all behavioral programming, but the reward-based approach is consistent with an enormous number of experimental studies. Earlier, we discussed Schultz's experiments using monkeys that suggest the use of reward discounting in a way that is consistent with reinforcement learning algorithms.[225] At the human level, studies of human behavior by Maloney and Landy show that the extent to which humans make such trade-offs is very refined.[226]

Formally, the task of each behavior is to map from an estimate of the relevant environmental state s to one of a discrete set of actions, $a \in A$, so as to maximize the amount of reward received. For example, the obstacle avoidance behavior maps the distance and heading to the nearest obstacle $s = (d, \theta)$ to one of three possible turn angles; that is, $A = \{-45°, 0°, 45°\}$. The *policy* is the action so prescribed for each state. The coarse action space simplifies the learning problem.

The approach to computing the optimal policy for a particular behavior is based on the standard reinforcement learning algorithm, termed Q-learning,[227] discussed in chapter 5. This algorithm learns a value function $Q(s, a)$ for all the state-action combinations in each behavior. The Q-function denotes the expected discounted return if action a is taken in state s and the optimal policy is followed thereafter. If $Q(s, a)$ is known, then the learning agent can behave optimally by always choosing as the policy $\text{argmax}_a\, Q(s, a)$. Figure 8.8 shows the table used by the litter collection behavior, as indexed by its state information.

Each of the three behaviors has its own two-dimensional state space. The litter collection behavior uses the same parameterization as obstacle avoidance: $s = (d, \theta)$, where d is the distance to the nearest litter item, and θ is the angle. For the sidewalk-following behavior, the state space is $s = (\rho, \theta)$. Here, θ is the angle of the centerline of the sidewalk relative to the agent, and ρ is the signed distance to the center of the sidewalk, where positive values indicate that the agent is to the left of the center, and negative values indicate that the agent is to the right. All behaviors use the logarithm of distance in order to devote more of the state representation to areas near the agent. All these behaviors use the same three-heading action space described above. The relative rewards for the behaviors are chosen such that collision avoidance is twice as important as picking up litter, which in turn is twice as important as staying on the sidewalk. These values are used to generate the Q-tables that serve as a basis for encoding a policy. Figure 8.9 shows a representation of the Q-functions and policies for the three behaviors.

When running the simulation, the Q-table associated with each behavior is indexed every 300 milliseconds. The action that is the policy is selected and submitted for arbitration. The action chosen by the arbitration process is executed by Walter. This in turn results in a new Q-table index for each behavior, and the process is repeated. The path through a Q-table thus evolves in time and can be visualized as a thread of control analogous its use in computer science.

8.4   Module Multiplexing

At this point, we need to address the issue of how the model might manage multiple programs at once, but before tackling this issue, it will be helpful

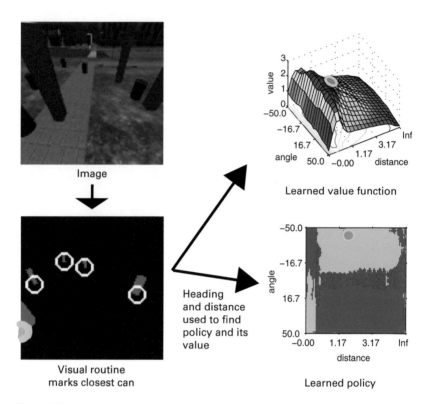

Image

Learned value function

Heading
and distance
used to find
policy and its
value

Visual routine
marks closest can

Learned policy

Figure 8.8
The essential portion of the litter cleanup module after it has been learned. The upper left inset shows the color image is used to identify the heading to the nearest litter object as a heading angle θ and distance d. Using this state information to index the table allows the recovery of the policy and its associated value. For the latter, shown in the lower right, green = −45 degrees, red = 0 degrees and blue = +45 degrees. The fact that the model is embodied means that there is neural circuitry to translate this abstract heading into complex walking movements. This is true for the graphics figure that has a "walk" command that takes a heading parameter. The upper right shows the value function $V(s)$ for the state, which can be computed from $Q(s, a)$ by choosing the action that maximizes it. These values are all-important as they govern the interactions between modules in several ways.

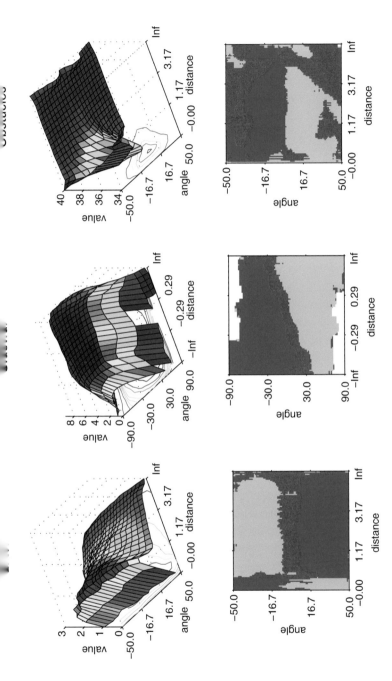

Figure 8.9

Q-values and policies for the three behaviors. The top row shows $\max_a Q(s, a)$ for the three behaviors: litter collection, sidewalk following, and obstacle avoidance. The bottom row shows the corresponding policies for the three behaviors. The obstacle avoidance value function shows a penalty for nearby obstacles and a policy of avoiding them. The sidewalk policy shows a benefit for staying in the center of the sidewalk $\theta = 0$, $\rho = 0$. The litter policy shows a benefit for picking up cans that decreases as the cans become more distant. The policy is to head toward them.

to review how silicon computers do this. The simplest strategy would be to put them in a queue and run them in the queue order, but the problem with that strategy is that a very long unimportant program may hog the processor, making more important programs wait. Thus, the standard silicon solution is to break processor time into quanta of about 100 milliseconds or so and work on multiple jobs a little bit at a time. A good analogy is a simultaneous chess exhibition where a grandmaster will play against several opponents. The grandmaster typically will have the opponents arranged in a rectangle around himself so he can easily transit the interior while studying the positions at each of the boards. What happens is that the grandmaster generally will know who the best players on the other side of the board are, and so will stop opposite them and spend more time thinking before choosing a move. In the same way, a processor transits the programs, executing the next set of the instructions in each program, and spends the most time on the important programs' instructions.

What sense does it make to think of neural circuitry as doing more than one thing at a time at this level of abstraction? As mentioned before, this obviously happens for the neural circuitry that controls the basic life support functions of the body, but crucially these only rarely have to be monitored cognitively. Speaking very loosely, for the brain they are a "background job." As for cognitive behaviors that require complex sensory-motor correspondences, the issue is very much open. Again, we can try and stretch silicon concepts for leverage; if a behavior has no run-time associations and all references are "hard-coded" in memory, then they can be run in the background, but if they need a run-time association (i.e., analogous to a variable in a subroutine), then that dynamic binding should deplete the "working memory" budget.

Certainly, if for some reason the brain was forced to execute single programs sequentially, it would face all the disadvantages that silicon computers do. This would also be exacerbated by the slowness of the neural circuitry, for as you know, in processing a single program, the smallest step usually requires of the order 200 milliseconds minimum, and usually much more, say seconds. These small differences are important, particularly for behaviors in athletics. The difference between expert cricket batsmen and duffers is that the experts are 100 milliseconds faster in anticipating the bounce point of the ball on the pitch.[228]

The facility with which the brain can do two things at the same time has received extensive study by psychologists. In particular, Ruthruff et al.[229] had subjects do a dual task where they had subjects map one of three tones and one of three letters visually displayed onto two key-presses. Subjects

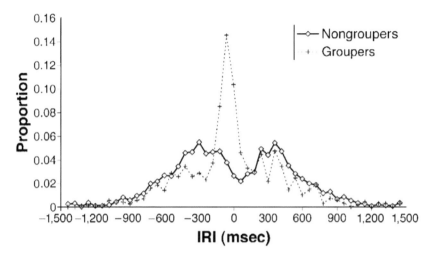

Figure 8.10
The Pashler laboratory study of multitasking. Subjects viewing a screen see one of three characters and have to respond by pressing a corresponding key. Simultaneously, they hear one of three tones and have to press a corresponding key. The plot shows that some subjects (the "Groupers") can treat the two tasks as a single complex task and press both keys simultaneously. Other subjects order the tasks so that one is handled before the other. Elaborate controls imply that the "Nongroupers" are dealing with a cognitive bottleneck. Reprinted from E. Ruthruff, "Dual-task interference with equal task emphasis: Graded capacity sharing or central postponement?" *Perception and Psychophysics*, vol. 65, p. 5, 2003, with permission from Springer Science and Business Media.

had to press correctly the appropriate two keys when the sound and letter were displayed simultaneously. The data are shown in figure 8.10. The subjects fell into two groups. In one group, subjects were able to see the task as a combined tone-and-letter task and therefore could simultaneously press the two answer keys. In another, one task was done either about 300 milliseconds before or after the other.

The authors relate these data to a well-known test of rapid serial visual representation (RSVP). When searching for two images one after the other in a very rapidly presented set of images, subjects have difficulty responding correctly when the second target is within 300 milliseconds of the first.[230] An example query would be "indicate when a chicken follows a telephone." Given what we know about working memory, it is hard not to suggest that it is the reason for the delays. In terms of programs, if the program has no variable arguments, then there is no issue, but if there is a variable argument, then apparently the neural circuitry needs time to handle the bookkeeping associated with its thread.

However, in most ecological settings, there is usually sufficient time to switch tasks if need be. Ongoing behaviors need to continue for several seconds and need to manage multiple tasks at once as illustrated with an experiment by Shinoda et al.,[231] who had subjects drive a virtual car. Driving was done by having subjects use a car simulator that had a steering wheel and pedals (figure 8.11A) while wearing a head-mounted binocular display that displayed a scene of a small town. Subjects were instructed to follow a lead car as well as obey traffic signs. This meant that when approaching an intersection, subjects had the dual tasks of following and looking for traffic signs at the intersection. What you can see from a typical subject's data shown in figure 8.11B is that the two tasks are managed by switching the gaze successively from one to the other. At one point in time, the subject is looking at the car and, at about 750 milliseconds later, the gaze is switched to the sign, and then back to the car, and so on. Remember that the visual cortical memory is retinotopic, so the major fraction of its neurons are sensitive to gaze position. Thus, their firing patterns are very different when gaze is changed. How are we to interpret what is going on here with respect to running multiple programs? One possibility is that the follow program is switched on when gaze is following the car, and the sign program is switched on when looking at the sign. But this seems very unlikely for two reasons. One is that, taking a cue from the dual-task experiments, there would be the overhead of turning the programs on and off. In addition, we know that from a standing start, it takes about 100 milliseconds to compute an answer,[232] so stopping and starting would have a temporal overhead. The second reason for positing simultaneously active programs is that in order to change gaze, the new gaze point for the next program must be computed while the current program is running—thus, for at least this part of the computation, the two programs have to be running simultaneously, a central assumption of the modules formalism.

8.5   Program Arbitration

If several modules are active simultaneously, an important issue that needs to be addressed with the modules approach is that concurrently active behaviors almost certainly have different state tables and also may prefer different actions. Some kind of arbitration mechanism must be available to choose among different modules' competing demands. It turns out

A

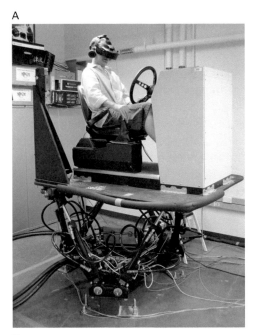

B

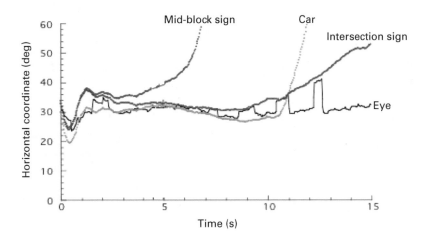

Figure 8.11

(A) Example of a virtual reality car simulator used to gather driving data. (B) Eye traces from a similar environment where human drivers follow a lead car and obey traffic signs show gaze (the black trace denoted as "Eye" in the figure) repeatedly sampling the intersection sign location and the lead car location. Modified from H. Shinoda et al., "What controls attention in natural environments?" *Vision Research*, vol. 41, pp. 3535–3545, 2001.

that RL's Q-learning leads directly to ways of addressing this issue, as noted by Humphrys[233] and Karlsson.[234] The specification of a program as a state-action table leads directly to two arbitration issues. One is the choice of action. It could be that the different modules recommend different actions. In general, how to resolve this dilemma is not settled, but there are two broad kinds of solutions. One occurs when the action space is common to the modules. This happens in our chosen example of walking where all the modules have a common action space of different headings. In that case, a logical strategy is to *average* the different heading recommendations. Another is to *select* one action over the others. In this choice, one logical strategy is to choose the one with the maximum Q-value for all modules. This happens in choosing a fixation location, as gaze can only point to one point in space at a time.

A separate issue occurs with the states of the modules. Typically, the way a module acquires information as to which state it is in is through sensing. If a module could continually update its state in this way, it would always have accurate state information; however, for several reasons it is impractical to have continuous sensing. Vision is the easiest illustrative example. Because visual information is acquired through routines and routines typically are monopolized by one module at a time, module states must sometimes depend on internal estimates. This reality, together with internal and external noise sources, means that the state of a module can become uncertain. Thus, computations that use module states much necessarily take uncertainty into account. Let's explore these issues in more detail.

The first issue to address is that of modeling uncertainty. In the simulation, noise is modeled in the growth of uncertainty over time. A simple assumption is that, if a state estimate is not updated by sensing, then its uncertainty growth can be modeled with a Kalman filter.[235] Basically, uncertainty is a normal distribution whose mean drifts and variance grow over time. With this model, we can describe the arbitration schemes.

Arbitration by Averaging: Heading
In the walking environment each behavior shares the same action space, so that Walter's heading arbitration can be handled by making the assumption that the Q-function for the composite task is approximately equal to the sum of the Q-functions for the component behaviors:

$$Q(s,a)^E \approx \sum_{i=1}^{n} Q_i(s_i,a),\qquad(8.1)$$

where $Q_i(s_i, a)$ represents the Q-function for the ith active behavior. Thus, the action that is chosen is a compromise that attempts to maximize reward across the set of active behaviors.

This formulation can incorporate uncertainty in a straightforward way. Because Walter may not have perfectly up-to-date state information, he must select the best action given his current estimates of the state. A reasonable way of selecting an action under uncertainty is to select the action with the highest expected return. Building on equation (8.1), we have the following:

$$a_E = \arg\max_a E\left[\sum_{i=1}^{n} Q_i(s_i, a)\right],$$

where the expectation is computed over the state variables for the behaviors. By distributing the expectation and making a slight change to the notation, we can write this as

$$a_E = \arg\max_a \sum_{i=1}^{n} Q_i^E(s_i, a), \tag{8.2}$$

where Q_i^E refers to the expected Q-value of the ith behavior. In practice, one can estimate these expectations by sampling from the distributions provided by the Kalman filter.

Arbitration by Selection: Gaze
Arbitrating gaze requires a different approach than arbitrating control of the body. Reinforcement learning algorithms are best suited to handling actions that have direct consequences for a task. Actions such as eye movements are difficult to put in this framework because they have only indirect consequences: They do not change the physical state of the agent or the environment; they serve only to obtain information.

To simulate the fact that only one area of the visual field may be foveated at a time, only one module is allowed access to perceptual information during each 300-millisecond simulation time step. That behavior is allowed to update its state information with a measurement, while the others propagate their estimates and suffer an increase in uncertainty. Simply put, as time evolves, the uncertainty of the state of a behavior grows, introducing the possibility of low rewards. With this framework, the selection strategy updates the module that has *the most to gain* by being updated, or equivalently, the most to lose by not being updated. Thus, the approach taken here is to try to estimate the value of that information. Deploying gaze to measure the state reduces this risk.

Estimating the cost of uncertainty is equivalent to estimating the expected cost of incorrect action choices that result from uncertainty. Assuming that the expected rewards for an action selection (coded in Q-functions) are known and that Kalman filters can provide the necessary distributions over the state variables, it is straightforward to estimate this factor, $loss_b$, for each behavior b by sampling, using the following analysis(see ref. 224). The loss value can be broken down into the losses associated with the uncertainty for each particular behavior b:

$$loss_b = E\left[\max_a \left(Q_b(s_b, a) + \sum_{i \in B, i \neq b} Q_i^E(s_i, a)\right)\right] - \sum_i Q_i^E(s_i, a_E). \qquad (8.3)$$

Here, the expectation on the left is computed only over s_b. The value on the left is the expected return if s_b were known but the other state variables were not. The value on the right is the expected return if none of the state variables are known. The difference is interpreted as the cost of the uncertainty associated with s_b. The maximum of these values is then used to select which behavior should be given control of gaze.

To summarize this use of gaze: Besides the executive and alerting facets of attention, the third important aspect is that of orienting. A module's resources must be focused to acquire the information necessary for its function. The reward uncertainty formulation places a burden on the orienting task of any given visual module, as a resource often necessary for its function, gaze, is competed for across the current module set. However, figure 8.12 shows that resolving this competition by allocating gaze that would reduce its reward-weighted uncertainty the most is a superior strategy compared to standard methods of gaze allocation.

Figure 8.13 gives an example of seven consecutive steps of the sidewalk navigation task, the associated eye movements, and the corresponding state estimates. The eye movements are allocated to reduce the uncertainty where it has the greatest potential negative consequences for reward. For example, the agent fixates the obstacle as he draws close to it and shifts perception to the other two behaviors when the obstacle has been safely passed. Note that the regions corresponding to state estimates are not ellipsoidal because they are being projected from world space into the agent's nonlinear state space. Subsequent tests on human subjects' driving in a simulator, show that the fixation interval probability functions for competing tasks can be fit almost exactly with the Sprague model.[236]

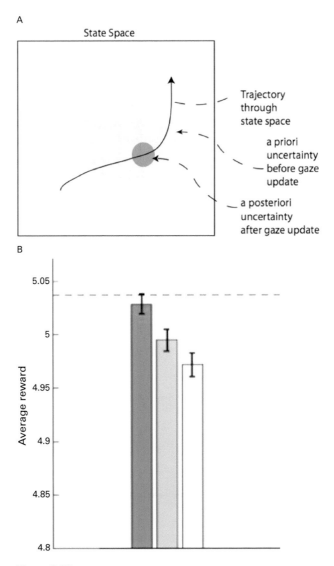

A

State Space

Trajectory
through
state space

a priori
uncertainty
— before gaze
update

__ a posteriori
uncertainty
after gaze update

B

Average reward

5.05

5

4.95

4.9

4.85

4.8

Figure 8.12
(A) The Sprague model of gaze allocation. Modules compete for gaze in order to update their measurements. The figure shows a caricature of the basic method for a given module. The trajectory through the agent's state space is estimated using a Kalman filter that propagates estimates in the absence of measurements and, as a consequence, builds up uncertainty (large, light-shaded area). If the behavior succeeds in obtaining a fixation, state space uncertainty is reduced (dark shading). The reinforcement learning model allows the value of reducing uncertainty to be calculated. (B) In the sidewalk venue, three modules are updated using the Sprague protocol, a sequential protocol, and a random protocol (reading from left to right). The Sprague protocol outperforms the other two.

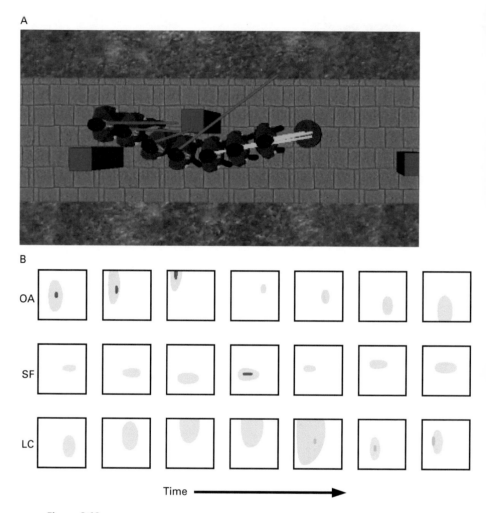

Figure 8.13

(A) An overhead view of the virtual agent during seven time steps of the sidewalk navigation task. The blue cubes are obstacles, and the purple cylinder is litter. The rays projecting from the agent represent eye movements: red corresponds to obstacle avoidance (OA), blue corresponds to sidewalk following (SF), and green corresponds to litter collection (LC). (B) Corresponding state estimates. The top row shows the agent's estimates of the obstacle location. The axes here are the same as those presented in figure 8.9. The beige regions correspond to the 90% confidence bounds before any perception has taken place. The red regions show the 90% confidence bounds after an eye movement has been made. The second and third rows show the corresponding information for sidewalk following and litter collection.

One possible objection to this model of eye movements is that it ignores the contribution of extrafoveal vision. One might assume that the pertinent question is not which behavior should direct the eye, but which location in the visual field should be targeted to best meet the perceptual needs of the whole ensemble of active behaviors. There are a number of reasons to emphasize foveal vision. First, eye-tracking studies in natural tasks show little evidence of "compromise" fixations. That is, nearly all fixations are clearly directed to a particular item that is task relevant. Second, results by Roelfsema et al. (see ref. 219) suggest that simple visual operations such as local search and line tracing require a minimum of 100 to 150 milliseconds to complete. This timescale roughly corresponds to the time required to make a fixation. This suggests that there may be little to be gained by sharing fixations among multiple visual operations.

8.6 Alerting

The foregoing exposition of module dynamics has focused on helpful modules that are succeeding in carrying out a predetermined agenda, but in an uncertain world there needs to be ways of adjusting the agenda. This is a very knotty problem in general and needs to be handled at different levels in the cognitive hierarchy. If things are completely out of kilter and elaborate reprogramming has to be done, in our line of reasoning the right thing to do is to punt this up to the debug level. What happens at this level is an open problem. However, for many situations the problem can be resolved by an existing module that happened to be inactive. The assumption is that the problem has come up before, and there is a module that can address it, but that module was not already running owing to working memory resource constraints. One way to handle this is to have an alerting module that uses a composite of the task, prior knowledge, and image features to trigger a module change.

As an example of the alerting module, consider the problem of dealing with exigencies during freeway driving, an example of which is shown in figure 8.14. Freeway driving is characterized by a very large motion stimulus, but for the most part that stimulus's main component is the radial motion produced by the parallel trajectories of the driver's and surrounding vehicles. Nonetheless, there are interrupts in this pattern that must be dealt with. In the case of a car in front encroaching on the driver's lane, the event must be detected by an alerting system, and then the executive must switch the requisite module into the active module set.

A

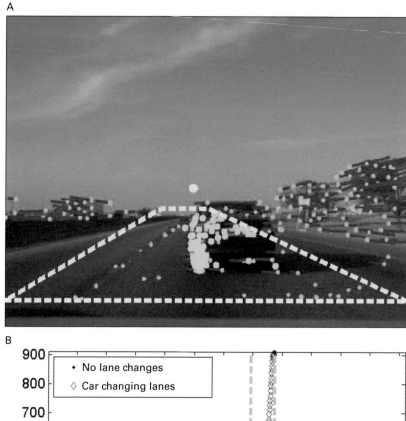

B

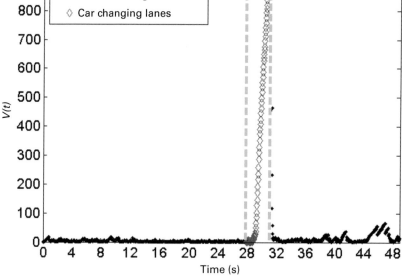

Figure 8.14

A potential job for an alerting module: Detecting unusual variations in optic flow while driving. (A) An encroaching car produces a pronounced deviation from background radial flow expectation. Radial flow can be dismissed as a normal expectation, but the horizontal flow of a car changing lanes signals an alert. (B) The timeline shows that this signal, as measured by a space and time-window integration, is easily detectable.

Although this example has only been tested in the case of driving, the expectation is that this approach might lead to a model of low-level attention that would test the environment for situations like this one that require a change of agenda. These changes span two levels of the cognitive hierarchy. If the change easily can be handled by another module, the suite of modules is updated. However if not, it must be handled at the level of conscious awareness, which can resort to simulations to diagnose a more elaborate response.

8.7   Program Indexing

The successful progress of Walter is based on having a running set of behaviors B_i, $i = 0, \ldots, N$ that are appropriate for the current environmental and task context. The view that visual processing is mediated by a small set of behaviors immediately raises two questions: (1) What is the exact nature of the context switching mechanism? (2) What should the limit on N be to realistically model the limitations of human visual processing?

Answering the first question requires considering to what extent visual processing is driven in a top-down fashion by internal goals, versus being driven by bottom-up signals originating in the environment. Perhaps somewhat optimistically, some researchers[237] have tacitly implied that interrupts from dynamic scene cues can effortlessly and automatically attract the brain's "attentional system" in order to make the correct context switch. However, a pure strategy of predominantly bottom-up interrupts seems unlikely in light of the fact that what constitutes a relevant cue is highly dependent on the current situation. On the other hand, there is a strong argument for some bottom-up component: Humans are clearly capable of responding appropriately to cues that are off the current agenda.

The model of the switching mechanism is that it works as a state machine as shown in figure 8.15. For planned tasks, certain behaviors keep track of the progress through the task and trigger new sets of behaviors at predefined junctures. Thus, the behavior "Look for Crosswalk" triggers the state "Near Crosswalk," which contains three behaviors: "Follow Sidewalk," "Avoid Obstacles," and "Approach Crosswalk." The bottom of figure 8.15 shows when the different states were triggered on three separate trials.

This model reflects the view that vision is predominantly a top-down process. The model is sufficient for handling simple planned tasks, but it does not provide a straightforward way of responding to off-plan contingencies. To be more realistic, the model requires some additions. First, behaviors should be designed to error-check their sensory input. In other

State Machine Diagram

Behavior List

- Follow Sidewalk
- Avoid Obstacles
- Pick Up Objects
- Look For Corner
- Look For Crosswalk
- Approach Crosswalk
- Wait For Light
- Follow Crosswalk
- Approach Sidewalk

On Sidewalk

- Follow Sidewalk
- Avoid Obstacles
- Pick Up Objects
- Look For Corner
- Look For Crosswalk

On Crosswalk

- Follow Crosswalk
- Approach Sidewalk

Near Crosswalk

- Follow Sidewalk
- Avoid Obstacles
- Approach Crosswalk

Waiting For Light

- Wait For Light

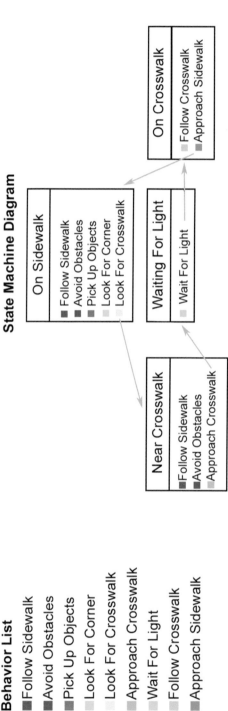
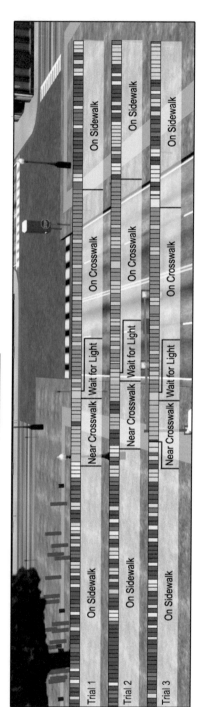

Figure 8.15

(Top left) A list of behaviors used in Walter's overall navigation task. (Top right) The diagram for the programmable context switcher showing different states. These states are indicated in the bands underneath the colored bars below. (Bottom) Context switching behavior in the sidewalk navigation simulation for three separate instances of Walter's stroll. The different colored bars denote different behaviors that are words, if a behavior's inputs do not match expectations, it should be capable of passing control to a higher-level procedure for resolution. Second, there should be a low-latency mechanism for responding to certain unambiguously important signals such as rapid looming.

Regarding the second question of the number of active behaviors, there is reason to suspect that the maximum number of them that are simultaneously monitored might be modest. That is the ubiquitous observation of the limitations of spatial working memory (SWM). The original capacity estimate by Miller was seven items plus or minus two,[238] but current estimates favor the lower bound (see ref. 214). The identification of the referents of SWM has always been problematic, as the size of the referent can be arbitrary. This has led to the denotation of the referent as a "chunk," which is jargon that postpones dealing with the issue of not being able to quantify the referents. by way of comparison, the thread concept for keeping track of independent state information in independently running behaviors is more utilitarian.

Although the number of active behaviors is limited, there is reason to believe that it is usually greater than one. Consider the task of walking on a crowded sidewalk. Two fast walkers approaching each other close at the rate of 6 m/s. Given that the main source of advanced warning for collisions is visual and that eye fixations typically need 0.3 second and that cortical processing typically needs 0.2 to 0.4 second, during the time needed to recognize an impending collision, the colliders have traveled about 3 m, or about 1½ body lengths. In a crowded situation, this is insufficient advance warning for successful avoidance. What this means is that for successful evasions, the collision detection calculation has to be ongoing. But that in turn means that it has to share processing with the other tasks that an agent has to do. Remember that sharing means that the behavior has to be simultaneously active over a considerable period, perhaps minutes. Several elegant experiments have shown that there can be severe interference when multiple tasks have to be done simultaneously, but these either restrict the input presentation time or the output response time;[239] the crucial issue is what happens to the internal state when it has to be maintained for an extended period.

8.8   Credit Assignment

The development up to this point has assumed the existence of internal rewards, but how should they be obtained? As mentioned at the outset, the brain has the central problem of rating the value of its behaviors in a way that is helpful to its survival and procreation. It is easy to see that something like caloric value could be handled, but what are plausible mechanisms for more abstract rewards? As noted in chapter 2, when the brain is in charge of coming up with its own estimates of the value of doing things, there is lots of chances to lose one's bearings. In some helpful cases, like food intake, the body provides helpful feedback, but the more abstract programs represent a challenge. There are, however, some things that can be done, and furthermore these lend themselves to a computational account. In particular, it turns out that the sets-of-modules formalism suggests two plausible ways of estimating rewards that are computationally tractable. One way follows directly from the multitasking venue. When multiple programs are simultaneously active, they can compare running estimates. If in addition they have access to a total reward estimate, it turns out that they can calibrate their own individual contributions. Another obvious way for humans and many other animals is by observing another's behavior. If the state-action description of the demonstrator can be mapped onto the observer's own internal representations, this allows for the observer's reward estimates to be modified efficiently.[240]

Calibrating Reward by Comparing Active Program Estimates

This setting simplifies the problem by assuming that individual reinforcement learning modules are independent and communicate only their estimates of their reward values. The modules can be activated and deactivated asynchronously and may each need different numbers of steps to complete. Each active module's program represents some portion of the entire state space and executes some part of the composite action, and, as an additional helpful condition, let's assume they have access to a global performance measure, defined as the sum of the individual rewards collected by all of the $\mathcal{M}$ active modules at each time step:

$$G_t = \sum_{i \in \mathcal{M}} r_t^{(i)}. \tag{8.4}$$

The central problem that can now be addressed successfully is how to learn each module's composite Q discounted reward estimates $Q^{(i)}$ $(s^{(i)}, a)$ when only global rewards G_t are directly observed, but not the individual values $\{r_t^i\}$ (figure 8.16).

The key additional constraint that we introduce is an assumption that the system can use the sum of rewards from the modules that are co-active

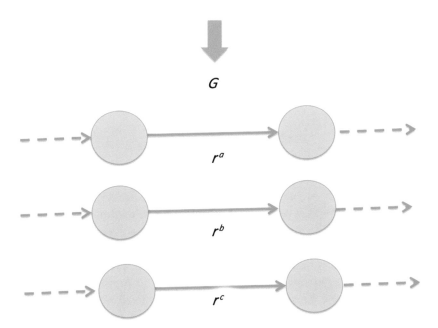

Figure 8.16
The credit assignment problem is a fundamental problem for a biological agent using a modular architecture. At any given instant, when multiple modules $(i) = \{a, b, c\}$ are active and only a global reward signal G is available, the modules each have to be able to calculate how much of the reward is their due.

at any instant. This knowledge leads to the idea to use the different sets to estimate the difference between the total observed reward G_t and the sum of the current estimates of the individual rewards of the concurrently running behaviors. Credit assignment is achieved by bootstrapping these estimates over multiple task combinations, during which different subsets of behaviors are active. Dropping the temporal subscript for convenience, this reasoning can be formalized as requiring the individual behaviors to learn independent reward models $r^{(i)}(s^{(i)}, a)$. The current reward estimate for one particular behavior, i, is obtained as

$$\hat{r}^{(i)} \leftarrow \hat{r}^{(i)} + \beta\delta_{r^{(i)}} ,\tag{8.5}$$

where the error on the reward estimates δ_r is calculated as the difference between the global reward received and its estimate from sum of the component module reward estimates:

$$\delta_{r^{(i)}} = G - \sum_{j \in \mathcal{M}} \hat{r}^{(j)} , \qquad\qquad\qquad\qquad\qquad (8.6)$$

which can be informatively rewritten as

$$\hat{r}^{(i)} \leftarrow (1 - \beta)\hat{r}^{(i)} + \beta \left(G - \sum_{j \in \mathcal{M}, j \neq i} \hat{r}^{(j)} \right). \qquad\qquad (8.7)$$

To interpret this equation: Each module should adjust its reward estimate by a weighted sum of its own reward estimate and the estimate of its reward inferred from that of the other active modules. When one particular subset of tasks is pursued, each active behavior adjusts the current reward estimates $\hat{r}_i$ in the individual reward functions according to equation (8.7) at each time step. Over time, the set of tasks that have to be solved will change, resulting in a different set of behaviors being active, so that a new adjustment is applied to the reward functions according to equation (8.7). This bootstrapping process therefore relies on the assertion that the subsets of active behaviors visit all component behaviors. Nonetheless, the formalism is elegant in that new modules can quickly learn their worth simply by being coactive with other calibrated behaviors.[241]

Calibrating Reward by Observing Behavior

Another important way the brain can calibrate the value of its programs is to observe the execution by another person. The important and tricky step that is needed is that the learner must be able to take the observations of the demonstrator and translate them into his or her own internal representation. This is not easy, but monkeys can do it, as seen by famous experiments by the Rizzolatti laboratory. A neuron that fires when the monkey is reaching for a food item will also fire when the experimenter reaches for it (see ref. 149). But at the level of abstraction of the operating system, the action has to be mapped into a states-and-actions formalism programmed by reinforcement. Once that can be done, the subsequent steps are very straightforward.

The data observed are going to be sequences of state-action pairs $O = \{(s_j, a_j), j = 1, \dots, N\}$, so it is easiest to work with the Q-value or action-value function $Q(s_j, a_j)$. So the observer sees a behavior and abstracts the sequence. Next, because he or she has learned a Q-table, the Q-values for that sequence can easily be accessed. Now what the brain would like to do is estimate the rewards R given O. This is a job for the Bayes rule:

$$P(R|O) = \frac{P(O|R)P(R)}{P(O)}.$$

(8.8)

Now for a big assumption. Estimate $P(O|R)$ using

$$P(O|R) = \frac{1}{Z} e^{\alpha E(O,R)},$$

where Z is just a normalizing factor to make sure the probabilities sum to unity. But the expected reward $E(O, R)$ of the observations and a given reward set is just $\sum_j Q(s_j, a_j)$. But when using sets of programs, the reward is just their sum, so we can write the sum as

$$\sum_i c_i \sum_j Q(s_j, a_j),$$

where to prevent the Q values from becoming arbitrarily large we use Q_1 as Q normalized to have a maximum of unity and additionally require the c_i to sum to one.[242]

The essential thing to note here is that given some observations, the only unknowns are the c_i, which reflect the relative values of the rewards. The intuition is that the form of the $Q(s_j, a_j)$ will not change. The sole thing at issue is their values *relative* to each other. Because the equations turn out to be simple, they are easily solved.

Figure 8.17 shows the results of testing this algorithm on both avatar and human data. Figure 8.17A shows the importance of recovering correct module reward estimates. With the correct estimates, an avatar can reproduce the behavior (figure 8.17A, top), but with very different estimates, the generated trajectory is very recognizably different (figure 8.17A, bottom). Figure 8.17B shows human data for three separate conditions: pick up litter and avoid obstacles, just avoid obstacles, and just pick up litter. Figure 8.17D compares the last two of these cases and shows that the case where litter is the goal can be reliably distinguished from the case where obstacle avoidance is the goal.

8.9 Implications of a Modular Architecture

The most important benefit of this kind of model is that it encourages framing experimental questions in the context of integrated natural behavior. From a psychological perspective, there are dramatic differences

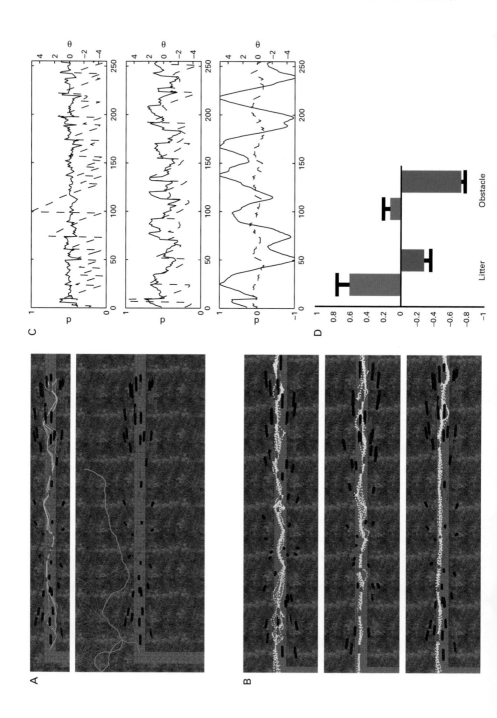

between this framework and traditional approaches to studying visual behaviors:

• The desired schedule of interrupts under normal behavior has a temporal distribution that is very different than worst-case laboratory situations. In the lab, subjects are typically in extremis with respect to reaction times, whereas natural behaviors typically allow flexibility in responding.
• In a multiple-task situation, the most important task facing the deployment of gaze is to choose the behavior being serviced. This problem is hardly considered in the search literature, which concentrates on within-task saliency of individual targets.
• The natural timescale for studying behavior components is of the order 100 to 200 milliseconds, the time to estimate state information. Below that, one is studying the process of state formation, a level of detail interesting in its own right but below the central issues in human behavioral modeling.
• The context for the deployment of visual routines is reversed from a laboratory situation. In that situation, the typical structure of a task forces a bottom-up description. The image is most often presented on a previously blank CRT screen. In a natural task, the particular test needed in a gaze deployment is known. Furthermore, this test is known before the saccade is made. Thus, in the natural case, the situation is reversed: The test can be in place before the data are available. This has the result of making the test go as fast as possible. The speed of tests may account for the fact that fixation times in natural situations can be very short. Dwell times of 100 milliseconds are normal, less than half those observed in many laboratory studies.

All of these observations suggest that the operating system level issues reveal a very different order of priorities for behaviors situated in their

---

Figure 8.17
(A) Testing the inverse reinforcement learning algorithm. The original data for an avatar walker is depicted in yellow. With the correctly recovered reward weights, the avatar can generate a trajectory very similar to the original, but with very dissimilar reward weights, the generated trajectory is correspondingly dissimilar. (B) Data from 10 human subjects show very similar navigation trajectories. Top: Pick up litter and avoid obstacles. Middle: Pick up litter only. Bottom: Avoid obstacles only. (C) Representative data sets for the individual modules' state spaces for one of the human subjects. (D) Weights computed for the human subject data assuming the modular format. Left: Pick up litter data. Right: Avoid obstacles data. (A–C) reprinted from C. A. Rothkopf and D. H. Ballard, "Modular inverse reinforcement learning for visuomotor behavior," *Biological Cybernetics*, vol. 107, pp. 477–490, 2013.

natural contexts. While the model assumes extensive structure, the choice of each of its components is motivated by a specific purpose, and the whole combines to direct the performance of human behaviors. A competing performance model might look very different but would have to address these issues.

The rule of abstractions is that one must stay within the boundaries of the chosen venue; however, from the complete perspective of the working brain system, the operating system constructs have to be represented neurally in some way. While this question is still unresolved, one promising direction is that of using neural oscillations at different frequencies as control signals (see refs. 13, 14). This idea was mentioned in chapter 1, but here we can be more specific in tying down putative functions. In managing a task such as making a sandwich, phase coding in the θ frequency is a way of keeping track of the task from beginning to end. However, to keep track of a step in the task such as spreading peanut butter, the β frequency can be used. And to keep track of the spreading itself, the γ frequency can be used.

8.10 Summary

The focus of this chapter was to introduce the issues associated with managing multiple behaviors. Although the system is complex, most of the constraints follow from the top-level assumption of compositional behaviors. Here are the main important features:

1. *Working memory is about running processes* Once one decides to have a set of running behaviors, the questions of how many and when are they running are immediate. Furthermore, they have ready answers in observations of human behavior in the classic observations of working memory and eye movements: Working memory suggests the number of simultaneous behaviors is small; eye movements suggest when a behavior is running as each fixation is an indication of the brain's instantaneous behavior being updated.

2. *There must be an operating system* The restricted number of active behaviors means that there must be a mechanism for making sure that a good behavioral subset has been chosen. Such a mechanism must interrogate the environment and add needed behaviors as well as drop behaviors if needed to meet the capacity constraint. Simulations using the OS lead to different perspectives than bottom-up analyses. For an excellent review and perspective of neural structures implicated in multiple competing behaviors, see Cisek and Kalaska.[243] However, their proposal has "attention" selecting

actions from a dorsal stream, whereas the stance herein prefers modules (aka behaviors) to be selected by "attention" and the selection of actions to be taken care of at a lower hierarchical level.

3. *Behaviors are abstracted in state-action tables* The key assumption has been on characterizing behaviors as decomposable into sets of independent modules; that each is responsible for some aspect of the overall behavior. The mechanism for programming behaviors is reinforcement learning. The essence of a module is captured by reinforcement learning's Q-tables that relate the states determined by vision to actions for the motor system. Indeed, the commands are in coded form, taking advantage of known structure in the body that carries them out. Assuming the existence of a table as is done at the reinforcement learning level finesses important details. Thus, a more detailed model is necessary to account for how the table index is created.

4. *Reinforcement drives gaze allocation* The reinforcement learning venue provides a different perspective on gaze allocation. One of the original ideas was a bottom-up view that gaze should be drawn to the most salient locations in the scene as represented in the image, where salience was defined in terms of the spatial conjunction of many feature points. However, recent measurements have shown that eye movements are much more agenda driven than that predicted by bottom-up saliency models. For example, Henderson has shown that subjects examining urban scenes for people examine places where people might be even though these can have very low feature saliency.[244] Walter's use of Q-tables suggests that to interpret gaze allocation, an additional level of indirection may be required. For example, the controller for sidewalk navigation uses gaze to update the estimate of the location of the sidewalk. To predict when gaze might be allocated to do this in our model requires knowing the uncertainty in the current estimate of the sidewalk location.

Arguably the most important theme in recent vision research is that no component of the visual system can be properly understood in isolation from the behavioral goals of the organism[245] (see also ref. 222). Therefore, properly understanding vision will ultimately require modeling complete sensorimotor systems in behaving agents. The computational cognitive hierarchy described here is certainly not true in all of its particulars, and it leaves many details unspecified. However, it does provide a framework for thinking about action-oriented human behavior.

IV Awareness

The most abstract levels in the brain's hierarchy of representations are those of awareness and social communication. These have been some of the most difficult to model computationally.

Conscious decision-making is being successfully characterized by reinforcement learning and Bayesian decision-making formalisms, and these results are highlighted in chapter 9. Decision making in the course of interacting with other agents is harder but is approachable with game theory techniques.

Despite a considerable history of experimental investigation and proposals, the brain's emotional system has only recently been understood at the level where computational models can be tried. One large precondition for this progress has been the separation of the brain's emotional state, which is a product of many interdependent systems, and the feelings that we are aware of. This distinction allows the emotional state to be connected to the underlying circuitry while postponing the issue of awareness. Consequently, one can pursue descriptions in terms of the lower-level computational abstraction such as reinforcement learning. In contrast, the study of feelings and their use in communication can be tackled without recourse to lower-level systems. Some current efforts to use affect use robot simulations, both to elucidate what are the essential features and as a complement to discourse. Chapter 10 describes this work.

The issue of consciousness itself was widely regarded as intractable until only about two decades ago but has gained acceptance. The Association for the Study of Consciousness held its fifteenth conference in Kyoto in 2011. In this developing field, there are many divergent views on how to proceed. Rather than attempt to be comprehensive, chapter 11 provides one view, biased toward being compatible with the brain's hierarchical structure.

9   Decision Making

Whenever you are conscious, and perhaps even whenever you are not, multiple computations are going on in your brain, which maintain and update current answers to some key questions: Is anything new going on? Is there a threat? Are things going well? Should my attention be redirected? Is more effort required? You can think of a cockpit, with a set of dials that indicate the current values of each of these essential variables.

—Daniel Kahneman, *Thinking, Fast and Slow* (Farrar, Strauss and Giroux, 2011, p. 59)

At this point, we have covered the basic neural machinery and its embodiment and now turn to what we are calling social issues. This level of abstraction explores the computational models that address a human being's behavior in making decisions that can be reported or observed by others. The abstract structure of neural programs for decision making was described earlier, covering the most important components of reward and certainty and discounting, but this chapter is focused on overt evidence for their importance. We have already discussed how these components are useful in describing computation "under the hood" but of course they are important in a human's overt behavior as well. It turns out that computationally motivated modeling is driving the exploration of these questions together with the tools of functional magnetic resonance imaging (fMRI) in humans and single-cell recording in monkeys. The emergent evidence shows that these primates both act in a way that can be explained by the reinforcement learning framework.

Of decision-making behaviors, the hardest to model are those when humans or monkeys are pitted against other primates, for the simple reason that the adversary need not be helpful, but can be acting for its own benefit at the other's expense. To model this more difficult venue, one resorts to computational models derived from game theory, but the principal result is still the same. Primates act in a way that is sensitive to these measures. To get started, let us revisit the central parameters of reinforcement learning that impact decision making.

• *Reward* Given the diversity of possible behaviors, there needs to be some common system of valuation. Internally, there is a great amount of evidence that dopamine is used for this purpose. In decision making, there is increasing evidence from several laboratories that primates work from a common system of valuation.
• *Uncertainty* Some things are easy to decide upon, but the ubiquity of noise in the world means that in many cases the information guiding choices is very uncertain. Nonetheless, choices still have to be made under those circumstances. Experiments show that classical models for dealing with noisy circumstances might be used by primates.
• *Discounting* Predicting the future is uncertain, and this affects choices because the uncertainty makes future rewards unreliable. Suppose you can have $200 now or $220 three months from now. You might wonder if you need to calculate investment opportunities to answer this question, but it turns out experimentally that individuals can be modeled as having a discount factor that scales their value for future rewards.

9.1 The Coding of Decisions

Representing Decisions
Before examining the evidence for the role of the major variables surrounding decision making, it is good to know that there are identifiable structures in the brain that can be implicated in this process. However, finding such evidence requires considerable inventiveness. A major difficulty in searching for cells in the cortex that represent decisions is knowing where to look, as cells in different parts of the cortex represent different information. As one example, Romo et al. examined an area termed the ventral premotor cortex (VPC), which is known to transform sensory information into action.[246] Their experiment had a monkey comparing two successive tactile vibratory stimuli of different frequencies delivered to the hand. The second frequency could be either higher or lower than the first, and the monkey had to signal the answer by pressing the appropriate of two keys with a finger of the other hand. It takes a while to train a monkey to do this, but they become very adept, and for frequencies 8 Hz apart in the range 10 to 34 Hz, they learn to perform at the level of greater than 90% correct. Figure 9.1 shows the basic experimental design.

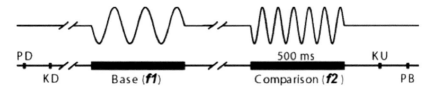

Figure 9.1
An experiment designed to explore the representation of decision variables in cortex. A monkey receives a vibratory stimulus to one hand of a certain frequency f_1. Shortly thereafter, a second stimulus of either a higher or lower frequency f_2 is applied. The monkey uses the second hand to signal the decision that the second frequency is higher or lower than the first by lifting up the finger (KU) and pressing one of two keys (PD). Reprinted from R. Romo et al., "Neuronal correlates of a perceptual decision in ventral premotor cortex," *Neuron*, vol. 41, p. 4, copyright 2004, with permission from Elsevier.

Cells in the VPC can be found that code for many different aspects of the task, but some reveal a definite correlation with the decision-making component. Two examples of these are shown in figure 9.2. The individual lines of spikes represent individual trials and are labeled with the pairs of frequencies tested. Many features of these spike trains are evident. The most important for the purposes of decision making are the ones in the interval where f_2 was presented. In the first cell, the spike rates are clearly representing the "f_1 greater" condition, whereas in the second cell, the same kind of information is represented, but in addition the response is correlated with frequency. What is also apparent is that there is significant activity during the interval before f_2 is presented, signifying that the cells are coding the first stimulus and anticipating the comparison.

You might be tempted to jump to the conclusion that the cells have an extraordinary amount of structure to produce all of these temporal features, but you have to remember to think of these cells as embedded in the larger context of the task. There is easily sufficient time for there to be several cortical–basal ganglia loops during the task. Thus, the likely scenario is that when the cortex codes f_1, this transitions the state to one where the code is being remembered but changing the state of the network that it is part of. Similarly, when f_2 is presented, the state changes again, and so on. You might also be wondering how a monkey's brain can program itself so quickly to do this task, but here again there is a ready answer. To do a task of this complexity, monkeys have to be given hundreds of examples, usually starting with a simpler version and gradually transiting to the final

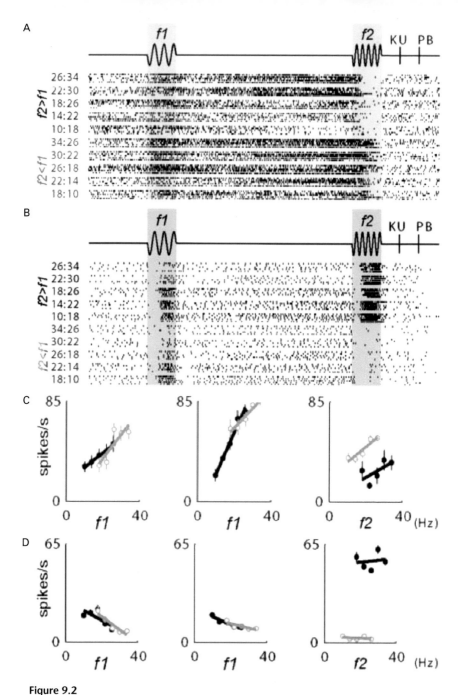

Figure 9.2
Two cells that are correlated with the decision-making task in the ventral premotor area of cortex. (A, B) Actual spike data laid out in correspondence with the task structure. Thus, first the spikes are lined up with the presentation of f_1, then the intervening interval, then the presentation of f_2, and finally the key-press response. (C, D) The spikes for the case where $f_1 > f_2$ are plotted in black and the converse is plotted in gray. Reprinted from R. Romo et al., "Neuronal correlates of a perceptual decision in ventral premotor cortex," *Neuron*, vol. 41, p. 4, copyright 2004, with permission from Elsevier.

configuration. Nonetheless, experiments like this one show that the neural correlates of decision variables can be found in individual cells, suggesting that the feature differences between the two variables has been abstracted away and that the essence of the decision is represented in a way that can be readily accessed. The next issue to tackle is the evidence for using a scalar value to evaluate this information.

The Coding of Value

It is not easy to find a way to probe the brain's use of a common value, but some experimenters have been able to gather suggestive evidence. Deaner et al. designed a setup whereby male monkeys had the choice of juice or looking at an image of something of interest to them with a smaller amount of juice.[247] The images varied between two kinds. One was that of a female monkey's perineum, or loosely, rear end, something of great interest to a male monkey with mating ambitions. The other image was of a lower-ranked cohort, which was of interest, but not as much. In one configuration, shown in figure 9.3, the monkeys chose between these two conditions by fixating one of two targets. By varying the juice values, the experimenters find the amount of times a monkey preferred one condition over the other. Of course, the monkeys vary their choices, but one can average the ratio of looks in any condition. Using this strategy, the experimenters could find the equilibrium values where they chose between the image and juice or just the additional juice the same number of times.

What this suggests is that they must have some internal measure that equates juice with looking at an image. Naturally, this could be handled by the "neuro" (aka dopamine). We cannot say definitively that this is how the bookkeeping is managed, but the experiment produces some very suggestive evidence.

These experiments are only a few of many that are providing increasing evidence that primates are extremely facile in making trade-offs when faced with decisions between outcomes of very different kinds of reward. Moreover, the evidence suggests that they can do this by being able to translate the different choices to a common internal value system.

9.2 Deciding in Noisy Environments

In the experiments that we have just reviewed, there was no uncertainty in the choice selection, in order to focus on the coding of features and use of a common internal value, but in the everyday world decisions are likely to be based on data that have an associated uncertainty. To isolate this issue,

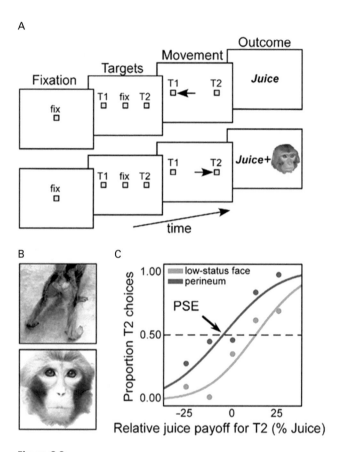

Figure 9.3
An ingenious method to test the trade-offs between items of different value from the Platt laboratory. Male monkeys fixate a central square and then are given two choices, T1 and T2. If they fixate the T1 target, they receive a juice reward. If they fixate the T2 target, they receive a smaller juice reward but view one of two images. One image is that of a low-status cohort and the other a female perineum. The monkeys prefer the perineum image. Over a sequence of trials with different relative juice rewards, the researchers can determine a relative value of one image over another in units of juice. From R. O. Deaner et al., "Monkeys pay per view: Adaptive valuation of social images by rhesus macaques," *Current Biology*, vol. 15, pp. 543–548, 2005.

experimenters create decision paradigms where uncertainty is the primary variation. One nice experimental situation, pioneered by Newsome et al.,[248,249] has monkeys making judgments as to motion coherence, as introduced in chapter 3 (see figure 3.10). A monkey subject fixates the center of a screen upon which moving dots are displayed. In the displays, some fraction of the dots are moving coherently in one of two possible directions, say to the right or left, and the rest are moving randomly. If most of the dots are moving randomly, deciding whether there is motion to the right is a very difficult problem, but as the coherently moving fraction is increased, the decision becomes relatively easier to make. In a common setup, the monkey makes the choice by making a saccade to one of two targets.

To gain purchase on modeling computation in this problem, consider flipping a coin. Someone tells you that a coin is fair, but you want to make sure. The obvious thing to do is to flip it a number of times. Just a couple of flips would not do, such as HTTT, but after HTHTTHTHHHTTH, you could start to be increasingly confident that the probability of getting heads was 0.5. If the coin was not fair, you could still estimate its heads probability by summing the number of heads and dividing by the number of throws. Where $\hat{p}$ is this estimate, one standard deviation in the estimate can be estimated by

$$\sqrt{\frac{\hat{p}(1-\hat{p})}{N}} \, ,$$

where N is the number of throws. So what this is indicating is a fairly obvious but nonetheless crucial point; namely, that the estimate increases in reliability with the number of throws.

In making decisions in noisy conditions, one can use this analysis to estimate that the probability is greater than a certain value. This is a classical model from statistics termed a *barrier model*. If we want to be sure that the probability of a decision has achieved a certain confidence level, and the decision itself can be broken down into subcomponents that are analogous to biased coin flipping, then we can use the binomial model to precompute the numbers of heads and tails that are needed and report "yes" when that number is exceeded. This is shown schematically in figure 9.4.

Researchers Cook and Maunsell used a Newsome-like paradigm in ventral inferoparietal cortex (VIP) that had been implicated as a site of such measurements.[250] Their experiment had the nice touch of using two areas of potential coherent motion that the monkey had to choose between with an eye saccade. As usual, the degrees of coherence of the moving dots

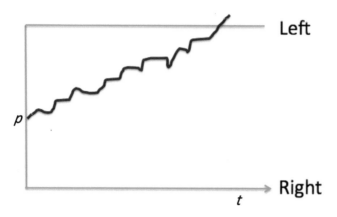

Figure 9.4
In a barrier model, evidence for a decision accumulates over time. In terms of the coin-flipping example, the problem can be seen as getting a certain level of representing the probability of heads exceeding a certain value. Each heads result moves one toward the barrier, and each tails result moves it away. When a sufficient level is reached, a decision is made.

were varied to produce problems of varying difficulty. Collecting the neural responses of neurons to many trials, they binned them according to the measured reaction times and summed the spikes in each bin. Figure 9.5 shows this result. What can be seen is that if a barrier is set at a level that works for all the successfully classified examples, then the individual summed responses are perfectly in register with the response times, indicated in milliseconds using the associated colored numbers.

You will note that the task setup is a little different from that of the vibrotactile experiment. The monkey starts by fixating a central cross, then two simultaneous random dot motion fields appear on the display, one of course at the location of a measured neural receptive field, and the other opposite so that the conditions can be balanced. Next at the decision time, one of the patches exhibits coherent motion for 750 milliseconds. Again to appreciate how this might be set up internally, one can appeal to the cortical–basal ganglia loop to step the cortex successively through the different states of the task.

Another computational issue that still needs to be addressed in this experiment is how the evidence accumulation process is actually implemented in neurons. If we assume that the state issue just discussed is settled, there is still the issue of measuring relative coherence in motion dot displays. Given the putative computational hierarchies, the logical abstraction for this step

A

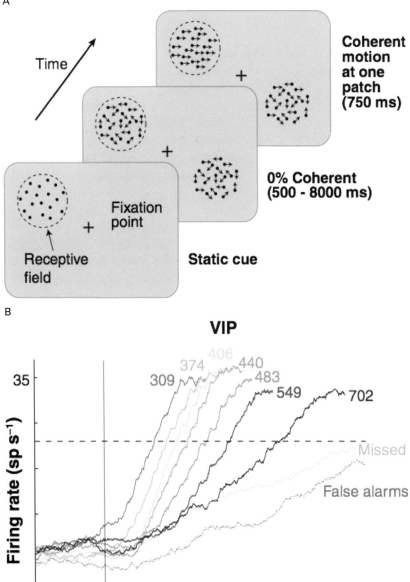

Figure 9.5
In a barrier model, evidence for a decision accumulates over time. When a sufficient level is reached, a decision is made. Reprinted by permission from Macmillan Publishers Ltd: E. P. Cook and J. H. R. Maunsell, "Dynamics of neuronal responses in macaque MT and VIP during motion detection," *Nature Neuroscience* vol. 5, p. 10, copyright 2002.

is that of visual routines. Cook and Maunsell assumed that neurons could integrate a coherence signal; that computation in turn appealing to a rate coding for spikes. Inspired by the coin-flipping example, another option is that of discrete measurements that probabilistically report either "coherent" or "random," but the issue is currently unsettled.

We don't really know if the monkeys are in a state of conscious awareness when they are making all these decisions, but we suspect they are with some confidence because when similar experiments are done with humans, similar trade-offs are seen.[251,252]

Learning Feature Distributions

Besides learning to make discrete choices, humans can also learn the distributions of features connected with different values. Trommerhäuser et al. had subjects learn to point at a blank plane in a way that maximized reward.[253] Points on the plane were valued according to superpositions of normal distributions of positive and negative reward values such that subjects had to learn to pick positively rewarded landing points and avoid negatively rewarded points. Despite the complicated reward boundaries of the superposed distributions, they were able to do this.

9.3 Social Decision Making

It's one thing to make decisions when playing against the environment but quite another to play against another animal, whether monkey or human. The big change of course is that the adversary has a mind of its own and may not always choose options with the best interests of its opponent in mind. Nonetheless, even though this situation is more complex, as we will see, it can still be modeled with computation.

One of the hardest features of animal behavior to explain is altruism because it seems to fly in the face of fitness-driven evolution. Simple forms of altruism can be explained by the usefulness of protecting one's genetic investment in relatives other than direct offspring. For more complicated models, we have to turn to game theory, which shows that altruistic behavior can be optimal but the conditions under which it is so are delicate. In any case, we do better when we can estimate average behavior over trials. One-shot decisions are risky and problematic.

Protecting Your Genes

The viewpoint of the gene provides a starting point for thinking about altruism. The crucial insight was the brainchild of William Hamilton. His idea was that individuals would invest in relatives to the extent of the amount of genes that they shared with that relative. Identical twins have identical DNA, but a child has ½ of the DNA of the parent, and the grandchild has ¼.

It is hard to imagine how confusing things were before Hamilton's clarifying work. Now things are not only much clearer, but we can appreciate a revolution in thinking. As Dawkins has eloquently portrayed,[254] in the long view we should think of the genes as in charge, using the phenotype as a way to get copies of themselves into the next generation.

But while the genetic view explains a lot of data, it does not explain all of altruism. A national news story from 2005 in India described how a girl age 12 and her younger brother had escaped the path of a speeding car when the girl looked back to see her neighbor's child still in danger. She ran back and pushed him out of the way but was hit by the car and lost her leg in the process. Because she and her neighbor did not share any genes, we have to look elsewhere for explanations of such cases. It turns out that we can begin to model these cases with computation, and the specialty that does so is called *game theory*. Such models do not settle all the complexities of social interactions by a long shot. But they do introduce the important idea that cooperation is a rational thing to do. And if that is true, then it is entirely possible that the need to cooperate is specified in the genes. It is extremely likely that we are hard-wired to be good—or at least most of us.

Social Intelligence and Game Theory

At the height of the Cold War between the United States and the Soviet Union, the specter of a nuclear holocaust prompted the best thinkers to try and come up with some sort of escape plan. Morgenstern and von Neumann's newly developed game theory proved a way of thinking about the problem.[255] In game theory, the negotiations between two sides are presented in terms of a table of the various combinations of choices open to both sides. There can be an arbitrary number of sides, but in all our examples we will only use two, with fictitious players Alice and Bob. In our first example, Alice and Bob each have options C and D. Figure 9.6 represents the value to each of the players for different combinations of choices. The way the game is played is that both players make their choices without knowing in advance what the other will choose. The choices are then revealed and the payoff matrix consulted to award points to each player.

In this example, Alice should choose option C and Bob should also choose option C, as that results in the highest payoff for both players.

A

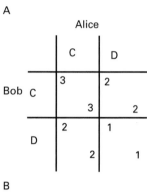

B

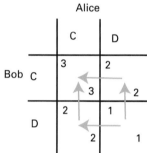

Figure 9.6
(A) A very simple game in which if Alice and Bob both choose "C," they each receive 3 points, and so forth. (B) If they know the payoffs, they can reason that they should each choose this option.

This is also a *Nash equilibrium* point, as neither Alice nor Bob can improve on these choices assuming the other player does what is best for him or her.

The model for the Cold War stalemate is more like our second example: the famous *prisoner's dilemma* game. In this game, summarized by the payoff matrix in figure 9.7, two prisoners each have the chance of receiving significant rewards for cooperating with the authorities, provided his confederate refuses. If they both rat on each other, the rewards are minimal, and if each refuses to turn in his confederate, the rewards are modest. The payoff matrix looks like this:

In the slightly confusing jargon of this game, not turning in your confederate is termed "cooperating" (C) and ratting on him or her is termed "defecting" (D). The numbers in the matrix are to be interpreted as years off their sentences.

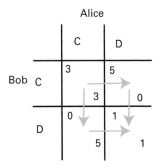

Figure 9.7
The prisoner's dilemma game. Alice and Bob are hardened criminals who are caught and separately asked to turn in his or her partner in return for years taken off his or her sentence. If they cooperate, C, that is, refuse to tell, then the sentence possible is 3 years below maximum. Alice's reward is shown in the upper left of each box, and Bob's reward is shown in the lower right. So if Bob defects, D, that is, turns in Alice, while Alice cooperates, he receives a 5-year sentence reduction and she none.

What should Alice do? If Bob is going to defect, then from figure 9.7, the best Alice can do is to defect as well. She'll at least get a 1-year reduction. If Bob is going to cooperate, then by defecting she'll receive a 5-year reduction compared to only 3 years off for cooperating. So no matter what Bob does, the rational thing for Alice to do is defect. Naturally, the same logic holds for Bob as well. Thus, it seems that the players inevitably must each choose "defect." Interpreting this example in terms of the Cold War, it seemed as though the right thing for the United States or Soviet Union to do was to "defect" and shoot their missiles first. A considerable number of people on both sides advocated doing just this at the time.[256]

Although this would not work for a one-time situation, the game analysis changes for repeated trials. In this case we have to imagine that Alice and Bob are recidivists. They play the game, see the choice that each has made, and then play again. This version of the game is known as the *iterated prisoner's dilemma*, or IPD. If they are going to play the game over and over again, there is the possibility that they could find the compromise strategy where they both cooperate. To see that there might be good prospects, we can analyze the outcomes where a defector and a cooperator each play against a player using the strategy "tit-for-tat," abbreviated TFT. The tit-for-tat strategist cooperates for one move and then chooses the opponent's most recent choice move for the subsequent turns. Owing to the uncertainty in the opponent's choices, let's use a discount factor γ that devalues future rewards. This allows the calculation of the relative payoffs by summing the series that would result from repeated games.

The cooperator playing against TFT will get

$$3 + 3\gamma + 3\gamma^2 + 3\gamma^3 + \cdots = \frac{3}{1-\gamma},$$

and the defector playing against TFT will get

$$5 + \gamma + \gamma^2 + \gamma^3 + \cdots = 5 + \frac{\gamma}{1-\gamma}.$$

By comparing these two values, you can show that when γ is greater than ½, the cooperator does better. Thus, if you value future rewards, it's rational to cooperate, but if you do not, you should defect.

The pioneering studies introducing IPD were done by Axelrod.[257] To see if cooperative strategies would appear, he had groups of human players play against each other. Each player was free to choose his or her own strategy. For example, you might tolerate two defections in a row, but then you defect also to send a message. Or you might cooperate most of the time and defect once in a while to get a profit. Each player was free to choose his or her own strategy against any player at any time. Surprisingly, it turned out that the straightforward tit-for-tat strategy beat all rivals.

However, the initial tests with human subjects did not allow for too much subtlety and left the thought that perhaps it might be possible to do better with a computational approach that was capable of remembering lots of information about the strategies of individual players. To test this idea, Axelrod turned to a genetic algorithm. The genetic algorithm was trained using a three-move history of the encounters with individuals. An element of the specific encoding can be visualized as shown in figure 9.8A.

To understand the encoding, consider the third line of figure 9.8A. This is interpreted as: if three turns ago I and my opponent both chose cooperate (C), and if two turns ago we also cooperated, and in the penultimate turn my opponent chose D while I cooperated, this time I should choose C. With all possible combinations of the last three turns, the "DNA" string gets quite long, as the different strategies need to be represented by different values in each of the 2^6 possible previous positions (shown in red letters on the figure). As a consequence of this encoding strategy, there is a vast number of *possible* individuals. The total number of such summary strategies, where each has a unique string, is 2^{2^6}, a large number indeed.

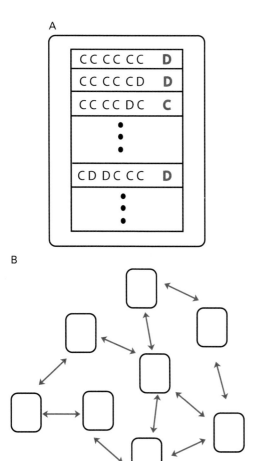

Figure 9.8
(A) The particular coding of one of the individuals in the population. The black letters on a row represent possible moves for both players in the most recent three turns. The red letters represent the policy for that individual. (B) The genetic algorithm finds good policies by having individuals play against each other to compute a fitness score for each individual.

The mechanics of tuning the genetic algorithm are to have a population of individuals with different strategies and let them play against data from human players. Individuals in the population each get a score depending on how well they did. Next, a new round of players is bred. Pairs of individuals are picked randomly with the best performers having the best chances of being picked. For each pair, random rows representing specific three-move policies are swapped to make a new individual. After many generations, very good artificial players have evolved.

Axelrod showed that the genetic algorithms (GAs) beat the TFT strategy. Because TFT was already bettered its human opponents, you might wonder how the GA did it. The secret is that while TFT is a best strategy against good opponents, more run-of-the-mill opponent strategies typically have weaknesses that can be exploited in ways that TFT does not take advantage of. If you know your opponent allows two defects before punishing you, you can sneak one or two in for a profit before behaving.

Snail Darters and Vampire Bats

When we see humans playing games that are advertised as being a model of their deep-seated altruistic behavior, computation or not, we tend to be skeptics. We are great at mastering abstractions and playing according to made-up rules, so perhaps this is something we can do but may not do as part of ingrained behavior. To show this view is shortsighted, we turn to the animals that play forms of IPD.

A snail darter is a small fish that schools. When a large fish appears nearby, there is always the possibility that the large fish is a predator. But also there is the possibility that it is not. The school wants to react appropriately to a real predator but not to waste energy on a false alarm. The solution adopted is to allow a pair of scouts to approach the predator. As they get close, if it is dangerous, the predator should reveal itself. But which of the pair should go first? The solution adopted is a form of IPD. Each fish advances by a discrete small amount in turn: one first taking the lead and then the other. The turns are of course like turns in IPD. Choosing to cooperate means going ahead. Turning tail is defecting. By behaving in this way, the two fish share the risk of information gathering.

In a similar way, vampire bats share the results of their food gathering. They suck blood from large animals, but having exploited a food source, they will fly back and share the blood with other bats in the nest. They sacrifice some of their food with the understanding that they might be the beneficiaries on another occasion.

Monkeys Playing Games

If snail darters and bats are genetically programmed to exhibit cooperative behaviors, then obviously primates should be, too. But the question is, do they? And if they do, what parts of the brain are implementing these algorithms? Recent work has been done at several laboratories in order to answer these questions.[258]

Glimcher has pioneered experiments using a game that requires a probabilistic solution strategy called *work or shirk*.[259] In this case, the computer plays the role of overseer of a factory that pays a worker a wage W. If the worker, played by the monkey, shows up, a product worth P is produced. The worker would like to skip, but if the worker does and is caught, the worker will lose the wage. If the boss inspects, it costs the boss I, and if the worker shows up, it costs the worker E. These data allow us to assemble the payoff matrix describing the different payoffs, and this is shown in figure 9.9.

Naturally, the worker would like to not show up and be paid while the boss would like to not have to pay the cost of inspection, but if either of them do this, they can be penalized by the other. The solution is to adopt a probabilistic strategy. It turns out the best values to each side are such that the probability of the boss inspecting should be E/W and the probability of the worker shirking should be I/W. Glimcher showed that monkeys not only can play this game successfully, but also that they are sensitive to shifting the payoff values. So if the ratio of I to W is changed, then the monkey adjusts its probability accordingly. If the cost of inspecting goes up, then there should be more shirking, and this is what happens.

Another simple game is *matching pennies*. Two players turn over a coin at the same time. They can choose the side to have face up. If the sides match, then player A gets a point and player B gets minus a point. If the sides do not match, the payoffs are reversed. This very simple game has some subtleties, however. If player A chooses more heads than tails, and player B notices this, then player B can win points by choosing more tails on average. To counter this, player A should randomly choose heads or tails. Then there is no strategy that can win. However, if player A does this reliably, then any strategy that player B picks is as good as any other. For example, if player B picks heads all the time, the average payoff is still zero.

Lee has trained monkeys to play a version of matching pennies.[260] Using real pennies with monkeys would be difficult, as a lot of time would be lost in training the monkeys not to throw the coins or try and eat them. The solution is to have the monkey use eye movements to indicate choices.

A

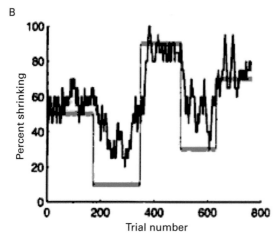

B

Figure 9.9

(A) The work or shirk game. If the boss inspects and the workers are at it, the boss receives the value of the product (*P*) minus the wages for the work (*W*) and the cost of inspecting (*I*). The worker gets the wages *W* minus the energy expended (*E*). (B) By manipulating *I/W*, one can change the equilibrium point for shirking. As the Glimcher lab showed, monkeys can track this shifting equilibrium. (B) from P. Glimcher, *Decisions, Uncertainty, and the Brain*. MIT Press, 2003.

What happens is that the monkey stares at a light spot in the center of a display and then, when the light goes out, looks to one of two lights displayed a short distance away on either side. The choice of left or right is equivalent to the heads/tails coin decision. The monkey is doing this for a juice reward. Before it looks, the computer has already selected left or right as the reward location. So if the monkey matches this by looking, then it actually gets the payoff, otherwise not and a new trial begins.

What is most interesting about this experiment is its control case. In this case, the computer selects reward locations from trial to trial at random. So as we just discussed, all strategies that the monkey can try are equivalent. This shows up in the data of the monkey's responses, which are arbitrarily biased to a side it prefers. But in the test trials, the computer keeps track of the monkey's responses and tries to minimize its payoff. So if the monkey frequently chooses "right," the computer chooses "left" more often and vice versa. In this case, the monkey must be random in its choices to maximize rewards, and that is exactly what it does.

So we have gone way beyond snail darters. The new factor exhibited by the monkeys is their use of probability in making their choices. To review, as we saw in the discussion of the standard neural model, when there are unknown payoffs with uncertainty in their estimates, the best thing to do is to sample the one you think is best increasingly often, and in the limit you'll have a deterministic strategy. This is the classical *bandit problem* result. It's when these payoffs are changing that it becomes trickier. If they do, you have to keep sampling what you estimate are the losing options more frequently. So you have a probabilistic strategy where your probabilities reflect the continuing uncertainty in the environment. Game theory can be seen as being at the most volatile end of the uncertainty strategy spectrum. Not only are the payoffs varying, but the reason they are varying is that they are being guided by another agent that is trying to exploit you! In these cases, the best defense is a purely probabilistic strategy that is tuned to your opponent's choices. The fact that monkeys play these games optimally is incredibly revealing. Not only does it say that they can do this, and are genetically programmed with these abilities, but also that they are tuned to the social demands of dealing with other monkeys in a computationally optimal way.

Payoff Matrices Can Be Learned
While we have covered the extensive evidence that primates understand and can play games, it of course also follows that they can learn to play these games. Learning in games has been extensively studied.[261] The work and shirk game just studied allows for a learning algorithm that uses its probabilistic strategies. The basic idea was first formally developed by Robinson[262] but refined to use gradient ascent by Singh et al.[263] and Bowling and Veloso.[264] The basic idea is for the player to use the history of the opponent's play as a guide to estimating their probabilistic strategy and then move the probability in a direction that would improve the player's payoff. This algorithm is very effective in finding Nash equilibria but needs some modification in solving harder problems where the desirable solutions are not Nash equilibria.

The main attraction of gradient ascent algorithms is that the machinery they need is well within the bounds of biological plausibility; however, the most straightforward formulation is unlikely to be the whole story. At least one additional factor that has to be dealt with is the brain's extensive reliance on prior estimates. This use has been pointed out in a nice study by Markman et al.[265]

Cooperating in IPD

While the monkeys can handle work or shirk if they use a probabilistic strategy, IPD still presents a formidable obstacle, as learning to cooperate is a much more delicate problem for the simple reason that the desired result is not a Nash equilibrium. Tit-for-tat is a good strategy for playing IPD, but all versions of it can get jammed if opponents pick too many "defects" or pick "defect" at the wrong time. For example, suppose two TFT strategies play against each other where one TFT strategy is coded as "do whatever your opponent did last time." If they both cooperate, that is great, but if the opponent ever picks "defect," then it's easy to get stuck in defect forever. One part of breaking out of this rut is to have probabilistic strategies. You do not have a rigid strategy, but on each round, pick your move with a given probability. But this still is not enough.

Zhu has shown one way of making an additional assumption that can get to the desired mutual cooperation point.[266] The basic problem is that left to their own devices, it seems that the best choice for both players is to defect, when they could earn more if they were able to cooperate. If they were able to do this, it would be all to the good, but after the first defect by either player, it is hard to recover. Zhu was able to fix this for the case where players are choosing their moves probabilistically. The repair starts by assuming a fraction of defects for each player and defining a *perturbed game* about that point. Within the small range of possible changes in fractional strategies, it is possible to limit the losses due to the proportion of changes to a small amount. This makes it possible to test the intentions of the opponent by upping the fraction of cooperation. A rational opponent also can test to see if the change in the fraction of cooperation is increasing. If this test reports in the affirmative, then that player can up his or her fraction but with a certain probability. In this way, two rational opponents can gradually walk their fractions of cooperation to near 100%.

This strategy works, but as discussed so far, it is incomplete, as the rational player also has to be ready to penalize an opponent who always cooperates. For that player, the best strategy is to always defect. The way Zhu handles this is to use a two-level hierarchical system. The top layer implements the probabilistic strategy and cooperates a large fraction of the time if the opponent is thought to be trying to cooperate as well. However, if the opponent seems to be a defector, or in the case of the all-cooperate strategy a sucker, then the lower level ups the fraction of defections. Based on these fractional measurements, it turns out that the hierarchical strategy is well-behaved and can steer a cooperator to the desired point as shown in figure 9.10A or exploit a die-hard cooperator as in figure 9.10B.

9.4   Populations of Game Players

Finally, let's change the venue and look at populations of individuals. This venue has special advantages in that we can explore the dynamics of group interactions as well as model the spatial locality of groups. To show the properties of probabilistic strategies in populations, let's use the setting of the familiar *rock-paper-scissors* game. As any school kid knows, both players start with one hand behind the back and on command produce a hand configuration that is one of *rock* (a fist), *paper* (hand open), or *scissors* (index and middle finger extended). Rock beats scissors, paper beats rock, and scissors beats paper. We can represent this as a game with payoffs of one for winning and minus one for losing, as shown in table 9.1.

Now think of a population of rock-paper-scissors players. We can model their behavior in terms of the fraction of the total number of players that are playing any particular strategy. What happens is that at any round, they randomly choose someone to play with and then use their strategy. Let's assume that the players operate as follows: If any particular strategy is preferred, other players will gradually change their choices to countermand that strategy. So if lots of players are picking "rock," other players will switch to "scissors," and so on. How do you model the dynamics of this behavior? One way do this is to write an equation that expresses a rationale for there to be increases in a particular fraction. For example, let's model the fraction of players choosing "rock" with the equation

$$\dot{r} = r(s - p).$$

In this equation, the rate of change of players choosing "rock" is denoted by $\dot{r}$ (it could have just as easily been written as dr/dt, but the "dot" notation is nice and compact). What the equation is saying is that if the scissors players outnumber the paper players, then choosing rock is a good thing to do, and the proportion of rock players will increase. Furthermore, the rate of increase is proportional to the difference between the fraction of scissors

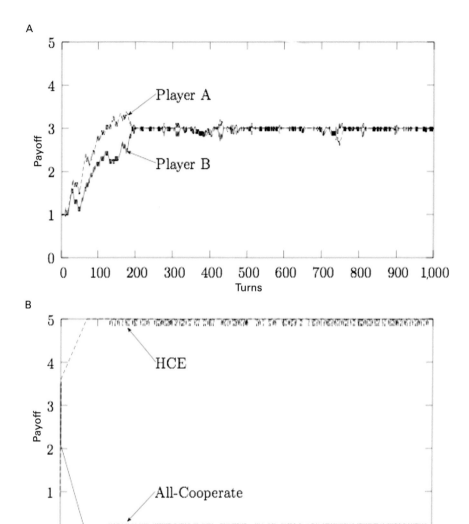

Figure 9.10
Zhu's model, called HCE, for detecting potential cooperators in the IPD works both
for (A) formidable opponents and (B) patsies. Basically, the hierarchical strategy implements a way to ask the questions "If I raise my fraction of cooperation, will you
do so also?" and "If I lower my fraction of cooperation, will you let me get away with
it?"

Table 9.1
The payoffs for the rock-paper-scissors game

		Alice		
		Rock	Paper	Scissors
	Rock	0	1	−1
		0	−1	1
Bob	Paper	−1	0	1
		1	0	−1
	Scissors	1	−1	0
		−1	1	0

players and the fraction of rock players. Similarly for the other players, there is

$$\dot{s} = s(p - r)$$

and

$$\dot{p} = p(r - s).$$

So the model predicts that the proportions will change unless the rates are all zero; that is,

$$\dot{r} = \dot{p} = \dot{s} = 0.$$

For which values of r, p, and s does this happen? You can see right away from these equations that one possibility is

$$r = s = p,$$

and because these fractions have to add up to unity, this occurs when they are all equal; that is,

$$r = s = p = \frac{1}{3}.$$

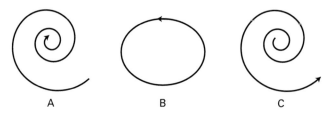

A   B   C

Figure 9.11
When a perturbation of some kind pushes the population off of an equilibrium point, the resultant dynamics can be stable as in cases A and B or unstable as in case C. It turns out that for rock-paper-scissors, the resultant trajectory will be stable in that it will oscillate around the equilibrium point.

Because once the fractions achieve these values they will not change, this set of fractions is termed an *equilibrium point*. If the fractions get to this point, they will stay there. Of course, this is an idealization of the real world, as one cannot expect that the fractions will remain perfectly equal. Thus, it is useful to analyze what would happen if they are somehow moved off of this point by a little bit. The possibilities are that the equilibrium point would be unstable or stable. In the former case, once the fractions drift off of the equilibrium point by even a little bit, they keep going. In the latter, they oscillate about the point or return back to it. It takes just a little mathematics, but it can be shown that this particular point is stable and that a slight perturbation will cause the fractions to oscillate around the point but not to move away from it, which is case B in figure 9.11.

Modeling a game as being played by a population of players might seem odd for rock-paper-scissors, but if the rules are changed to another game, you will see that the methodology can lead to important understandings. For that, we will turn to a *public goods game* studied by Hauert et al.[267]

In the public goods game, workers can opt to either participate or not. If they do choose to work, they receive their wage minus an expenditure; but if they show up for work and do not actually do anything, they can do better than the actual worker when it comes to receiving their wage. The wage is a based on the number of people who actually worked, divided by the number of workers (whether they worked or not). So you can see that this game has a version of the prisoner's dilemma in it, in that it is attractive to "defect," or not work at all and still get something. However, the Hauert et al. variant makes this more interesting, and that is to give workers the option of sitting out for a fixed sum. Basically, a person can choose to go on welfare and still have an income. What this does, under certain parameter settings, is set up a cycle like the one in the rock-paper-scissors game. Initially, workers find it profitable to defect, as the remaining workers make enough of a sum to be profitably shared. But as the number of nonworkers increases, the total value of their goods produced dwindles to the point where it is better to sit out, which workers start to do. But as the total number of workers in the wage pool decreases, then a person who chooses to work can do better than sitting out, so he or she will choose to work. Thus, there can be a repeating cycle, such as shown in figure 9.12A.

The stability of the cycling is even stronger if the game is played out on a two-dimensional grid that models local communication. Thus, if the situation is modeled as a two-dimensional array of instances of the game and the strategy adopted is the best of those in a 3 × 3 neighborhood, then different kinds of behaviors result depending on the relative values of the different payoffs. For intermediate values for wages, the best strategy at a location is dynamic, changing over time, but for larger values, the simulations show that the welfare recipients go extinct, and defectors can live at the boundaries of cooperators, as shown in figure 9.12B.

9.5 Summary

The study of overt decision making allows us to link the machinery that was first posited at the neural level. To choose between one course of action and another requires having a common basis for comparison. Chapter 3 presented the basic measurements that determined that the likely currency of value was the neurotransmitter dopamine. Using this scalar, a neural circuit that has the right connections can organize its transitions to follow paths of high expected discounted value. More recent work has shown that this organization extends to overt decision making that can be measured with functional magnetic resonance imaging in addition to single-neuron recording. All these experiments are basically consistent with the reinforcement learning model.

Additionally, the decision-making apparatus can be probed with high levels of stimulus noise. These experiments seem consistent with barrier models that accumulate evidence until a criterion is reached.

The most difficult kind of decision making occurs in the social context of making decisions in concert with an adversary who may or may not cooperate. In this context, the theory of game playing has been a great help in elucidating the basic parameters and trade-offs. Some situations are easy, and as a result it is simple to choose strategies where players can cooperate for mutual benefit. However, very difficult games, such as the iterated prisoner's dilemma, have to be tackled with probabilistic strategies as well as ancillary strategies. One is that of Zhu, which explores cooperation via

A

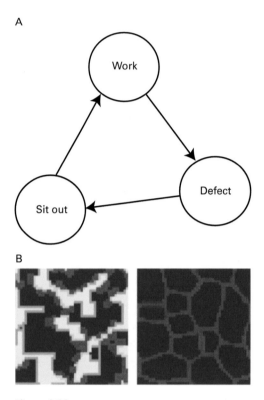

B

Figure 9.12

(A) The representation of the cycle of choices made in a public goods game. If many are working, workers may choose to defect. But if too many defect, it may be more profitable to sit out, decreasing the total number of workers. If the working number is very small, it becomes profitable to choose to work. (B) The game can be played on each square of an $N \times N$ grid where players adopt the most successful strategy in a 3 $\times$ 3 grid neighborhood. Blue = workers, red = shirkers, yellow = sitting out. Toroidal boundary conditions apply; that is, at the edge of the grid, the calculation of the 3 $\times$ 3 grid neighborhood wraps around. The leftmost panel shows a snapshot for the case where parameter values cause the choices to exhibit a dynamics with waves moving across the grid. The rightmost panel shows a case for substantial worker pay values. In this case, sit-outs die away, and a stationary pattern of workers and shirkers is observed. From C. Hauert et al., "Volunteering as red queen mechanism for cooperation in public goods games," *Science*, vol. 296, p. 5570, 2002. Reprinted with permission from AAAS.

low-risk fractional bets. Another, by Hauert et al., provides a third option of opting out in addition to the standard cooperate and defect choices.

All of these efforts are extraordinarily significant, as they sketch the possibility that the primate genome provides an inherent ability for its phenotypes to find cooperative solutions to problems instead of the most destructive options. Furthermore, this basic methodology has a rational mathematical basis.

Of course, a huge question that is not addressed by all the formalism covered is related to that of "awareness." How does the brain parse the circumstances of instances of a decision making interaction in novel social circumstances? All of the formalisms used in chapter 8 and earlier could in principle be done automatically. However, the kinds of decision making studied here ultimately seem to need some symbolic binding capability to associate the problem at hand with the internal machinery in a flexible way. While we won't be able to solve this problem in any complete way, the intent is to try and eat away at it in the remaining chapters.

10   Emotions

A discussion of emotion ... requires a mention of reward and punishment, drives and motivations and, of necessity, feelings. [It] entails an investigation of the extremely varied devices of life regulation available in brains but inspired by principles that were anteceded to brains and that, by and large, operate automatically and somewhat blindly, until they begin to be known to conscious minds in the form of feelings.
—Antonio Damasio, *Self Comes to Mind* (Pantheon, 2010, p. 108)

Emotions are so basic to human existence that they are often taken as the sine qua non that distances us from computation as done by a silicon computer:

It is clear, however, that, without the preferences reflected by positive and negative affect, our experiences would be a neutral gray. We would care no more what happens to us or what we do with our time than does a computer.[268]

This quote implicitly appeals to the aspect of emotions that is enshrined in our consciousness, which is the sensations, or feelings that they engender. We can laugh or feel elated when happy and cry when sad. However, it turns out that the experienced feelings of emotion are just the tip of the iceberg of a vast set of underlying components of neural and visceral anatomy that produce them. Thus, of course, present-day computers do not have feelings and people do, but this distinction misses the central question addressed here: whether human emotions can be seen to have a purpose that can be described by computation.

So things are turned around; humans have emotions, but perhaps like almost every other attribute of the brain, our premise is that they are best seen as part of computations that are essential to their owner. A vital sign for a robot is that it is low on battery power and must plug itself in. This was a problem tackled by early mobile robots and is still studied.[269] In a crude analogy, the state of the robot plays the role of a human emotional state. When a state that is important to well-being or survival is reached, the robot must adjust its activities and take care of the signaled issue. The charging state plays the role of a sleep state in humans; you might have heard the human sleep metaphor "charging one's batteries."

Even given this basic idea of a system usefully monitoring its body state, attempting to elaborate the emotional computations in humans is not an easy task, partly because the components of emotion are only partially understood, being grounded in complicated interactions between the viscera and neurochemistry. These complex interactions can be further obscured by the semantics of emotions as feelings, which prevents the deconstruction of the term "emotion" into functional components. The goal of this chapter is to attempt to wade through these complications and focus on a computational hierarchy for emotions, but, before setting out on this course, it is very necessary add a caution to what is going to follow. Even if the discussion can be situated helpfully in the forum of human brain models, it must be conceded that, although the puzzle of what are emotions has been extant for a long time,[270–273] most scientific advances in the understanding of human emotions are fairly recent, and an integration of them is still very unsettled. Setting computation aside for the moment, there are many different ongoing arguments as to just how emotions, particularly conscious feelings, are grounded in the brain.[274] The view represented here attempts a synthesis of many different elements and is biased toward hierarchical neural organizations. Adding to that is the fact that the computational characterizations of hierarchical functions of the emotional system are for the most part completely unsubstantiated. However, the point of discussing them, despite their gossamer support, rests on the belief that the problems they address are vital and need to be solved one way or another.

Despite these challenges, a flood of recent research on emotional systems has made the task of dissecting putative computational contributions of emotions considerably easier, and several recent books make essential reading.[275–278] A review material from all this work will set the stage. The upshot is that fairly recent clarifications of the deeper aspects of emotions change ways of thinking about them dramatically. One huge change derives from information about how different brain structures contribute to an emotional response. Another derives from tracking the temporal course of neural emotional signals. These vantage points allow us to speculate on their computational uses.

10.1 Triune Phylogeny

Let's now turn to the evolutionary time course of emotion-related components. This course might at first blush seem like a digression, but in fact
it turns out to be essential, as it leads directly to the emotional system's
hierarchical architecture. A primary feature of emotions is that they have a
distinct time course through the body, with conscious awareness of them
only at one end of the transit. Because this transit can recapitulate the
body's phylogenesis, it is essential to understand it. Ekman summarizes the
legacy of emotion this way:

> What is unique is that when an emotion occurs we are dealing with current funda-
> mental life tasks in ways [that] were adaptive in our evolutionary past. This is not to
> deny that our own individual past experience will also influence how we deal with
> these fundamental life tasks, but that is not what is unique to emotions (p. 56).[279]

One proposal is that the modern brain's emotional repertoire is thought to
have evolved in three broad stages, and as a consequence, the neural representation can be parsed into three successive layers, as depicted in figure
10.1. Cognoscenti will immediately recognize this figure as reminiscent of
the three stages in MacLean's evolutionary triage[280] and be quick to point
out that animal data subsequent to the initial development do not support
the clear-cut correspondences of reptiles, birds, and mammals that he tried
to make; the actual picture is messier.[281] However, for our purposes it is
important not to throw the baby out with the bath water. The trichotomy
of emotional development into components at different *abstractions* in
mammals is essential.

In the first stage, the brain is composed of the forerunner of the brainstem. Nuclei in the brainstem such as the pre-hypothalamus and the pre-
mammillary bodies control the body via a servo-like signaling mechanism
based on complex neurotransmitter molecules. There are many of them.
Including the important monoamines discussed in chapter 2, there are
about 50 in all that together provide a huge range of regulatory functions.
This type of control is very direct. If an animal is threatened, then heart
rate is increased in anticipation of the response. This level of structure can
also learn associations, such as a given stimulation being associated with a
negative reward, but the timing of the pairing of stimulus and response has
to be appropriate; the stimulus has to appear before the response, and the
response must arrive soon enough afterwards. Besides the chemical signaling capabilities, there is a huge armamentarium of neural signals that code
exteroceptive sensations, such as touch, smell, taste, vision, and hearing, as

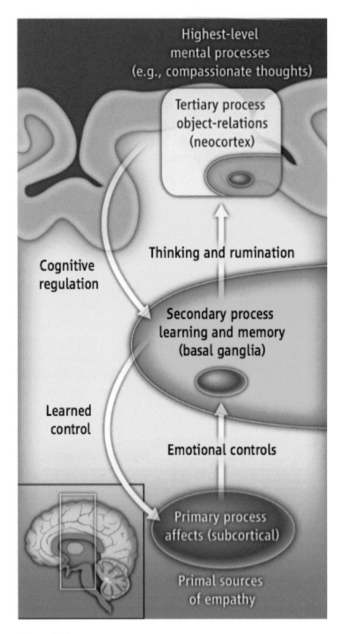

Figure 10.1

The three-level organization of affect from Panksepp.[282] Level 3 is the most abstract: conscious awareness of affect through the explicit coding of emotional coordinates. Level 2 is a precursor stage. Chemical formation of programs at this level use serotonin and dopamine to shape choices through reinforcement mechanisms. Level 1: Brainstem processes control levels of arousal in part through neurotransmitters acetylcholine and norepinephrine to modulate basic drives.

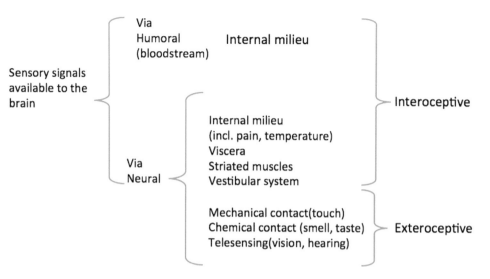

Figure 10.2
Damasio's succinct summary of the body's internal signals.[283]

well as less appreciated interoceptive signals including the viscera, striated muscles, and the vestibular system, as summarized in figure 10.2.

With the arrival of stage 2, the situation changes dramatically. A complex of forebrain structures is added that are associated with the perception of emotions. These are the limbic cortex, hippocampus, amygdala, and thalamus. These structures are intimately connected to the older hypothalamic system but extend its functioning in a crucial way. Their addition allows the brain to remember states. Remembering the discussion of the forebrain from chapters 2 and 4, the main function of the cortex is to add huge amounts of content addressable memory (CAM), and the main function of the hippocampus is to program that memory with new experiences. Thus, the value added in stage 2 allows more elaborate behaviors that respond to rewards that may be delayed or depend on complex contingencies. In computational terms, the new structures provide the machinery needed to learn by reinforcement as described in chapter 5. Neurotransmitters associated with reward are used to score the cortical states, which in turn represent state-action coding of complex behaviors. Thus, all the mechanisms associated with programming are in place in stage 2. However, the focus is on the internal world of the animal's states.

The arrival of stage 3 hallmarks an outreach program for the brain. The focus on internal states is augmented hugely by an enormous amount of cortex devoted to models of external states of the world in the sensory cortices and ways of dealing with them in the motor cortices. Primates in particular have large recent expansions of the cortex that deal with hands, especially fingers, reflecting their greatly increased manual dexterity. From this vantage point, one common interpretation of the advance of stage 3 is that of reasoning.

However, based on the earlier observations in chapter 1, it is extremely unlikely that there is enough compute power for the slow brain to do elaborate formal reasoning literally. Furthermore, there are no new structures in stage 3, only more cortex—the additional cortex being focused on interactions with the external world. So based on the anatomic evidence, it seems more likely that the mechanisms of stage 2 were preserved and extended to dealing more elaborately with environmental challenges. Support for this more gradualist view comes from recent experiments with rats that seem to show that even these lowly mammals have at least some of the elements of a developed stage 3 emotional system.[284] The experiments suggest that rats have empathy for other rats in distress; when a rat had access to another that was tied to its cage, the rat would opt to free the captive by gnawing through its bindings, even at the expense of an alternative chocolate reward. Care was taken to make sure that the rats had no relationships with each other or that there were no ancillary reward inducements. What this experiment indicates is that the emotional system, complete with the triune hierarchical structure, is in place in the rat, if in a more reduced form.

To summarize the three-stage development, in the first place neural control systems are put in place to regulate the body's subsystems. Second, these neural control systems are given an abstraction in limbic cortex. Finally, these abstractions are connected to the cortex's abstractions of the external world. Given our focus on computational abstraction, you already can intuit that the goal of this chapter is to define computational levels of abstraction that will elucidate the value of organizing emotions into a hierarchy. But before we can arrive at a point of sketching such a provisional information-processing hierarchy, we need to elaborate the relevant details of emotion-processing circuits substantially.

10.2   Emotions and the Body

For animals in a competitive milieu, it pays to be able to respond quickly, so it's of little surprise that one of the most important functions of the emotional system is to get the body to react quickly to events. Thus, circuit wiring in the brainstem, particularly in the reticular system, can quickly get the body's systems going. Thus, a primary use of emotions concerns patterns

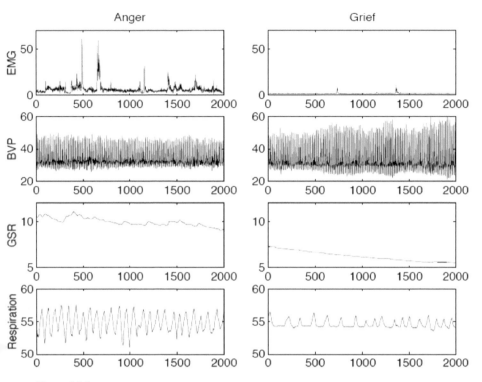

Figure 10.3
Comparing the emotions of anger and grief, we see significant differences in measurements of the electromyograph (EMG), which measures muscle neural activity, the galvanic skin response (GSR), which measures skin conductivity, blood venous pressure (BVP), and respiration rate and amplitude.

of activation of the body's states. These are multifaceted, as a single emotion will typically have many associated body correlates. For example, figure 10.3 shows this basic connection with some of the body's subsystems for two emotions, anger and grief. Anger shows elevated activity on the muscles through the electromyograph, elevated levels of galvanic skin response and respiration, and lower blood venous pressure compared to those of grief. Consequently, the different levels of activation of the viscera effectively can provide a code for these emotions. The different patterns of activation, which result from the need to implement the animal's basic drives, can also provide helpful cues for the recognition of the emotions themselves.

Such a recognition should happen at the next level up, where the limbic system is involved. Here, the system needs to be quick also. Thus the motivation for the dopamine/serotonin scoring system for actions, which allows their rapid evaluation. In particular, Damasio has stressed that the emotional circuitry provides a fast, chemically based scoring system, grading plans based on the utilities of previous experiences, which have pre-stored evaluations. And what is the genesis of these evaluations? As he points out, the body has a vast complex of systems whose homeostatic regulation has the goal of well-being. So, when the body needs a judgment as to the goodness or badness of something, it can turn to the momentary state of the body (see ref. 275).

The direct connections between the body and emotions seemed to produce a chicken and egg problem. Did the body's reaction to stimuli as reflected in the viscera produce emotional feelings or did the analysis of complex stimuli into affective components initiate the body's reaction? William James held the former (see ref. 270) and Cannon the latter (see ref. 273). Today, we would have to say that the answer is either or both. Some of the confusion comes from the fact that the question confounds different timescales. The feelings of emotion are produced within seconds and, as LeDoux has shown, can manifest themselves after the basic mechanisms have been fired up. Furthermore, the basic mechanisms are much faster than perceptions. It can be the case that by the time perceptions are involved, the vast network of circuitry that deals with emotion has all been engaged and done extensive processing. However, the dynamics of the interconnected circuitry, which spans all three levels, can easily allow it to be excited from either a low-level or high-level context. As discussed, peripheral sensors can record potentially dangerous stimuli, triggering fight-or-flight circuitry. In contrast, high-level deliberative thought can reveal potentially dangerous flaws in abstract plans and, as a consequence, excite lower-level circuitry. Support for the top-level control of emotions comes from recent work classifying emotional states from the fMRI signal,[285] two examples of which are shown in figure 10.4.

Emotion Codes

Regardless of the level, it is helpful to organize the emotional circuitry into categories. This can be seen as an indexing strategy that can allow competing circuits to be constructively separated from one another. Level 1 uses a basic set of codes for behavior. Here, Panksepp[286] identifies the set

{SEEKING, RAGE, FEAR, LUST, PANIC, PLAY},

as basic classifications of behavior that can be elucidated with electrical deep brain stimulation. As a motivation for the value of such a categorization,

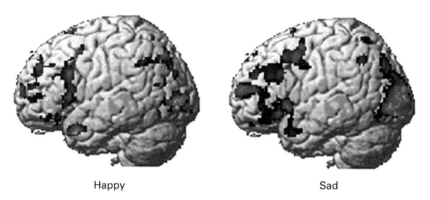

Happy   Sad

Figure 10.4
Comparing the emotions of happy and sad we see significant differences in measurements of the fMRI signal when subtracted from a control. Courtesy of Marcel Just, Carnegie Mellon University.

keep in mind the need for fast responses. If a particular behavior is called for, it would be a great help if all its variations somehow could be indexed so that alternatives could be accessed rapidly. Furthermore, inappropriate behaviors need to be excluded so that they do not steal resources. The emotional triage into characteristic states is a way of speeding things up.

The value of codes associated with "emotions" at level 2 centers on their use in the formation of programs, and in that role the most important identification is of emotions with the mechanisms learning through reinforcement. In that vein, Rolls has put forth an interesting hypothesis that ties emotions to reinforcement and changes in reinforcement.[287] His proposal is that emotions can be seen as related to reward outcomes. The overall picture is shown in figure 10.5. Positive emotions, such as happiness—here given a scale of pleasure, elation, and ecstasy—are associated with expected positive reward that was actually obtained. Fear, here elaborated as apprehension, fear, and terror, is associated with negative reward actually received. An omission of a positive reinforcement or its termination also can produce negative emotions on a scale from frustration or sadness to anger or grief to rage. Finally, the omission or termination of a negative reinforcer can lead to relief.

Of course, what is known about reinforcement learning algorithms allows one to raise the objection that it's the expectation of the reward that should be the genesis of the feeling. We do not have to wait until the outcome to be afraid; we can see what is about to happen. Rolls, of course,

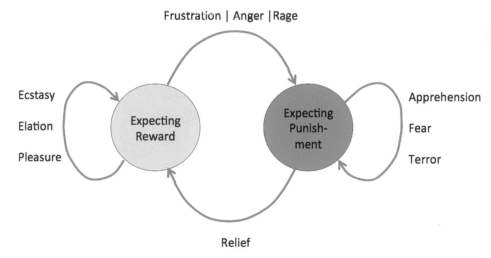

Figure 10.5
Rolls's proposal for summarizing emotion (see ref. 278) reorganized by the author.
If the agent is in a state of receiving reward and this continues, feelings of pleasure,
elation, and ecstasy in increasing order of stimulus duration or strength are experienced. However, if a positive stimulus was unexpectedly discontinued or an expected
positive stimulus was not delivered, the feelings of frustration, anger, and rage may
be experienced in transiting to an "expecting punishment" state.

knows that "emotional states can be produced by remembered reinforcing
stimuli," but of course it's all memory in terms of stored cortical program
states. Rolls's contribution is the idea of associating certain emotions with
the consequences of reward outcomes, rather than the absolute value of
reward itself. Again reverting to computational terms: In the temporal difference reinforcement learning algorithm, the adjustment is made based on
the difference between the expected reward and the actual reward received.
The association of emotions with reward dynamics harbors another benefit
in that as secondary reward is a universal currency for the brain, emotions
acquire a universality in that they can evaluate arbitrary situations and are
not tied to specific behaviors. Compare this to the more focused organization of the hypothalamus.

   The emotional codes considered at the two lower levels speak to the
need for fast *internal* behavior indexing schemes, but we now proceed up
one level of increased abstraction in order to discuss use of emotion codes
that have an *external* focus. Many animals live in groups wherein communication among members of the group is important, but nowhere is

that more so than with human primates. For us, the role of the face is paramount. Six basic emotions {angry, happy, sad, fear, disgust, surprised} are preserved across cultures and signaled facial muscles. Paul Ekman, the pioneer of quantitative studies of such systems, showed that the facial muscles used to express this set are used in well-defined groups. Figure 10.6 shows these groups collaborating in different parts of the face to produce common emotional signals. The emotional set here does not readily map onto the Panksepp categories, probably reflecting the fact that the latter are adapted to the demands of communicating in groups.

Even though the face codes might not map onto the more primitive codes exactly, they are demonstrably coupled to the rest of the supporting circuitry. The French physiologist Duchenne working in the mid-nineteenth century was the first to bring to light the close relationship between emotions, the muscles of the face, and the body. By stimulating particular facial muscles in a subject, he could produce emotional expressions such as happiness. Furthermore, he concluded that these expressions produced genuine emotional feelings. Much subsequent work has confirmed these ideas. For example, biting a pencil produces a pseudo-smile. If subjects read a text in this pose and are asked questions about it afterward, they will produce more positive interpretations than a control group. This experiment, which might appear a bit strange, contains a huge lesson for us when we allow ourselves to speculate in computational terms. The facial muscles signal affect, and as they are part of the motor system, they make use of extensive somatosensory feedback to register that the affect signal has been carried out. But here is a case of putative blowback. Because the feedback signals that positive affect has been achieved, that signal apparently leaks into the overall system and biases the mechanism that would be used to evaluate the signal into an overall positive affective state. A subject's internal neural state apparently biases itself toward the conclusion "I smiled so I must have been happy." This mental observation in turn biases subsequent conscious processing. So the muscles of the face are connected to the underlying brain processes that produce emotions. Furthermore, these emotions are connected to the visceral changes, too. Thus, we have relationships between the face muscles, the viscera, and the emotions.

Up to this point, one might be tempted to conclude that the three-stage emotional circuit necessarily is tightly coupled. However, there are separate neural pathways that seem to respect the stage 2/stage 3 distinction. For communicating emotional expressions, there is a pathway from the motor cortex and brainstem that projects onto systems of neurons that indirectly or directly drive the muscles. This contrasts with felt or

A

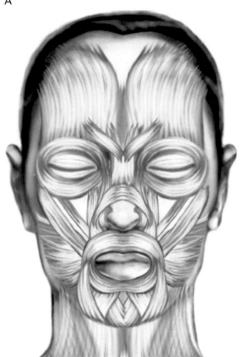

B

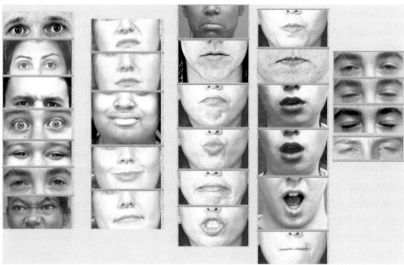

Figure 10.6

(A, B) The muscles in the face that evolved to control eye and mouth opening and closing have been co-opted for use in social signaling. Pioneering work by Ekman has established distinct muscle groups that are used in the construction of facial expression for each of the six basic emotions. Adapted from http://www.cs.cmu.edu/~face/facs.htm.

spontaneous emotions, where there is a separate pathway that involves the limbic system and hypothalamus. Each of these two systems also have separate motor cortical areas associated with them. An easy way to remember this dichotomy between communicated and genuine emotions is to think of job interviewees. They no doubt want to project happiness via the former system, but may be feeling anxiety, represented by the latter system. Because signaled and felt emotions are predominantly driven by separate cortical structures, it should come as no surprise that they can be selectively damaged, say by a small cerebral stroke or a tumor. In the case of the volitional system, damage to motor cortex in the right hemisphere will cause an inability to use the left-side muscles in an emotional expression like a smile. For damage to the spontaneous system, that person can smile voluntarily but will fail or have a very muted facial response to a humorous comment. Figure 10.7 shows these cases.

Of course, to signal a genuine emotion, the motor system must be engaged as well, but the main point is that in this case, the limbic system is the prime driver, whereas during the faux emotion used for communication, the limbic seems either bypassed or reduced to a subsidiary role.

10.3   Somatic Marker Theory

The use of the face for emotional signaling in itself signals a change of processing for the forebrain. This ability exposes the use of emotions in the abstract. They may or may not be felt and in their abstract cortical state can be reasoned about in a way that is disconnected from the rest of the body. This disconnection is the central subject of Kahneman's book *Thinking, Fast and Slow*. Human behavior can be very different depending on whether humans use the fast system for generating a response or whether they take time to be more deliberative. However, for the purposes of understanding emotions, it is helpful to adopt Damasio's focus and distinguish between thinking and acting. We can all think of cases where procrastinating can have costs, but mainly thinking about things is less risky than acting on them. According to Damasio, the brain's design has this cost very much embedded in its architecture in that acting has to go through an emotion-based scoring system. When it is dysfunctional in some way, bad actions result. Thus, the cortical system for generating actions has to be connected in some way to the emotional scoring system. While there are possibly many cortical regions that play a role in the generation of emotions, clinical studies have revealed a cortical area that seems to be necessary for the proper functioning of emotions. This general area is part of the prefrontal

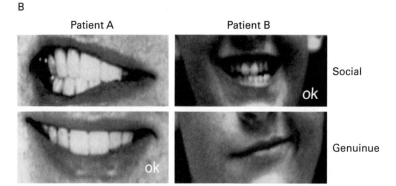

Figure 10.7
Are your emotions genuine or are you faking it? It matters in the cortex (ctx), as two
different areas govern these two options. (A) If emotions are genuinely felt, limbic
cortex is in charge. If you are signaling a desired affect in a social exchange, the
motor cortex is the primary supervisor. (B) This dichotomy is supported by patients
that have selective damage to one or other of these areas. Damage to motor cortex is
indicated by a skewed smile. Damage to limbic cortex results in a muted expression.
(A) from A. R. Damasio, *Descartes' Error,* copyright (c) 1994 by Antonio R. Damasio,
MD. Used by permission of G. P. Putnam's Sons, a division of Penguin Group (USA)
LLC. (B) from *Neuroscience*, 3rd ed., ed. D. Purves et al., 2004. Reprinted from Trosch,
R. M. et al., *Journal of Neurological Science*, vol. 98, pp. 195–201, 1990, with permission
from Elsevier.

cortex. As confirmed by several studies, patients that have selective damage to this area exhibit irrational decision making and insufficient risk aversion. Probably the most famous case of a person with this damage is that of Phineas Gage, who survived a railway construction accident that focally and irrevocably damaged his emotional system, as shown in the reconstruction in figure 10.8.

A more recent prototypical case is Damasio's patient "Elliot," who had a tumor resected from the prefrontal area. Before the operation, Elliot was a successful businessman and husband. After the operation, which unavoidably damaged his prefrontal cortex, he recovered but was in some way a different person. He was no longer risk averse and made a series of bad decisions that cost him his job and marriage. Notably, when tested on moral judgments, he scored very highly; the problem was that although he could articulate the right things to do, he could not convert this knowledge into actions. What went wrong? Elliot passed a variety of intelligence tests and had no obvious deficit to working memory or the ability to focus but was singularly lacking in emotional responses (figure 10.8). But does this inability to interpret emotional signals impair judgment? There are a number of ways this can be tested.

One such test is the Wisconsin Card Test. Subjects draw cards that have positive and negative numbers on them from one of four decks. If the card has a positive number, the experimenter pays the subject that value in some redeemable "money." If the number is negative, the subject pays the experimenter from his supply. Unbeknownst to the subjects, two of the decks are safe and boring. On the average, the payoff from these decks is positive, but the deviation of the values is relatively small. The other two decks are unsafe but exciting. The average payoffs are negative, but the deviations in their payoffs is relatively large.

When normal subjects play this game, they typically sample all the decks initially, but when they discover the payoff situation, they focus their draws on the decks with the small but positive average payoff. An interesting finding is that galvanic skin response, the measurement of an electric potential on the skin that signals anxiety, is present when the normal subjects reach for a risky deck, before they can articulate the rules that describe the average payoffs of the four decks. It's as if their bodies "know" that the risky decks are risky at a very early stage in the game. In contrast to the normal subjects, the prefrontal patients never acquire the safe deck rules. They continue to sample the risky decks even though doing so will lead to penury. Furthermore, their galvanic skin response is absent, suggesting that there is a disconnect from the helpful body system that signals risk.

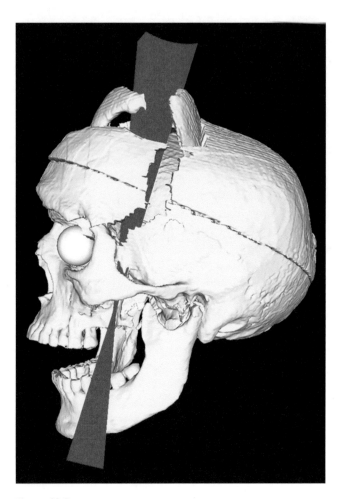

Figure 10.8

Damasio's patient "Elliot" is a clinical case of the modern day. However, a more famous precursor case is that of Phineas Gage, who had a serious accident while supervising blasting operations during construction of a railroad bed in hilly terrain. A meter-long rod he was using to tamp explosive charge shot through his skull when the charge accidentally exploded. Amazingly, he was conscious and lucid when taken for medical care. He recovered fully in a sense, but his emotional system was permanently impaired and his subsequent life went downhill. The red bar in the figure shows the reconstructed path of the rod, which putatively damaged the parts of the forebrain used to "read" emotional state. From J. D. Van Horn et al. "Mapping connectivity damage in the case of Phineas Gage," *PLoS ONE*, vol. 7, p. e37454, 2012.

Damasio's central argument is that humans do not have the ability to reason through all the possibilities of any given situation but have to have some quick way of evaluating potential choices. One example he gives is that of deciding whether or not to enter into a business relationship with the rival of a close friend. The contingencies in this decision are complex, and it would take an impossibly large amount of reasoning to explore them all. Hence Damasio's proposal that this decision is handled by appealing to prestored similar situations. This account stops short of being computational, but at this point we have the means to evaluate it in computational terms.

From the standpoint of computation, one important problem that somatic marker theory addresses is that of computing a correct score for alternatives. This is not easy for the brain, as any cigarette smoker who has tried to quit knows. However, for the most part we do it amazingly well. We engage in behaviors that have distant rewards such as pursuing a higher education or writing a book. At the same time, we can juggle these against more immediate goals such as getting to the market or brushing our teeth. Of all of these, only buying food is directly related to something the brain can calculate easily. Nonetheless, there is a huge repertoire of things that do have computable scores, and the part of the cortical memory that monitors this part of the body functions as a CAM that represents the result. Furthermore, given the reinforcement learning model, we know that if the computation depends on some history of states—or, more colloquially, memory—then a putative cortico–basal ganglia circuit can handle that also. Thus, it stands to reason that the brain would want to co-opt this machinery for more abstract calculations such as the value of helping a friend when you have chores to do. The limbic system neurons already represent estimates of what is good and bad for you; all you have to do is associate them with the new abstract experience. Of course, saying this belies that difficulty of specifying the circuitry in detail. But we know in which direction to head.

That making decisions based on previously scored situations is compatible with reinforcement learning theory Damasio himself notes. But escaping the combinatoric costs that doom straightforward logical approaches is not easy, and there are potential pitfalls here also that must be dealt with. The principal difficulty is that a situation like the business-deal example would not have been seen in exactly the same form before, so it has to be compared to a similar situation. The backgammon net does this with a decomposition. Individual neurons recognize features of the situation, and each such feature contributes a portion of the total score. It is hard to imagine the more general case being handled differently, because if there was no way to decompose the decision, then the number of all possible decisions would be insurmountably large in the same way that unchecked reasoning options are insurmountably large. Thus, the decision of the moment, to be practical, has to be broken down into components that each can have stored scores that are then added up. Walter can simultaneously negotiate the sidewalk, pick up litter, and avoid obstacles because he can compose the scores for each of the tasks done separately. If he were not able to do this, them the combinatorics of all possible tasks would be overwhelming.

Another point to remember is that there are only a few neurotransmitters, notably dopamine, that are responsible for this signal. Doing the actual dopamine math is another biological step that must be accounted for, too, but in the shadow of looking up the correct decomposition, it's just a detail.

Somatic marker theory attempts to replace reasoning, but we know that we can "reason" and in fact that is compatible with reinforcement learning models. If you remember from our discussion of reinforcement learning algorithms in chapter 5, if there is a *model* of the situation, then it's computationally practical to run the model backwards from future goals. You might wonder what could be gained by doing this, as the score at the starting point already reflects future rewards, but we could have information that something might have changed. You have a plan for going to the supermarket that is scored based on its normal hours, but a friend tells you that it is closed owing to some special circumstance. So you can compute a new score in the moment using the new data.

10.4   The Amygdala's Special Role

Although we tend to give all the emotions the same status, one of them is special, and that is *fear*.[288] Fear is associated with the amygdala, so much so that rats lacking an amygdala have no fear response. Figure 10.9 shows that this connection extends to humans as well. A patient missing an amygdala is asked to sketch faces exhibiting the basic emotions. All the attempts are reasonable with the exception of the illustration of fear; the rest of the emotions are spared the consequences of amygdaloid damage.[289]

The amygdala is a separate subsystem that is ideally placed between the cingulate cortex and the "abstract end" of the hippocampus (figure 10.10). The association with fear may cue us to focus on the amygdala's role in selecting new programs based on their importance in dealing with the

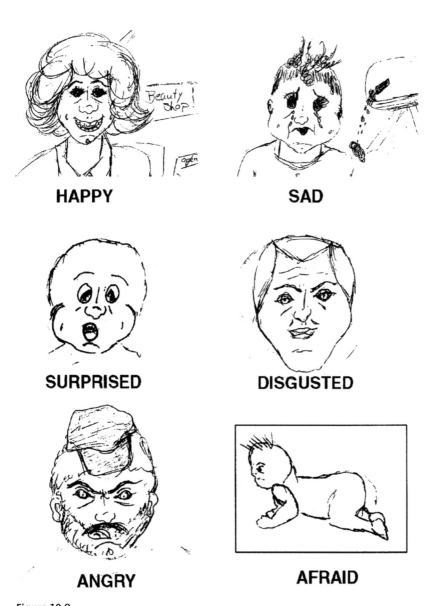

HAPPY   **SAD**

SURPRISED   **DISGUSTED**

ANGRY   **AFRAID**

Figure 10.9
A patient who has had amygdala damage was asked to sketch faces depicting different emotions: *happy, sad, surprised, disgusted,* and *angry* are rendered appropriately, but *fear* is not. One possibility is that the emotion labeled "Fear" is actually a composite of feelings associated with the uncertainty of dealing with the external world, whereas the other emotions are fundamentally a more controlled response generated from internal factors. From R. Adolphs et al., "Fear and the human amygdala," *Journal of Neuroscience*, vol. 15, pp. 5879–5891, 1995.

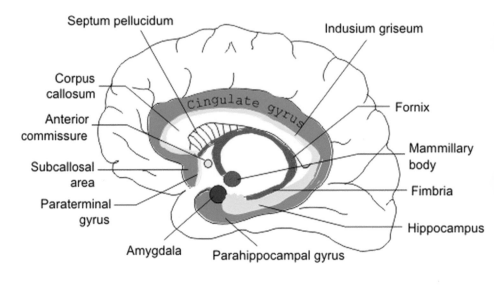

Septum pellucidum

Indusium griseum

Corpus callosum

Anterior commissure

Subcallosal area

Paraterminal gyrus

Cingulate gyrus

Fornix

Mammillary body

Fimbria

Hippocampus

Amygdala   Parahippocampal gyrus

☐ Limbic Gyrus   ☐ Intralimbic Gyrus   ☐ Fornix & Inner Arc

Figure 10.10
Anatomically, the limbic cortex is structured to be next to the amygdala-hippocampus subsystem. Thus, the brain's internal model of the viscera (cartooned in purple on the left) is placed adjacent to the structures that make new programs. From R. R. Edelman et al., eds., *Clinical Magnetic Resonance Imaging*, 3rd ed. Copyright © Elsevier-Saunders, Philadelphia, 2006.

external world as opposed to the other emotions, which have their focus on our *internal* states. Extensive subsequent experiments have extended the characterization of the amygdala's response as one of *importance*. As introduced in chapter 2, the external importance concept describes the amygdala's basic computational role: rating the value of permanent modifications to the brain's repertoire of programs.

The external/internal distinction takes getting used to. Inside the body, the various mechanisms have to cooperate for homeostasis and, by evolutionary design, work together for the animal's successful function. It would be a huge mistake to think that the parts in any way "know" what they are doing, but nonetheless the overall result is positive in terms of the survival of the animal. In contrast, the neocortex deals with the outside world, which, in a competitive environment on an evolutionary timescale, involves threats from various agents and other exogenous hazards. Thus, any outside characterization of the effects of these sources would tend to be negative. The amygdala is triaging to extract the most important "threats" that have to be dealt with, and the rest of the forebrain is incorporating these responses into its repertoire.

Some support for this view comes from the experiments by Markman et al. briefly mentioned in the previous chapter (see ref. 265). They had subjects play a game that contained wins and losses in dollars. There were two groups of players. One group was given a $50 ticket and told that if they achieved a certain level of performance, they could keep it. Another group was told that if they achieved the same level of performance, they would *get* a $50 ticket. The game was structured so that some amount of exploring was necessary to be successful. In the trials, the group that started with the ticket did more poorly than the group that started without it. Even though the stakes were identical, the group that started with nothing took more risks and as a result learned the game structure more thoroughly. Why should this be so?

Let's bring the amygdala into play and consider the following ecological scenario where a person is wandering in the desert with the need for water. If that person starts with a canteen, the strategy may be to ration water and search effort, whereas when starting with nothing in a more desperate situation, more risk taking may be called for. The contrast is that in the first case, the person's evaluation is more focused on *internal* features of competence, whereas in the second, the uncertain *external* world is the decision driver. From the perspective considered here, where the amygdala evaluates external factors, the game-playing result can be interpreted. Having the resource indexes into the internal system, and not having it indexes into the external system.

10.5 Computational Perspectives

James' original explanation of emotions linked them explicitly to the body's signals. Rising heart rate signaled anxiety or anger. Thus, the causality between felt emotions and these indicators was direct. But the developments just discussed reveal that the structure of emotions must be much more complex. A central new concept is that of a cortical *model* of the circuitry in the limbic system. The body itself has the liver, heart and lungs, and so forth, but the limbic system has abstract models of their structure. This is of course a great advantage in that if the body is to take action, it is important to have an idea as to its working order. The rest of the cortex is characterized as creating summary abstractions of vital fight-or-flight decision making to be run by the hypothalamus, so obviously it would be important for an internal abstraction that weighed the extent to which the body is onboard with this evaluation. Thus, the body provides an important primal source of evaluation of good and evil. Good is what is good for the body and bad is what is bad for the body. But remember that while the limbic system has the requisite body information, it's in the form of an abstract model. This information must be summarized and communicated to the hypothalamus, which then calculates the appropriate amounts of neurochemical currencies to the forebrain to mediate action.

To revisit the basic question, it is easy to see why the notion of "emotions" is difficult to analyze: the word describes a very complicated system that was developed in stages over evolutionary timescales and now serves diverse functions in the forebrain's computational repertoire. Let's deconstruct the emotional system a bit to examine some of these functions.

Speed of Response

The most important function has been introduced by Damasio: a fast mode of computation. Animals that can make fast decisions have a survival advantage over those that cannot. However, mammals burdened with a slow neural response are at a potential disadvantage. The solution is to ask the body. The body maintains an ongoing representation of its state in the limbic system that can be accessed quickly compared to laborious stage-by-stage reasoning. Once this state is queried, the parameters for a response can be quickly relayed by the dopamine-serotonin-noradrenaline-histamine system.

Social Communication

A second major function can be seen in the communication of emotions. As shown in figure 10.7, different cortical areas are most active in the signaling of genuine and socially motivated emotional states. In each case, a cortical program is active, but in the second case, the purpose is to communicate the higher-order bits of a social communication. It is easy to see how important the essential nature of affect is in this process. The sentence "I'm glad to see you" has very different meanings when the affective content is anger, sarcasm, or happiness. The operating system abstraction layer focuses on the tactical issues related to multitasking in the moment. But emotional abstractions have a role to play in a longer timescale, signaling strategic decisions.

Several groups, notably those of Breazeal[290] and Picard,[291] have studied the role of emotions in this regard, focusing on the roles of emotion in communication and signaling body state, respectively. Figure 10.11 shows

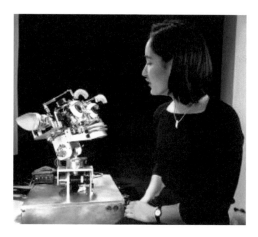

Figure 10.11
One of Breazeal's first experiments had a human interacting with a robot designed with an affective interface. The robot has no language ability but uses vision and audition sensors to read and respond to the human's perceived affective state with affective signals of its own.

a session with one of Breazeal's early robots, which can only communicate with affect changes. Newer versions have speech capability and are breaking new ground in terms of coloring speech communication with affect.[292]

Studying emotions with robots can seem a long way away from decoding their neural structure in humans, but it is precisely our premise that getting a handle on the mechanics of affect at this level can provide important hints of what part of affect is especially important in normal social communication. It also helps that humans are experts at judging the genuineness of affect: subject-robot interactions provide valuable data that allow the tuning of the model system.

Emotional Tagging

A third and very speculative function of emotions is in something we will term *tagging*. Recall that the operating system abstraction drew the analogy between items in the brain's "working memory" and points in silicon computers. There is tremendous advantage in being able to refer to complex objects deictically. Pointing to "this one" or "that one" in communication can save a laborious process of characterizing them on the basis of distinguishing collections of features. In silicon, this process is implemented by virtue of the structure of the computer memory. A very large list of features can be referred to economically by using the address of the memory location where the description begins. While this solution is not available to the brain in any direct way owing to the very different architectures, an interesting possibility that would capture these features may be available by using the limbic system.

To motivate the tagging hypothesis, let's examine the data from Ramachandran's patient records. A patient with Capgras syndrome was unable to recognize his own parents when he saw them, declaring them impostors.[293] Ramachandran's conclusion, endorsed here, is that the patient damaged his emotional system. Because his meeting of his parents was unaccompanied by any feelings about them, he concluded that the only way this could be possible would be if they were in fact not his parents. At first blush this seems incredible, but less so in recalling the experiences of Ramachandran's phantom-limb patients. Despite being able to see their amputated arm, a phantom-limb patient will maintain that the arm is still there. From a normal perspective, the patient's observation seems exotic, but from a computational perspective, it can all make sense. If the brain's programs combine data within a Bayesian framework, which summates evidence in a democratic (probabilistic) way, then all information represented in the cortex's abstract neural models is combined and adjudicated systematically and mechanically. For normal subjects, the result almost always jibes with expectations, but for patients with brain traumas, their reports, which may not jibe with that of an external viewer, can make perfect sense internally. For the phantom-limb patient, there is visual evidence for a missing limb, but the internal proprioceptive system reports it is still on line. The latter evidence apparently turns out to be more compelling. For the Capgras patient, there is positive visual evidence for the parents, but the internal emotional system reports negative emotional saliency. In this case also, the internal evidence turns out to be more compelling.

Now let's interpret these results in terms of a putative method of realizing silicon references. The brain's option to denote things deictically is to *use the limbic bits*. When executing a program that is about your parents, you can reference their inclusion in that program explicitly by linking components of the program to the active limbic neurons that can denote them. How can you distinguish thoughts about what your parents are thinking and thoughts about yourself? Simply use a special pattern in the limbic system that is active when thinking about your parents. This strategy is what we are calling tagging. It would be hard to underestimate its potential utility or generality. How could your brain keep track of what you are thinking about what Sam is thinking? Just use different emotional affect patterns to augment each state. Thus from a CAM standpoint "thinking about me" becomes differentiated form "thinking about Sam." It's not so hard to imagine the implementation because those bit patterns will naturally occur in each of the possibilities. In effect, labeling states with the affective system enormously adds power to the brain's reasoning by introducing a form of notation for what would be characterized in silicon as variables.

The virtues of tagging can be further appreciated by comparing it to possible alternates. One method that might be used is to tag another person with his or her location. After all, this is used regularly by deaf people in the process of signing. After having made a complicated sign or set of signs in a position in space, they can refer to it without signing it again by just pointing to its location in space. That works in conversation because the sign cannot move, but not for people, who may move around or even disappear from view. Other sensory candidates are likely to also suffer from the same impermanence. In contrast, the feelings for a person are far more stable.

10.6   Summary

For a summary description of emotional state and conscious feelings, one is hard put to better that of LeDoux and Damasio[294]:

Conscious feelings facilitate learning about objects and situations that cause emotional responses. Thus feelings enhance the behavioral significance of emotions and orient the imaginative process necessary for planning of future actions. In brief, unconscious emotional states are automatic signals of danger and advantage, whereas conscious feelings, by recruiting cognitive abilities, give us greater adaptability in responding to dangerous and advantageous situations. Indeed, both emotions and feelings also play a major role in social behavior, including the formation of moral judgments and the framing of economic decisions.

The emotional system is extraordinarily complex, and an enormous amount of work will have to be done to sort out a comprehensive computational model of its function. Nonetheless, at this point we can attempt a summary of some of its important features. The overall organization is shown in figure 10.12. To recap, the hippocampus has the function of parsing experience into a form that can be incorporated into the cortex. The set of experiences chosen for parsing is selected by the amygdala. A crucial step in this process is the computation of value. A major candidate for this computation is the limbic cortex. Limbic cortex has extensive connections with lower brain regions that deal with emotional state, which in turn is a reflection of body state. Thus, provisionally limbic cortex fills the function of turning emotional state into an explicit index that can be read by

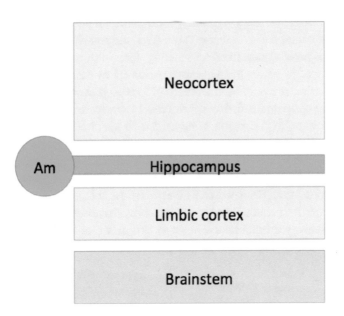

Figure 10.12
Anatomically, the limbic cortex is structured to be next to the amygdala-hippocampus subsystem. Thus, the brain's internal model of the viscera is placed adjacent to the structures that make new programs. Am, amygdala.

the hippocampus to make appropriate connections to neocortex. In short, the amygdala has to make value choices between episodic events from the outside world; limbic cortex provides the rating data needed to make those choices.

The above description focuses on a midlevel abstraction of emotions and feelings, but we can also take a step backward and view the entire system in relief as is done in table 10.1.

The initial stages can been seen as elaborations of the most basic needs of the body such as hunger and thirst together with its large array of interoceptive and exteroceptive sensations (figure 10.5). These are summarized by the hypothalamus at the brainstem level, which can elicit basic whole behaviors that can be evoked through deep brain stimulation. The circuitry up to the hypothalamus can be modeled as a first stage in the organization of emotions.

At the limbic level, these get both refined and abstracted so that they can be interfaced to the neocortex. Emotions are elaborated into approximately six or seven basic categories that are cross-cultural. From a computational standpoint, their importance may be that they constitute a more

Table 10.1
A proposal for the brain's use of emotional computational hierarchies

Level	Computational Model
Language	Emotions have abstract "feelings" in consciousness
Signaling	Emotions mapped onto facial muscle codes
Cortex	External-world representations communicate with basic drive abstractions
Limbic	Abstractions of behaviors allow finer-grain responses
Hypothalamus	Organized nuclei of drives
Basic	Basic body needs (e.g., thirst, hunger) communicated via neurotransmitters

generic characterization of the value of behaviors and their dynamics that can be applied to arbitrary behaviors. At the end of chapter 8, we saw the need for some kind of alerting mechanism that the operating system could use to modify the agenda. The emotions as coded by the limbic system can potentially be part of this role.

At the next level above, it is not easy to differentiate from the limbic level. The neocortex contains vast and sophisticated representations of the outside world. One differentiation is captured by Kahneman's dichotomy of fast and slow,[295] where an interpretation would be that an older, faster cortico-limbic system can be conceptually separated from a newer, neocortically dominant system. The earlier system has been championed by Gigerenzer.[296]

At the neocortically dominant level, social needs get met. Whereas genuine emotions signaled through facial muscles have one cortical pathway, symbolized forms of emotions have another distinct pathway. This development opens up the communication of emotions abstractly and ultimately through language as well.

Throughout this hierarchy, there should be loops that refine their internal states. Thus, in Kahneman's "thinking fast" scenarios, the emotional states are allowed to direct behavior, whereas in "thinking slow" scenarios, abstracted emotional states are allowed to propagate into the future and may refine action choices and their consequences. However, in this process, presumably new value scores must be obtained, so that proposed behaviors must somehow be sent down to the drive center for evaluation.

11 Consciousness

Imagine entering a theater just before the beginning of the show, noticing the stage, the chatting audience, and a few side doors leading backstage. As the house lights begin to dim and the audience falls silent, a single spotlight pierces the descending darkness, until only one bright spot shinning on stage, remains visible. You know that the audience, actors, stagehands, and spotlight operators are there, working together under invisible direction and guided by an unknown script, to present the flow of visible contents onstage. As the house lights dim, only the focal contents of consciousness remain.

—Bernard J. Baars, *In the Theater of Consciousness* (Oxford University Press, 1997, p. 41)

Baars's image of consciousness as a theater has been appropriately criticized by Dennett for its projection of Cartesian unity,[297] but for introductory purposes on this difficult topic, the description of the support team of stagehands, which for a movie theater can be augmented to include scriptwriters, makeup artists, costume designers, directors, stunt people, and many others, is central. From the previous chapters, you can appreciate the vast amount of machinery in the abstraction pyramid that must be in place before a construction of a conscious narrative of any kind is possible. Thus, consciousness is a very abstract level in a computational hierarchy. And given that, as Damasio points out,[298] it also must be based on Darwinian principles, and all the lower abstraction levels have an inherent utility, it is hard to imagine that being conscious does not somehow confer additional utility for its owner.

By now, we have gotten used to the difficult but essential idea that the brain has not access to ground truth about the world and people in it but instead runs a model of these things. Most of this model is driven by previous experience, but at least for humans, an additional component is the ability to simulate situations in the future. Naturally, to be successful in this endeavor has enormous survival value. However, to run this simulation requires additional mechanisms that are not covered by purely memory-driven algorithms. These mechanisms are the candidates for a substrate that produces the feeling of conscious experience.

Before we try to pin down some attributes of consciousness, it might be helpful to visit the feeling of being unconscious. We have all been unconscious in a dreamless sleep. Nonetheless, it is a very gentle unconsciousness because, as you know, while you are asleep the brain has a lot to do to save the experiences that were selected form the last period you were awake. In addition, before you went to sleep, you had expectations about what things would be like when you awakened, and for the most part these are met. Perhaps a better experience of unconsciousness is obtained by those of us who have been under anesthesia during a medical operation. In these cases, the drift into unconsciousness and awakening is abrupt, and the interim experiences are a blank. Without family and hospital staff filling in the details, the time we were under anesthesia would be just seamlessly missing.

The experience of being conscious compared to being unconscious, with the latter as measured when consciousness returns, is easy to relate to because we all have it. But while we all have the feeling of being conscious, to try and jump into a satisfactory explanation of its holism is fraught with difficulty. The main reason is that "consciousness" is a summarizing word that binds together a lot of underlying structure. Thus, the job of this chapter is to deconstruct the experience of consciousness and examine its components and their respective functions together with the rationale for those functions. You have to be prepared for a disappointing experience at the end of this process. Think of a car. One can take it completely apart and lay out all of its pieces on the garage floor with their attendant functions labeled, but the dissected and labeled "car" is not the same thing as the assembled vehicle.

11.1 Being a Model

In beginning the deconstruction of consciousness, the most important of its component concepts is that of a model. The brain's programs of course do not have access to all the information about the world but only receive a small shadow of that information through the process of sensing and acting in it. As a consequence, the brain's internal programs have to estimate the essential state information about the world needed to direct promising behaviors. The consequences of working with state estimates rather that the real thing are profound, and we can only really appreciate them when something goes wrong, as when a brain is injured.

As noted in the discussion of emotions, a spectacular divergence of a brain's estimate and ground truth occurs with phantom-limb patients, explored by the psychologist and physician V. S. Ramachandran (see ref. 293). For most people who have the misfortune to lose a limb in an accident, that's it. The limb is gone, and they learn to adjust to life without it. But a small fraction of amputees have an unusual condition where they swear the limb is still there. They of course have very conflicting data in that they can look at the empty space where the limb used to be. Yet they still are quite convinced that they have it. You would think the issue would be settled when Ramachandran asks them to do something with the limb, like pick up a cup, yet the patients remain unfazed and come up with an excuse like "Oh, I can't do that right now because my arm is too tired."

Ramachandran's hypothesis is that the cortical circuitry that handles the face is near that which is used to handle the arm (which you can verify by glancing at figure 7.6) and gets co-opted into service by the programs that were running the arm. Thus, the neural circuit that handled the arm is satisfied. It needed inputs and it has them.[a] Naturally, there has to be some combination of evidence that reconciles the visual input with the ersatz haptic and proprioceptive input, but we can assume this happens and that the latter trumps the former. If you think in terms of your phenomenological experience of vision, this is hard to swallow, but if you think in terms of tests by routines that have varying results, it becomes much easier.

The astonishing phantom-limb data contain two very important concepts. One is that the model running the arm that uses the cortex is in a certain state and at a lower level of abstraction than that of conscious report. Furthermore, conscious report has no alternative but to summarize the situation down in the engine room as best it can. Hence the rationalization and equivocation. The other point is that conscious report, interrogating this less abstract state, uses time and necessarily comes after that state is generated. The idea that consciousness comes after subconscious processing has appeared in numerous other contexts.[299, 300]

The idea of running a model is not limited to just injured brains but of course is a general feature of working brains as well. A nice demonstration of model use comes from experiments done by Sekuler et al.[301] Subjects viewing a display saw two circles move behind a rectangular occluding surface as shown in figure 11.1. There were two different conditions. In one, as the circles met behind the surface—hidden from the observers—there was the sound of a "click." When this happened, subjects perceived the circle on the right as having collided with the circle on the left, bouncing off it and emerging on the same side. The left circle behaved oppositely. However, without

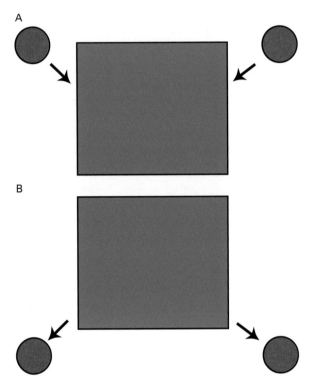

Figure 11.1
Two frames from an experiment to test perception. (A) A moment before two moving circles disappear behind an occluding surface. (B) A moment after they reappear on the other side.

the click the perception was reversed. The circles were seen to pass by one another and emerge on opposite sides. This is a stunning demonstration, as it reveals clearly that the brain has embedded its visual and auditory inputs in some kind of model. The visual histories are identical in both cases; it's just that in the first, the sound is associated with the physical act of colliding and changes the course of the internal simulation.

The ability to use models to interpret the world is so seamless and effective that we do not readily appreciate that it is acquired during child development in a lengthy staged process that spans the years from birth to puberty. The study of this process has been enormously productive and has revealed many interesting features of the different stages. We will just limit ourselves to a single observation that conveys some of the complexities involved.

DeLoache showed 2½-year-olds a drawing of a living room scene with a teddy bear in the drawing behind a couch.[302] When these children are subsequently taken into the real room for which the drawing was a model, they have no trouble finding the teddy bear. They have understood that the drawing represents the room and make the correspondences necessary to locate the teddy bear. It's what they cannot do that is especially interesting. Children of the same age are shown a dollhouse that contains a replica of the room that they will go into to find the teddy bear. In the replica, the bear is placed behind the couch. But when these children are taken into the real room, they are clueless. Somehow they cannot appreciate that something already in the real three-dimensional world (but just smaller) can also be a model for something else in the three-dimensional world. But at the same time, they can understand that something that is not three-dimensional at all and just contains a facsimile of things in the three-dimensional world is a useful representation of that world.

In some sense, this result might seem counterintuitive to us. Why does a child not find the three-dimensional problem easier to solve? As adult observers, we can appreciate that the difference between the dollhouse and the house they are in is just one of scale, and once that simple notion is understood, the problem is trivial. However, the genes have a different take on the problem, as the abstract encoding where the picture codes the problem is easier to solve. The paper drawing's abstractness tips off the child that the information on the paper contains symbols for something real. Evidence for the idea of symbolic representations comes from very early stages in child development. Wagner et al.[303] showed that 11-month-olds think an intermittent tone is more similar to a dotted visual line than a continuous visual line is to the dotted line. From the earliest moments, the abstraction trumps the concrete.

Agency

Up to this point, there have been two crucial ideas, the first being that the brain runs models and the second being those models prefer abstract currencies in their depictions. As you saw, the ability to abstract from two-dimensional images is highly developed, even in children, and we use it effortlessly.

What is also very developed in humans is the use of abstraction for characterizing agency. On an evolutionary timescale, anything that moves substantially is most likely to be an animal. Thus, this programming extends effortlessly to symbolic representations that move. As an example, we showed subjects movies of simple diagrams with movies of simple moving

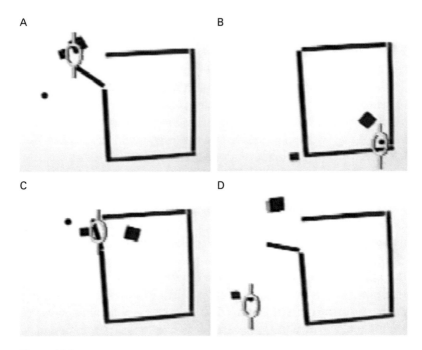

Figure 11.2

Four frames from a movie that has a simple scenario involving a circle and two squares. Normal subjects effortlessly attribute agency to the moving geometric figures. The eye position is shown on each frame as a white token with an oval center. Note that the eye position reveals information about the viewer's running model. In frame A, the fixation is in the middle of the confrontation. In frame C the fixation is leading the "running" figures. Thus even though the accompanying narrative mentions the big square, the focus is on the escapers.

tokens represented with filled squares, and circles, copying the format originally used by Heider and Simmel.[304] Figure 11.2 shows frames from one such movie. The movie contains only geometric shapes in motion, but when asked to describe such scenes, subjects vividly bring the scenes to life attributing sex and elaborate motives to the moving tokens. In the figure, the eye fixation point is also indicated as a white marker. The box is naturally a house seen from above. The small figures are seen as a pair of animates that have to deal with the large square, which is also treated as an animate. What is interesting is that normal subjects inevitably have huge overlaps in their descriptions of the scene. They see the movie as one of conflict between the large square and the other two figures. In the frames shown in figure 11.2, note that the depiction in the subject's narrative and gaze location is consistently from the viewpoint of the small figures, suggesting identification with their point of view.

After reviewing the evidence for how we test the environment in chapter 6, perhaps we should not be surprised that we just use the barest of elements to define complex events. We are not worried by the absence of evidence of enfleshed characters. The story is a little different when there is positive evidence for details that are not quite right though. Jack Loomis made studies of human interaction with virtual human figures, or avatars, in virtual reality (VR) environments. He was after just this question: What cues about the virtual figures need to be present for the subjects to treat them as real? Subjects wear a head-mounted display and see the virtual figures that may be engaged in various activities or just standing around. In one experiment, he instructed subjects to remember a number printed on the front of the avatar and a name printed on the back.[305] As the avatar was standing still, to do this they had to walk around it. Loomis had two different ways of describing the avatar: in some runs, observers in VR were led to think that the avatars were driven by humans, and in other runs they were led to think that they were driven by a computer. The interesting result was that the subjects were very sensitive to these details and behaved toward the avatars differently under the different assumptions. When the avatars were non-human driven, subjects did not respect the avatars' personal space in the course of getting the name/number information. But when the avatars were thought to be human operated with great fidelity, the subjects gave them agency and respected their personal space, walking around them at a close but polite distance. Figure 11.3B shows another variant. Subjects gave staring avatars a wider berth than nonstarers, even though the state of the avatars was known to be identical in every other respect.

In another experiment, male subjects are playing blackjack in virtual reality for a while when two avatar players appear beside them and also start playing. Because the subject has been playing for a while, he has established a risk level in terms of the average amount that is bet on each hand. It turns out that social pressures are such that in a group, a male would like to bet a little more than the average of the group, because being a male risk-taker has status. The delightful result is that when the avatars appear and start betting more, a male will raise his bet to impress them. Remember, they are not real, just graphics figures!

Autism and Mindblindness

It is rather astonishing that we endow avatars with an agency. We know that they are not humans of course, yet in behaviors we accord them a

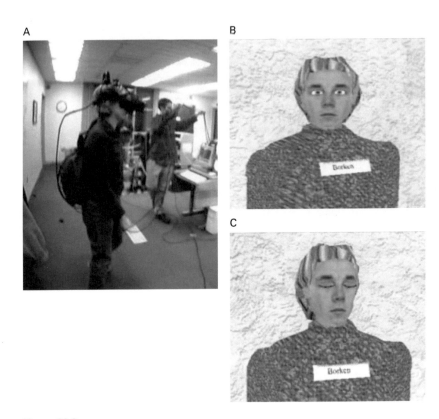

Figure 11.3
(A) A walking environment for virtual reality experiments. (B, C) Part of an experiment designed by Bailenson et al. (see ref. 252) to test agency. Staring avatars' personal space is respected more than that of nonstarers. From J. N. Bailenson et al., "Interpersonal distance in immersive virtual environments," *Personality and Social Psychology Bulletin*, vol. 29, pp. 1–15, 2003. Copyright ©2003, Reprinted by permission of SAGE Publications

measure of human respect. However, now that you have the picture of how the brain handles things in the cortex, it is perhaps a bit easier to appreciate. The brain uses tests to index behaviors, and the avatars pass enough of them to be given human perquisites.

The extensive social structures that we have to navigate, together with their accompanying linguistic demands, would seem to make it impossible that we could ever lose accoutrements of this sense of others' agency, yet in the special case of autism, it appears that that is exactly what happens. Severely autistic subjects have enormous problems reading the minds of other people. Baron-Cohen, whose research interest is in studying autistic subjects, has termed this "mindblindness."[306]

The classical experiment, done with young children, has an autistic child, say Ann, in a room standing in front of two boxes, one of which contains a candy box of "Smarties." The child's friend Fred leaves the room, and the autistic subject sees the experimenter take the candy from its box, let's say box A, and put it in the other box, box B. Ann is then asked: "When Fred comes back, which box will he say has the candy?" Of course, she should say box A because Fred did not see the transfer and would infer that the candy is in its original box. But the interesting observation is that the response is box B. Why? The thought is that the autistic subject cannot make a distinction between what she knows and what Fred knows. Because she saw the transfer and knows that the candy is in box B, then Fred must think so, too.

From a computational perspective, having trouble with agency in this way is not a big leap. We have seen that the important features of our own agency are intimately tied up with our body machinery. When we experience emotions, we are interpreting our momentary body state. In this light, appreciating the agency of others means that we have to simulate them using our hardware. Thus, we have to allow our readouts of their state to direct our internal simulation. Even normal subjects doing this often get it wrong, as we can all testify. At any rate, you can appreciate that this might be a technically difficult fragile program that could be damaged. Amazingly, autism is due to a genetic abnormality. The genes seem remote in the sense that they work at a very low level of abstraction to make proteins that construct the phenotype. Yet here the long arm of the gene reaches up to produce a profound and seemingly abstract defect.

Baron-Cohen's thesis is that a necessary precursor in the normal developmental pathway is that a child learns to use a caregiver's eye movements as a pointing tool. The caregiver has a pre-linguistic way of instructing the child just by looking at things. If a child fails to pick up on this indicator, it signals the beginning of trouble distinguishing the intentions of another person and his or her own. There is a lot of evidence to support this idea. Yu has shown that human subjects can learn words from a foreign language much more easily when they see eye movements.[307] In his initial experiments, adults listened to a child's story read in Mandarin. None of the subjects had any experience with Chinese. There were two conditions: in the first they heard the story read and saw the pages of the storybook turned at the appropriate points; in the second condition additionally they saw an indicator of the reader's eye-gaze position on the text from moment to moment as the story was read. Afterward, the subjects heard an animal's Mandarin code and had to pick one of five possible animals that it referred to. Figure 11.4A shows the result. Subjects who had seen the gaze position during the reading of the story did spectacularly better than those who had not. Subjects who had seen the reader's gaze got more than 60% of the tests correct, while subjects who had not performed no better than the 20% obtainable by guessing. Yu's tests have been supplemented by a graphical model that can gain leverage by utilizing the visual-auditory associations in learning. The implication is that as long as the child learner can detect that a caregiver is sending these signals, then they can be used effectively. This result has implications for autistic subjects also, as in development, they can have difficulties recognizing another's gaze direction.

Needless to say, having autism leads to all kinds of challenges for the person who has this difficulty, but the point here is about agency. A concept that we take for granted does not come without a struggle and in the case of autism can be side-tracked. You should also take notice that this deficit is not necessarily about intelligence, as autistics can have very strong intellects. It's about agency.

Conscious Will

If there is a jewel in the crown of consciousness, it's our so-called free will. We have the distinct sensation that we can choose among actions. I have the sensation now that I could go into the kitchen and make either coffee or tea. It seems that I make up my mind as to which one I want. But the thesis of this book is that it's all computation. So in that light, we ask what does it mean in a program to make choices?

Making choices in the exercise of our free will is not always so easy. We just saw that a phantom-limb patient does not have the ability to believe that his or her arm has been amputated, despite huge amounts of evidence. In virtual reality laboratories, you can have people wear a head-mounted display and show them a huge virtual pit in front of them. If you ask them to step into it, a large percentage of subjects refuse. They know that the ground in front of them is the solid laboratory floor, yet the faux visual evidence of the pit is overwhelmingly compelling. So at best, free will consists of weighing evidence in the choice making. If there is too much evidence mediating one choice, others are not made. I might muse that I have the free will to rob a convenience store, and perhaps we could create circumstances such that it would be an option, but the fact is that under any normal set of circumstances, I would not be able to do it. I do not have the freedom or free will to make such a choice.

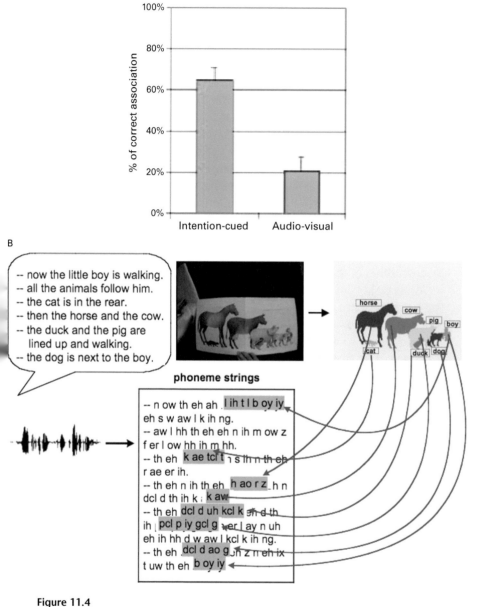

Figure 11.4

(A) Experiments show that human adult language learners are much better at this when they can see the teacher's eye-gaze position. (B) A basic problem in learning a language is that the learner has to associate parts of the speech stream with referents. A graphical model that associates visual and auditory features provides an explanation of this advantage.

Nonetheless, despite these arguments, at least some readers are still thinking to themselves: "Maybe that is the way your brain works, but I still feel that I have free will." Daniel Wegner has tackled precisely this question in a wonderful experiment that is able to separate when we feel we do something from when we actually do it, not an easy thing to do. The experiment makes use of a setup that is reminiscent of an old parlor game played with a special board—the Ouija board, so we will take a little detour to introduce its usage.

The Ouija board is a parlor game introduced in 1890, a more much sedate time with respect to today's video games. The board itself is a hardwood surface with "yes" and "no" alternatives marked on alternate sides of the board. Two people sit with the board on their laps and manipulate a *planchette*. This object is a raised surface that they can comfortably grasp from either side. The planchette is on pedestals designed to make it slide easily on the board. The players cooperatively slide the planchette about on the board. One of the players thinks of a question and then the subsequent motion of the planchette is supposed to guide it toward the answer.

What makes the Ouija board fun is that for some pairs, there is a compelling experience that the planchette is moving randomly and is not under the control—or wills—of the two players. Thus, the answers to the questions seem to come from an external source.

Incidentally, from what we learned in chapter 9, we can make a suggestion as to why this works—when it does. Let's focus on two important aspects of the game: (1) the players are trying to be cooperative, and (2) it's not fun if the planchette moves deterministically. As a consequence of these two desiderata, they are trying to be random. Because each person's movements are made in short segments and the net results of these movements can be sensed by both players, it could be the case that generating random movements is easier in the cooperative game. The players can each estimate the movement of the other player by subtracting one's own contribution and try to generate a new movement that is not in that direction, a fairly easy thing to do. At the conscious level, the movement of the planchette seems random because, owing to the success of coupled two-player game, it is convincingly close to being random.

In Wegner's experimental design, two people manipulate a small planchette-like structure over a table as shown in figure 11.5A. The planchette controls a computer cursor that glides over a cluttered scene containing many common objects. The participants hear audio signals that instruct when they are to stop moving the cursor and place it over an object. The experimental cleverness is that one of the participants is actually a confederate who hears precise instructions as to when and where to stop moving the cursor. In the actual runs, a trial consisted of 30 seconds of movement followed by 10 seconds of music indicating that they should make a stop. The subject heard words over the headphones that were billed as being used to provide a distraction. The confederate was of course supposed to be hearing distracting words also but instead heard instructions on what to do. The interesting focus of the trials was forced stops. On these, the confederate was instructed to move toward a specific object and given a countdown to use to stop on that object. The purpose for all this structure was that the subject heard a priming word at specific times with respect to when the confederate stopped. The subject thought the two of them were stopping, but in fact the confederate was controlling the stop. Both of them were asked to rate all the stops as to how intentional they were on a 14-point scale ranging from "I allowed the stop to happen" to "I intended to make the stop." The priming word in the forced stop was at 30 seconds, 5 seconds, and 1 second before the action and 1 second after the action. The results are shown in figure 11.5B. The interval rated the highest as to intentionality was 1 second before the action was made. Keep in mind that for none of these data is the subject actually controlling the cursor. One second after gets a low score as does 30 seconds prior. Wegner's model for this can be seen in figure 11.6. The idea again appeals to the abstraction gulf between the machinery that is directing the selection of actions and that privy to conscious report. The data suggest that hearing the word primes the low-level circuitry that actually generates the action. In reinforcement learning parlance, priming is part of the indexing state that will guide the action, and stopping is action selection. So like the phantom-limb patients, here the normal experience of conscious will can be interpreted as the product of running an internal model. This ingenious experiment allows the manipulation of the particulars of that model in the subject. The idea is that the perception of conscious will depends on a timing relationship between the computation of the state that guides the action and the execution of the action itself. The experiments suggest that normally this is about 1 second or less before the action occurs. We might be skeptical that the brain's programs are modifying the temporal relations that we are aware of, but the truth is that they are doing this continually (see ref. 29). When you touch your nose with your finger, the perception is that the finger and nasal skin meet simultaneously. However, the signal from the finger has a much longer transit to get to the brain and arrives later. The brain's programs take this into account and make the necessary adjustments.

A

B

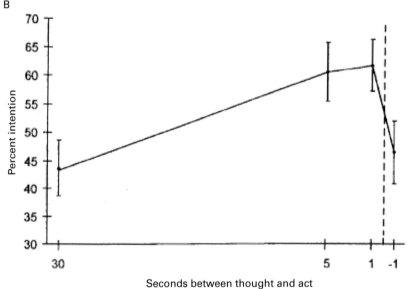

Seconds between thought and act

Figure 11.5
(A) The confederate and subject in the "I Spy" experiment moving a Ouija-like cursor to targets on a screen. (B) The ratings of willfulness produced by the subject pool depend on the time between when a prime word was heard and the actual stop of the cursor (see ref. 277).

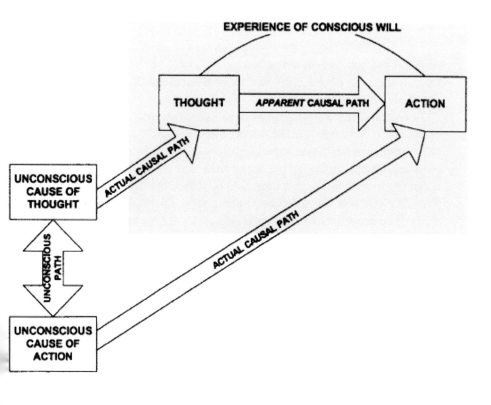

Figure 11.6
In Wegner's model, the perception of conscious will is produced as a by-product of an expected 1- to 0.5-second delay between the appearance of the representation of the action and the execution of the action. Lower levels in the hierarchy, depicted on the extreme left, process the stopping data and determine the interval between thought and action in a model simulator, depicted in green. From D. M. Wegner, *The Illusion of Conscious Will*. MIT Press, 2002.

11.2   Simulation

At this point, you have seen the basic components, so that it's time to put them together to obtain a description of consciousness. To do this, let's revisit unconscious behavior.

To compensate for slow (compared to silicon) circuitry, the brain stores behaviors in tabular form. The cortex is a vast storehouse of millions of prescriptions for the tiniest aspect of behavior, which includes the quanta of seeing, speaking, and feeling. These behaviors are rote; we don't think about them when we execute them. Think of driving a car along a familiar route. You probably can remember times when you have been unconscious; that is, you drove for miles without being aware in any deep sense of transiting the route. When you finally become aware, you can wonder whether you ran a red light or stop sign, but you are clueless. Of course you didn't. Your detailed set of programs for driving that familiar stretch directed eye movements, head and foot movements, stopped at red lights, started at the onset of green lights, signaled turns, and performed almost countless other small but necessary tasks. This example is just one of the myriad of behaviors that your brain stores. In fact that's what it's for; to execute a stored repository of behaviors that allow it to accomplish the goals laid in via an architecture refined by millions of years of evolutionary programming. Furthermore, those programs are constantly being fine-tuned everyday by loads of repetitions. But none of this is conscious.

So what is the role of consciousness? Almost everyone agrees that the number-one value of having a model is the ability to predict the future. Because the brain runs models, it's easy to imagine a reward system that places a great premium on brains that are slightly better at doing this. However, predicting the future is not easy because it has not happened yet. Thus, there are two crucial aspects that have to be dealt with.

1. The model must be able to run ahead of the present, constructing representations of possible worlds. This ability introduces a difficulty in that modeling the present and near future is easier because either is a vast source of sensory and motor cues of just how the present *is*. When the model is projected in the more distant future, it loses access to this vast sanity check.
2. The model must deal with potentially one-off situations that are not explicitly represented in memory. There are some situations that one could expect the memory to handle, and those are where the exact situation has not been seen, but it can be interpolated from behaviors that have been seen. However, after we remove these from consideration, there will always be situations that have not been encountered. They might have a compo-

nent that can be looked up in the table, but there are still some unique elements left over.

These two elements in conjunction mean that the programs that can run into the future have to have special functionality in order to deal with them. Let's run the risk of producing another jargon word that needs to be deconstructed and call these programs collectively *the simulator*. The simulator's primary value is that can run in the future, but it can also be tripped by either of the two conditions just mentioned. For example, you can be doing something in the present and stumble into a novel aspect not covered by the tables. The simulator kicks in and can run the tables with hypothetical assignments.

Let's imagine your ancestor of 10,000 years ago hunting the world's mammals. He forms a plan: He will distract the mammal while the rest of the group pelts it with spears and or stones. In this construction, it is vital to not get confused between the distractor and the attackers. The brain has to represent both parties with their hypothetical duties. In the brain, it is very unlikely that it is done with a few "grandmother" neurons in a specific place. As you saw previously, a much more likely candidate is the emotional system. For each person, the cortical imprint of the body's different functions provides a signature that means "ME." From this perspective, you can see the technical demands of denoting a person that is not yourself. You have the mechanism: You could just attribute its signature to another, but that would leave you severely autistic. What you have to do is mark the differences in that signature in a way that you have created another distinct individual. In effect, the simulator is doing a second-order simulation. You are not just simulating yourself in the future where you can tag items with your signature, but you have to abstract the signature itself to create an independent person "symbol."

One very important concept that can help here is that of working memory. Working memory is needed precisely in the case where an unfamiliar referent has to be spliced into a familiar program. Thus, it is a correlate of the running of the simulator. It helps keep track of the fact that you have left the realm of everyday existence and are running in hypothetical mode. The simulator uses this structure to adjust the value of itself—the plan-ahead program—against the standard background cacophony of routinized behaviors.

Consciousness Meets Computation

If consciousness is just going to be a program, then we had better recap a bit about programs, in particular, how programs are organized. As Newell pointed out, any complex system that has admitted of a description has incorporated the use of the notion of hierarchy, and the brain is unlikely to be an exception.[308] Hierarchies have two prominent properties:

1. As the description becomes more abstract, its components necessarily run slower than those of the lower level and take up more physical space.
2. The more abstract description omits details from the lower level.

A ready example from silicon is that of digital circuitry. The dependence of current on voltage across a gate is continuous, but digital circuitry abstracts that into two levels. The speed of such circuitry is governed by the time it takes for the levels to change between one level and another. Here, the continuous voltage value is abstracted away into a binary code. In a similar way, computational models of the brain are different at different levels, as the primitives at each level are correspondingly different, too. Just like silicon computing, brain computation has to be organized into hierarchical levels, each of which groups primitives from the level below into equivalences. The average cortical neuron receives input from two to 10,000 other neurons and yet at any moment summarizes those inputs into a single spike or silence. Higher levels make similar kinds of abstractions. So a key question is: If consciousness is a program, how does that program project into the various computational abstraction levels?

The brain's computational framework has sensitized us to the fact that most of the brain is about memories. The slow neural circuitry, more than 1 million times slower than silicon, means that at a basic level, behavioral responses have to have been precomputed and looked up by a fast indexing technique. To resvisit our metaphor, before Google, we might have been skeptical that this was possible, but now we can appreciate that, in a way analogous to Web crawling and indexing, the brain works over its lifetime continually to sort its behavioral programs so that they can be rapidly brought to bear on the situation at hand.

How do the memory circuits get formed? A computational answer for the peripheral visual circuitry has found wide acceptance, and this views those areas as constructing a code for natural images. Any good code should represent the image to be coded, but computational models have introduced a special spin; the image indeed should be coded, but a desirable property of the code is that its neural substrate should use an economical signaling strategy. Neurons signal in discrete spikes, and half of the metabolic cost is in spike generation, so codes that use less spikes and thus save energy are to be preferred. Stunningly, those codes when created in computational simulations appear very similar to the experimental observations from neural recordings.[309,310] This methodology implies that the brain has refined *Occam's razor*. Simple codes are preferred, but simplicity must balance the accuracy of the code against the complexity of its description.

How are memory circuits used? The visual codes that we have just been talking about allow the description of visual data, but they do not include a prescription for what to do with it. For that we must jump to a higher abstraction level that explicitly represents functional routines that manipulate sensory motor abstractions. Recall that human retinas have pronounced foveas so that the resolution at the point of gaze is 100 times increased over the periphery. As a consequence, almost all the time, primate eyes use rapid ballistic movements termed saccades to orient the high-resolution gaze point over interesting visual targets, and programming such quantile tests takes about a quarter of a second. We saw in the Roelfsema experiments how such tests could be concatenated into a sequence.

The above two examples are united by their use of computation as the core concept that guides the interpretation of data, yet they describe experiments that span two separate abstractions. When studying memory formation, the use of the memory is postponed to a higher level. And when studying the use of memory, the existence of the memory at a lower level is assumed. And these two abstractions are just two of several that must be assumed to handle the richness of human brain computation.

What about consciousness? Could it be the sole *deux ex machina* that escapes a computational description? Computation is not an all-powerful theory and has well understood limitations, but all of these concern mathematical infinities and are unlikely limitations for a satisficing animal brain. The odds-on bet is that when consciousness is understood at some mechanical level, that level will be isomorphic with computation. Furthermore, as some kind of program, it must fit in the computational hierarchy. Table 11.1 shows a more complete candidate computational hierarchy that posits consciousness at a separate *debugging* level in the brain's computation hierarchy. This is going to need a significant amount of elaboration.

Abstraction Hierarchies Transgressed
Abstraction hierarchies are so fundamental that discussions of brain organization that eschew them tend to be complicated. Searle, arguing against artificial intelligence models, and by extension computation, used a now-famous example of the Chinese room.[311] A person is in a room where Chinese sentences in characters appear at an input slot. While the person knows nothing of Chinese, he or she has access to a set of instructions in English that are of the form "if you see these characters output these

Table 11.1
A proposal for the brain's use of computational hierarchies with two additional
levels over table 1.2 to manage strategic issues and the demands of awareness

Level	Computational Model
Operating system	Schedule all behaviors; trade-off debugging and runtime behaviors: "if all the operations are succeeding, stay in runtime mode"
Debugging	Analyze unusual events using off-line mode: "why isn't the jelly sticking to the knife?"
Runtime [Scheduler]	Pick a suite of behaviors to handle the current situation: "take the lid of the jelly jar," "pick up the knife"
Behavior [Programs]	Sensory-motor coordinated actions: "inset the knife into the jar and gingerly bend and remove it"
Input/Output [Routines]	Task-specific tests of the environment: "locate the purple jelly surface"
Calibration [Data Abstraction]	Encode environmental statistics in specific circuits; filling-in phenomena: color constant neural circuitry
Neural	Models of specific neurons in circuits; excitation and inhibition
Synapses	Models of a neuron's components; roles for different neural transmitters; neural spike signaling; basic units of memory

Note: In the task of making a sandwich, different subtasks are described at different
levels of abstraction.

characters." Supposedly the person is a model for the computer, mimicking linguistic behavior but understanding nothing. But here is where the trouble starts: Is the person in the room actually a human or a computer (or a neuron)? There are different logical consequences depending on how one answers this question, and therein lies the difficulty.

The danger of confusing abstractions is easier to understand if we change venues for a moment and talk about computers and their abstractions. One can program in MATLAB, assembly language, or microcode, but each language is cast at a different level of abstraction, and each involves a different level of familiarity with the underlying computer that runs the programs. MATLAB requires no knowledge of the machine at all, whereas assembly language requires an elemental model of basic random access architectures, and microcode requires an intimate knowledge of a particular machine's low-level hardware instruction set. Recall the multiplication $x * y$. At the level of MATLAB, it is a set of symbols that can be manipulated. At the level of microcode, these are translated into binary representations one of which is successively shifted and added to the other. The point of these levels is that they cannot be mixed. Once an abstraction level is picked, one must stay within it: MATLAB and microcode cannot be mixed.

What seems so clear when couched in silicon computational terms somehow does not always survive the translation into philosophical arguments about human computation. To return to our example, Searle's attempt to gain purchase depends on transits across abstraction boundaries. Is the occupant in the room a human or computer? It has to be one or the other; to not insist on a choice leads to what I would term *the abstraction con*, which comes up repeatedly in discussions about consciousness. Consider Escher's famous drawing of hands drawing each other. Is it just a drawing or is it really a picture of something that could happen? Perhaps when you don't look, the hands transform into versions of "Thing" of television's *Addams Family* and take a few pencil strokes. But let us not let this kind of magical thinking displace the magical thinking (aka the abstraction con) upon which the drawing's kick depends: "It's a pair of hands" oscillating across the abstraction boundary separating "It's a drawing of hands."

Consciousness as a Modeling Level of Abstraction

We started the discussion of consciousness with Baars's very imaginative theater metaphor; however, its unitary nature has been criticized as the *Cartesian* theater by Dennett (see ref. 297). He points out through a series of examples that the nature of consciousness is not an exact depiction of spatiotemporal events but one that gets edited to produce a composite narrative, and that this editing process is ongoing. To focus on the editing process, he proposes a "multiple drafts" description of consciousness. Many different processes are actively deconstructing experience, and consciousness is just the perceived digest of this cacophony. Philosophers don't like to share concepts, so there would be resistance to seeing the multiple drafts model as a refinement of the Cartesian theater, but computational abstractions encourage this interpretation. At the embodiment level of abstraction, there *are* multiple drafts, as many different threads work on different aspects of a problem. But at the very removed top level of the conscious narrative, we access a much larger spatial and temporal scale that projects a more coherent description. We must remember that being a model of reality, it is incredibly useful to make generalizations that make the memory load more compact, and in this compression process, it pays to combine fragments of different experiences into a composite story that takes liberties with individual fragments of experience. Think how chaotic consciousness would be if this *didn't* happen.

Let's refocus on the specific computational consciousness issue and now wonder how consciousness gets represented in an abstraction hierarchy. Could it be that (a) consciousness is an abstraction that appears at a given level but has no discernible trace at the level below? The alternatives are that (b) consciousness is only manifested at lower levels or else (c) it has a trace that spans all levels. Many researchers would opt for at least option (a) given our phenomenological experience of hearing our own internal directive voice, but there is obviously also enthusiasm for option (c) as well. Koch for one has advocated a search for the *neural correlate of consciousness*.[312] Of course, unless one is a dualist, there will be a sense in which the answer must be option (c). But the huge cautionary note from the example of silicon computation is that one needs to understand the various abstraction levels in their own terms as well as the process of translating between them. Failing to keep them separate and mixing abstraction levels only results in confusion. This assertion can be seen as in sympathy with Lamme's view.[313] He points out that the various claims as to consciousness can seem confusing if they do not respect basic neural organizations.

Perhaps the best articulation of consciousness having its own abstraction level comes from Graziano.[314] He takes a neuropsychological stance and differentiates between an *awareness* level and an *attentional* level. Graziano takes great pains to differentiate between these two levels as they are often confused with each other in psychology. In a beautiful treatise, he points out that the awareness level treats the items at the attentional level as symbols that can be manipulated. This distinction is perhaps easier to appreciate given a sparkling metaphor from *The User Illusion*.[315] On a laptop computer, one can delete a file by dragging its icon to a trash can. Of course, that action bares no resemblance to what is happening inside the machine, but it is very satisfying nonetheless. In the same way, awareness does not have to be strongly tied to the huge substrate of neural circuitry that produces it. The perspective of this chapter resonates more or less completely with this view but stresses a computational perspective. Thus, we concur with Graziano completely that the two levels of awareness and attention should be treated as different levels, which we have denoted the debug level and runtime level, respectively.

As Graziano notes, the division in hierarchical levels also provides a ready solution to the distinction between awareness and sensation. You can be aware of feeling a headache or be capable of ignoring the sensation. But in the latter case, the feeling is still there, ergo it must have its own reality. This fact has been given a huge amount of attention in philosophical circles. Block takes Damasio to task in his review of *Self Comes to*

Mind because in Block's view, sensation has been improperly subjugated to self-awareness, whereas they should be two independent entities.[316] They should, of course, but from the computational perspective they are represented at *different levels of abstraction* and thus are incomparable.

It is when we get to the "attentional" hierarchy that the computational perspective displays its usefulness. Looking back at chapters 6 through 8, we can see that to realize the most basic of behaviors requires an enormous amount of sophisticated bookkeeping. While there is competition between modules, that computation takes place at the level of reinforcement learning circuits that represent behaviors at an intermediate level of abstraction. In this view, the settings for biased competition models, while informative, come across as pitched at too low a level of detail. The bottom line is that the attentional/runtime level is best thought of as a complex of neural machinery for implementing overlearned behaviors that Graybiel terms "habits" that, in favorable circumstances, are able to proceed without conscious monitoring. Naturally if a hierarchical description is going to have traction, it will have to have descriptions that specify how to translate between levels. But the difference between embodied levels running autonomously and being tweaked by awareness context could be very nuanced. In the same way, attention models at the neural level that skirt with the notion of awareness face the same issue. At this even lower level experimentalists catch circuits being modulated by an animal's attentional demands, but this may be most economically explained by the descriptions at the embodiment level just above.

However, the current state of knowledge of the brain provides only an outline of plausible abstraction hierarchies, thus leaving an enormous amount of work to do. At this point, we will content ourselves by elaborating the differences between what we have termed in table 11.1 the "Runtime" level and the "Debugging" level, the latter being the level where the contents of consciousness are most evident.

Will Work for Dopamine

Although what we are calling the debugging level of abstraction will be the most important for understanding consciousness, to appreciate this level it is important to revisit the abstraction level beneath it. How do programs get established in the first place? We know a bit about the programs themselves in that, in broad outline, the cortex stores elaborate states of the world, and the basal ganglia sequences through those states and triggers actions. While silicon computing sequences through states at an incredibly high 2-GHz rate, the basal ganglia probably works a about 0.3 Hz, or a billion times slower than silicon. Of course, the bandwidth of the cortex in terms of its parallel processing more than makes up for the slowness. We also know a bit about how programs are formed in that the job of the amygdala is to filter out what is important, and the job of the hippocampus is to slice and dice those components into pieces that are compatible with what the cortex and basal ganglia have stored. So although we cannot say too much about the details, the big picture of what programs are is sketched.

Given that we have programs, who—or rather *what*—is the programmer? Answering this question requires understanding at a basic level what programs do for us, and that is prediction. Ultimately, we need to be able to mate successfully and along the way to survive, and the way to do that is to be able to predict the future. And one hugely important way to do that is to save what has happened in the past along with its value so that if the situation recurs, one can estimate its outcome and value before its conclusion. The past is prologue.

In this process, it helps to keep in mind that the brain's programs are not literal parts of the external world, but just internal models of that world. Ramachandran[317] and Sacks[318] have beautifully described cases where, owing to some kind of brain or body injury, that model is divorced from the true picture of the world, but the brain's conscious owner is unaware of the schism. The point is that these cases of brain injury tell us about the healthy brain's structure, and that is one of building and maintaining representations of the world and programs for extracting reward from it. This last point needs elaboration, for although we act in the world to survive and mate successfully and along the way achieve measurable rewards in terms of behaviors that satisfy us in one way or another, to accomplish all this the brain's *models* need "pretend" rewards, or secondary rewards that stand in for the real thing. The main neurotransmitter that signals secondary reward is known to be dopamine,[319] which we have termed the "neuro." Its power is very much experienced by cocaine addicts who engage in behaviors that trigger dopamine release. In fact, most addictions can be conceived in terms of behavioral shortcuts to dopamine release. Most of us though are calibrated, in the sense that we can engage in socially acceptable behaviors that the brain can translate into a dopamine reward estimate. How can one choose between behaviors A and B? Simple. The brain can retrieve their dopamine estimates and pick the most rewarding.

Zombies

Characterizing programs as mechanisms for chasing secondary reward doesn't seem to leave room for consciousness, but of course we have it.

To create a conceptual space for it, we turn to the featured non-persona of philosophers and B-movies, the *zombie*. The zombie brain has an enormous raft of programs that negotiate to the hypothalamus to be valued in neuros. With this common currency, the brain can pick the most valuable for execution; no conscious thought required.

What is the zombie state? The earlier car driving example is perfect here. One drives miles of a familiar route, remembering nothing about a huge driving segment wherein one was guiding the car, obeying traffic lights, avoiding pedestrians, and handling numerous other small exigencies. During this period, for the driving behavior, the zombie programs were running. There was no use of any special monitoring because the states that were directing behavior had precoded expectations of consequences that were constantly being met; it is only when this does not happen that something non-zombie has to be done. Of course, in philosophy there always seems to be someone on the other side of the fence. Searle, again (see ref. 311):

It is true, for example, that when I am driving my car "on automatic pilot" I am not paying much attention to the details of the road and the traffic. But it is simply not true that I am totally unconscious of these phenomena. If I were, there would be a car crash.

This was written before the 2005 DARPA Grand Challenge that had five vehicles successfully complete a complicated desert trail loop autonomously. Hopefully, we can all agree that those vehicles were not conscious. The point is that the vehicles are essentially limited to zombie driving and did not crash.

David Foster Wallace captured the zombie state brilliantly in his essay "How Tracy Austin Broke My Heart,"[320] where he asserted that professional athletes have difficulty describing their feats precisely because *the descriptions are no longer accessible* when the overlearned skills are compressed in the zombie brain:

The real secret behind top athletes' genius, then, may be as esoteric and obvious and dull and profound as silence itself. The real, many-veiled answer to the question of just what goes through a great player's mind as he stands at the center of hostile crowd-noise and lines up at the free-throw that will decide the game might well be: *nothing at all* ... there's a cruel paradox involved. It may well be that we spectators, who are not divinely gifted as athletes, are the only ones able truly to see, articulate, and animate the experience of the gift we are denied. And that those who receive and act out the gift of athletic genius must, perforce, be blind and dumb about it— and not because blindness and dumbness are the price of the gift, but because they are its essence. (pp. 154–155)

Make our inner zombie the athlete and our conscious self the specta-
tor, and you have one of the best compact descriptions of the relationship
between the two ever written. The quote also highlights another point. The
zombie is usually disparaged as inferior, but the metaphor shows off its true
relevance: Zombie skills are those that have been honed to near perfection
through experience. Moreover, the Grand Challenge robot driving success
shows that, while the detailed programs that the brain might use for its
runtime abstraction level have yet to be pinned down satisfactorily in the
neural substrate, the computation that does the job is fairly well understood and moreover, is now the subject of standard texts.[321,322]

11.3   What Is Consciousness For?

Zombie programs depend on their model of the world being a very good fit:
All the contingencies that can occur have been seen a significant amount
of times, and their responses are coded. Thus, the search for alternatives
has been done and remembered. But before that happens, as models are
being constructed, the statistics of the model need to be gathered. This is
the role of consciousness. The best analogy for this is the process of debugging a program on a silicon computer. Once a program has been debugged,
it can be used in a zombie state where it is not tampered with further, but
before that state is reached, the programmer must stop the program, try
alternate versions of it, and test them to see if they behave according to
expectations. Like consciousness, this debugging process is much slower
and has substantial off-line portions of time.

However, an important contrast is that, unlike debugging, there is no
programmer in the human consciousness, just a neural search program
that is trying to fit a model to data. In this fitting process, it is central to
distinguish between effects that the human agent produces and effects that
the rest of the world produces. For this point, we return to one of the best
characterizations of consciousness, that of Wegner (see ref. 277). In his "I
Spy" experiment described earlier, the interval of time between when subjects think of an action and when they do it is manipulated. When asked
to rate their actions on a 14-point scale from "I caused it" to "It just happened," subjects rate the period that has the thought preceding the action
by 1second as the most causal. The stunning suggestion is that the act of
consciousness itself is just one more model that the brain uses, and its "fit"
depends on the temporal relationship between the brain's machinery and
the body's actions in the external world. Wegner characterizes that result of

a good fit between the two as "an illusion of authorship," to which I would add: produced and required by the neural program in its search process.

The search process of consciousness has a special and remarkable technical problem to deal with, which is that it must share the same neural hardware used by the zombie programs. This was brought to light by measurements of the firing patterns of cortical "mirror cells."[323] Rizzolatti et al. discovered when recording from a monkey's cortex in motor area F5 that the cells would respond when the monkey picked up a raisin but also when another monkey did so or when the experimenter did so. The profound implication of this is that there is finally concrete evidence that the monkey and by extension humans use one set of neural "hardware" for representing all these different events. What this suggests is that a large part of the experience of consciousness may be generated by mental simulation using the same neural circuitry that is used for everyday action, a point made much earlier by Merleau-Ponty[324] and amplified recently by Barsalou.[325]

Tagging

Searching by simulation using existing knowledge has a lot of advantages, the principal one being that, in the act of exploring the effect of changing one variable, the rest of the program can be used as is because it is already in place. Thus, the search process can systematically try out slightly different variations of the simulation program without extensive reprogramming of all the components.

The computational role for consciousness is to be the mechanism that does this bookkeeping, but this point requires quite a bit of elaboration. To give it a name, let's import a concept from the previous chapter and call this ability *tagging* in the sense that we can tag the actions of the brain according to the particular agent whose activities are being represented. We use the ability to tag in different ways:

1. We may need to reflect on the past or the future. In these cases, the same brain hardware handles the visual perception of the state of the world in the past and future, but, as they are not in the present, these venues must be tagged. The state of other animates is especially complicated and requires more elaborate tracking, as noted in the following items.
2. We may need to reflect on another's motives. In this case, the other person must be simulated on our brain hardware but of course is not us and must be tagged to keep the distinction apparent.

3. It becomes even more complicated if we have to consider what another person is thinking that we are thinking or when we must imagine the other person's future actions, but the overall ability still rests on tagging or the bookkeeping to record that the brain's activity is a simulation of reality, not reality itself.

4. Multiple personality disorders are a failure of the tagging system. Often, in order to protect itself from the consequences of remembering early abuse, the brain will adopt an alternate personality that did not experience it. In this case, the tagging is dropped to protect the user.

However, this boon comes with an attendant disadvantage, which is that some bookkeeping must be done to keep track of what is simulation and what is reality. To go back to Rizzolatti's example, the monkey's brain must somehow keep track of the difference between the experimenter doing the action—debugging mode—or the monkey itself doing the action—runtime mode. Let's make a speculative suggestion as to how this might be done at the neural level. What is needed is some kind of neural "tag" that can differentiate the references to oneself and other people. One possibility is position. Different people can only be in one place at a time so that spatial coordinates could be a "tag." People learning Mandarin are told to associate unfamiliar characters with places in a room to help their recall. Auditory differences could work, too. However, the suggestion that we would make would be to use the limbic system. How we feel about a person would be used as a unique identifier that wouldn't require hearing or seeing the person. Recall Ramachandran's patient who, when seeing his parents, declared them to be impostors, but had no trouble talking to them on the phone. A possibility is that he is using emotional tags to identify them, and through a cortical stroke this circuitry has become separated from his visual system circuitry, whereas the link to the auditory system has been preserved.

To return to the functional issue, we can appreciate the usefulness of tagging with an analogy to program debugging. When a computer program is running it is in the "zombie" state, where it sequences through a set of instructions until the end of the program. But when something goes wrong, the programmer interrupts this sequence, slows down the execution, and interrogates the values of particular variables, seeking to explain an unexpected part of the execution. In the analogy, the programmer is the conscious homunculus.[326] In runtime mode, the zombie mindlessly executes a set of responses to coded states of the world, and in doing so its programs get translated into all the more concrete models of the neural substrate. In debugging mode, the identical neural hardware is used, and most of the operations are indistinguishable from their runtime mode instantiations with one huge exception: the system "knows"; that is, it has the machinery to understand that, for portions of the debugging mode program, it is simulating the results. What can complicate the matter even further is that debugging mode may only be required for small portions of a large sequence. These observations have a huge implication for understanding consciousness through psychological experiments. Because the conscious mode (debugging) is utilizing the same underlying neural substrate as the unconscious mode (runtime), the former is actually being used as a probe to discover information about the various neural circuits rather than necessarily revealing information about itself, a point also made by Block.[327]

Many readers versed in philosophy of mind issues will recognize that tagging is a version of *simulation theory*. When one wants to make judgments about what another person is thinking, one just has to simulate that person's model on one's own neural hardware. Simulation theory is very attractive on a number of counts, but the most important factor from a computational standpoint is the sheer weight of structure at lower levels of abstraction for executing behaviors that form a foundation for the forebrain's abstract representations. This makes it compellingly easy to appreciate a proposal that posits a small amount of additional structure.

Another aspect is that the two foremost alternative accounts are relatively mute on how their primitives would be implemented. *Theory theory* holds that humans build "folk" theories of their world analogous to physics and can handle something like causal reasoning. However, attempts to handle reasoning with the field of artificial intelligence, while having made enormous strides in understanding the technical aspects of the problem, as we discussed in chapter 1, suggest that it is an unlikely candidate for human reasoning. *Rationality theory* holds that there are principles that humans share that allow reasoning about moral judgments. This tack is harder to dismiss outright, as there are various ways such an approach could be implemented. For example, the principles could be implicitly coded in generic game-playing algorithms such as the ones discussed in chapter 9. To paint the issues in relief, let's take a look at an innovative experiment from the Saxe laboratory.[328]

Different subjects read scripts with four different storylines. While taking a break during a chemical factory tour, Grace makes tea for her companion. In one scenario, she accidentally mistakes a poisonous powder for sugar and poisons her companion. In another she wittingly adds the poison. In two other cases she adds real sugar and in these cases she adds powder from a container labeled as hazardous. So in the last case she means to poison her companion but does not succeed. Subjects read these stories while in an fMRI scanner, and the scans are compared to the nominal case as shown in figure 11.7.

Perhaps one would have liked the different cases to have been more differentiated than they are in figure 11.7. Nonetheless, the clear distinction between the normal case—neutral belief and neutral outcome—and the case where malice didn't work implies that the cortex is sensitive to different moral choices. But how is this being done? In simulation theory, one runs the scenario using the brain's machinery and "reads off" the outcome. Rationality recognizes that the sets of rules that describe the two situations are different. However, a rapprochement is possible if we cast simulation theory and rationality theory at separate levels in the brain's computational hierarchy. Rationality theory is represented at what we are calling the debug level and consists of somehow-coded principles that are summary abstractions of previous simulation runs.

This division into separate description levels finds resonance in lots of other work, as shown in table 11.2.

This distinction is made by many other researchers as well. Kahneman (see ref. 294) dichotomizes thinking into "fast" and "slow" categories, but these are perfectly compatible with the simulation and debugging, respectively. Similarly, Graziano (see ref. 313) makes the distinction between referencing an item with the brain's attentional system, as would be done in simulation, and referencing an item in an awareness system, as would be done in debugging. Similarly, Dennett's notion of the "intentional stance" is pitched at the level of rationality theory,[329] whereas his multiple drafts model of consciousness is compatible with simulation theory (see ref. 297). Of course, we are far from decoding the neural minutiae of just how consciousness is represented at the level of spiking, but the confluence of abstract thinking is suggesting helpful overall directions in which to proceed.

11.4 Summary

Our principal thesis has been that much previous work has concentrated on finding a mechanistic explanation of how it *feels* to be conscious and, specifically, in trying to trace that feeling to the neural substrate. While it might be possible to do this, it also might be an extremely difficult task, equivalent to questioning why the mass of a proton has the specific value that it does? In the same way, consciousness, to work, is a useful added piece of computational neural circuitry that can perform abstract modeling

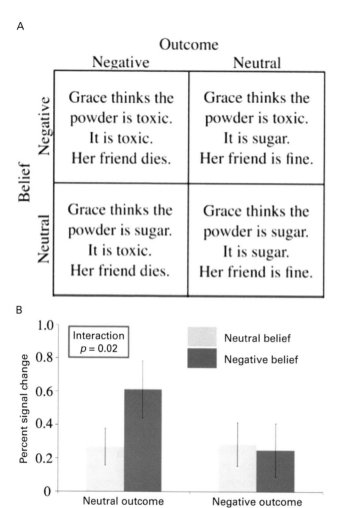

Figure 11.7

An fMRI experiment shows a difference in activation for the consequences of an action. If a subject has a belief in a bad outcome that does not materialize, the activation is significantly greater than that for an actual bad outcome. From L. Young et al., "The neural basis of the interaction between theory of mind and moral judgment," *Proceedings of the National Academy of Sciences U.S.A.*, vol. 104, pp. 8235–8240, 2007.

Table 11.2
Distinctions made in brain theory from several different vantage points can be seen as referencing different abstraction layers

Level	Theory of Mind	Emotions	Law
Debugging	Rationality theory	Social communication	*Mens rea*
Runtime	Simulation theory	Genuine	*Actus reus*

(see ref. 314). In this process, it may have to produce a signature that distinguishes the agent from the surround. The feelings that we can report indicate that the programs are running, which should be reassuring.

Why it produces the feeling that it does as opposed to another may not be a particularly useful question, however compelling it is to ask it. Ramachandran cannot resist and makes it one of his central unanswered questions about consciousness, phrasing it another way: How can we distinguish the consciousness that we readily accept that others have, which he terms third-person consciousness, from the consciousness that we have, which he terms first-person consciousness? Parenthetically, it's not apparent that we cannot do this to some extent in feeling empathy to another's tragic situation, and to the extent that we cannot is an indication of trouble.[330] Moreover, one should be careful what one wishes for. Imagine the mental chaos if you could feel another person's feelings in their raw intensity. However, the principal difficulty in resolving this distinction can be appreciated by recalling the all-important respect for abstraction boundaries in our discourse. Thinking about another's consciousness is at a more abstract level than living our runtime (zombie) existence. In the same way, if we want to characterize our own consciousness, we have to "tag" it so that now we are effectively debugging ourselves as a third-person simulation. To engage *simultaneously* the runtime environment and experience our own consciousness requires that we blur an abstraction boundary, which we have declared taboo.

Of course, taboos can be violated, and even celebrated, as they are in *Gödel, Escher, Bach*.[331] By violating an abstraction boundary, one can create an aptly named "strange loop," where one ends up unexpectedly at a lower level of abstraction, with the result that strange things happen. It's not that you cannot do these things, but, from a computational standpoint, you should not.

The question of why consciousness exists may have a ready answer when compared to its useful partner, the unconscious. Unconscious programs represent repeated and reliable interactions between the agent and the world that can be coded invariantly for each case. Consciousness is used to direct the search for new programs, and in that search it becomes essential to distinguish the agent from the surround. In a strong sense, the views in this chapter have advanced nothing particularly new, as its core ideas have been pioneered by many other researchers. Hopefully, its main value might be in steering the quest for understanding consciousness toward more accessible computational questions.

Notes

Chapter 1

[a]Modern computers have become increasingly complex and have moved toward Marr's characterization of the brain as "a whole lot of information processing devices."

[b]What we will characterize as memory turns out to be closer to the concept in computer science than in psychology.

[c]In the jargon, O is called "Big O" or simply "O."

Chapter 2

[a]However, the basal ganglia's dopaminergic circuitry is so complicated and extensive, and together with the fact that there are different kinds of dopaminergic receptor types, it is quite likely that dopamine is used for other distinct purposes in this subsystem in addition to its central reward function. However, in this overview we will sidestep this issue.

Chapter 3

[a]*Ligand gated* means the channel can be opened by a molecule, in this case N-methyl-D-aspartate, or NMDA.

Chapter 4

[a]As the peak of an iceberg above water is ⅛ of the total, by analogy at 1% you are looking at something less than the peak of the peak.

Chapter 7

[a]PhaseSpace Inc., San Leandro, California.

Chapter 11

[a]Experiments with monkeys have provided evidence that supports this conjecture. If monkeys lose a finger, then the cortical circuitry that monitored that finger gravitates its function to an adjacent finger.

References

1. P. H. Lindsay and D. A. Norman, *Human Information Processing: An Introduction to Psychology*. Academic Press, 1972.

2. D. Marr, *Vision: A Computational Investigation into the Human Representation and Processing of Visual Information*. W. H. Freeman, 1982.

3. P. S. Churchland and T. J. Sejnowski, *The Computational Brain*. MIT Press, 1992.

4. R. Penrose, *Shadows of the Mind: A Search for the Missing Science of Consciousness*. Oxford University Press, 1994.

5. J. R. Searle, "Minds, brains, and programs," *Behavioral and Brain Sciences*, vol. 3, no. 3, pp. 417–457, 1980.

6. S. Piker, *How the Mind Works*. Norton Press, 1999.

7. T. H. Cormen, C. E. Leiserson, R. L. Rivest, and C. Stein, *Introduction to Algorithms* (3rd ed.). MIT Press and McGraw-Hill, 2009.

8. C. S. Sherrington, *The Integrative Action of the Nervous System*. Yale University Press, 1947.

9. D. Marr and T. Poggio, "Cooperative computation of stereo disparity," *Science*, vol. 194, pp. 283–287, 1976.

10. A. Newell, *Unified Theories of Cognition*. Harvard University Press, 1990.

11. V. B. Mountcastle, "Modality and topographic properties of single neurons of cat's somatic sensory cortex," *Journal of Neurophysiology*, vol. 20, pp. 408–434, 1957.

12. L. Valiant, *Circuits of the Mind*. Oxford University Press, 1994.

13. G. Buzsáki, *Rhythms of the Brain*. Oxford University Press, 2006.

14. W. Singer, *Dynamic Coordination in the Brain*, ch. 11. MIT Press, 2010.

15. S. Lloyd, *Programming the Universe: A Quantum Computer Scientist Takes On the Cosmos*. Alfred A. Knopf, 2006.

16. H. T. Segelmann, "Computation beyond the Turing limit," *Science*, vol. 268, pp. 545–548, 1995.

17. E. Nagel and J. R. Newman, *Gödel's Proof*. New York University Press, 2001.

18. D. R. Hofstadter, *Gödel, Escher, Bach: An Eternal Golden Braid*. Penguin, 2000.

19. E. Schrödinger, *What Is Life? Mind and Matter*. Cambridge University Press, 1944.

20. R. Dawkins, *The Selfish Gene*. Oxford University Press, 1976.

21. M. Manto, J. M. Bower, A. B. Conforto, J. M. Delgado-García, S. N. da Guarda, M. Gerwig, C. Habas, N. Hagura, R. B. Ivry, P. Mariën, M. Molinari, E. Naito, D. A. Nowak, N. Oulad Ben Taib, D. Pelisson, C. D. Tesche, C. Tilikete, and D. Timmann, "Consensus paper: Roles of the cerebellum in motor control—the diversity of ideas on cerebellar involvement in movement," *Cerebellum*, vol. 11, 457–487, 2011.

22. S. M. Morton and A. J. Bastian, "Prism adaptation during walking generalizes to reaching and requires the cerebellum," *Journal of Neurophysiology*, vol. 92, pp. 2497–2509, 2004.

23. W. Schultz, P. Dayan, and P. R. Montague, "A neural substrate of prediction and reward," *Science*, vol. 275, pp. 1593–1599, 1997.

24. L. W. Swanson, *Brain Architecture*. Oxford University Press, 2012.

25. S. M. Sherman, *Handbook of Neuroscience for the Behavioral Sciences*, vol. 1, pp. 201–223. John Wiley & Sons, 2009.

26. D. H. Ballard, M. M. Hayhoe, P. K. Pook, and R. P. N. Rao, "Deictic codes for the embodiment of cognition," *Behavioral and Brain Sciences*, vol. 20, 723–767, 1997.

27. K. Kveraga, A. Ghuman, K. Kassam E. Aminoff, M. Hämäläinen, M. Chaumon, and M. Bar. M. "Early onset of neural synchronization in the contextual association network." *Proceedings of the National Academy of Sciences*, vol. 108, pp. 3389–3394, 2011.

28. A. L. Yarbus, *Eye Movements and Vision*. Plenum Press, 1967.

29. R. R. Llinás, *I of the Vortex*. MIT Press, 2001.

30. J. Hawkins, *On Intelligence*. Times Books, 2004.

31. M. Kimura and A. M. Graybiel, eds., *Functions of the Cortico-Basal Ganglia Loop*. Springer-Verlag, 1995.

32. J. D. E. Gabrieli, J. Singh, G. T. Stebbins, and C. G. Goetz, "Reduced working-memory span in Parkinson's disease: Evidence for the role of a frontostriatal system in working and strategic memory," *Neuropsychology*, vol. 10, pp. 322–332, 1996.

33. E. B. Baum, *What Is Thought?* MIT Press, 2004.

34. W. B. Scoville and B. Milner, "Loss of recent memory after bilateral hippocampal lesions," *Journal of Neurology, Neurosurgery & Psychiatry*, vol. 20, pp. 11–21, 1957.

35. D. Ji and M. A. Wilson, "Coordinated memory replay in the visual cortex and hippocampus during sleep," *Nature Neuroscience*, vol. 10, pp. 100–107, 2007.

36. A. Karni, D. Tanne, B. S. Rubenstein, J. Askenasy, and D. Sagi, "Dependence on REM sleep of overnight improvement of a perceptual skill," *Science*, vol. 265, pp. 603–604, 1994.

37. B. Kuipers, "Drinking from the firehose of experience," *Artificial Intelligence in Medicine*, vol. 44, pp. 155–170, 2008.

38. J. E. Lishman and N. A. Otmakhova, "Storage, recall, and novelty detection of sequences by the hippocampus: Elaborating on the Socratic model to account for normal and aberrant effects of dopamine," *Hippocampus*, vol. 11, pp. 551–568, 2001.

39. J. E. Lisman and A. A. Grace, "The hippocampal-VTA loop: Review controlling the entry of information into long-term memory," *Neuron*, vol. 46, pp. 703–713, 2005.

40. J. A. Ainge, M. Tamosiunaite, F. Woergoetter, and P. A. Dudchenko, "Hippocampal CA1 place cells encode intended destination on a maze with multiple choice points," *Journal of Neuroscience*, vol. 27, pp. 9769–9779, 2007.

41. E. Tulving, "Episodic memory: From mind to brain," *Annual Review of Psychology*, vol. 53, pp. 1–25, 2002.

42. M. T. Rogan, U. V. Stäubli, and J. E. LeDoux, "Fear conditioning induces associative long-term potentiation in the amygdala," *Nature*, vol. 390, pp. 604–607, 1997.

43. J. M. Allman, *Evolving Brains*. W. H. Freeman, 1998.

44. F. Rieke, R. de Ruyter van Steveninck and W. Bialek, *Spikes: Exploring the Neural Code*. MIT Press, 1997.

45. P. Dayan and L. F. Abbott, *Theoretical Neuroscience: Computational and Mathematical Modeling of Neural Systems*. MIT Press, 2001.

46. D. Serratt, B. Graham, A. Gilles, and D. Willshaw, *Principles of Computational Modeling in Neuroscience*. Cambridge University Press, 2011.

47. J. O'Keefe, *The Hippocampus as a Cognitive Map*. Clarendon Press, 1978.

48. J. O'Keefe and L. Nadel, "The hippocampus as a cognitive map," *Behavioral and Brain Sciences*, vol. 2, pp. 487–533, 1979.

49. M. E. Hasselmo, *How We Remember: Brain Mechanisms of Episodic Memory*. MIT Press, 2012.

50. S. Sreenivasan and I. Fiete, "Grid cells generate an analog error-correcting code for singularly precise neural computation," *Nature Neuroscience*, vol. 14, pp. 1330–1337, 2011.

51. C. Reid and J-M. Alonso, "Specificity of monosynaptic connections from thalamus to visual cortex." *Nature*, vol. 378, pp. 281–283, 1995.

52. B. Olshausen and D. Field, "Emergence of simple-cell receptive field properties by learning a sparse code for natural images," *Nature*, vol. 381, pp. 607–609, 1996.

53. S. Mallat and Z. Zhang, "Matching pursuit with time-frequency dictionaries," *IEEE Transactions in Signal Processing*, vol. 41, pp. 3397–3415, 1993.

54. M. Rehn and F. T. Sommer, "A network that uses few active neurones to code visual input predicts the diverse shapes of cortical receptive fields," *Journal of Computational Neuroscience*, vol. 22, pp. 135–146, 2007.

55. P. Lennie, "The cost of cortical computation," *Current Biology*, vol. 13, pp.493–497, 2003.

56. T. Gollisch and M. Meister, "Rapid neural coding in the retina with relative spike latencies," *Science*, vol. 319, pp. 1108–1111, 2008.

57. D. H. Ballard and J. M. F. Jehee, "Dual roles for spike signaling in cortical neural populations," *Frontiers in Computational Neuroscience*, vol. 5, p. 22, 2011.

58. C. Börgers , G. Talei Franzesi, F. E. N. LeBeau , E. S. Boyden, N. J. Kopell, "Minimal size of cell assemblies coordinated by gamma oscillations." *PLoS Computational Biology*, vol. 8, p. e1002362, 2012.

59. T. Womelsdorf, B. Limac, M. Vinck, R. Oostenveld, W. Singer, S. Neuenschwander, and P. Fries, "Orientation selectivity and noise correlation in awake monkey area V1 are modulated by the gamma cycle," *Proceedings of the National Academy of Sciences USA*, vol. 109, pp. 4302–4307, 2012.

60. *M. Vinck, B. Lima, T. Womelsdorf, R. Oostenveld, W. Singer, S. Neuenschwander, and P. Fries, "Gamma-phase shifting in awake monkey visual cortex."* Journal of Neuroscience, *vol. 30, pp. 1250–1257, 2010.*

61. H. K. Turesson , N. K. Logothetis, and K. L. Hoffman, "Category-selective phase coding in the superior temporal sulcus," *Proceedings of the National Academy of Sciences*, epub ahead of print, 2012.

62. Adrian, E. D. *The Basis of Sensation.* Christophers, 1928.

63. J. Y. Lettvin, H. R. Maturana, W. S. McCulloch, andW. H. Pitts,"What the frog's eye tells the frog's brain," *Proceedings of the I.R.E.* vol. 47, pp. 1940–1959, 1959.

64. A. Cobas and M. Arbib, "Prey-catching and predator-avoidance in frog and toad: Defining the schemas," *Journal of Theoretical Biology*, vol. 157, pp. 271–304, 1992.

65. D. B. Chklovskii, T. Schkorski, and C. F. Stevens, "Wiring optimization in cortical circuits," *Neuron*, vol. 34, pp. 341–347, 2002.

66. A. L. Hodgkin and A. F. Huxley, "A quantitative description of membrane current and its application to conduction and excitation in nerve," *Journal of Physiology*, vol. 117, pp. 500–544, 1952.

67. H. Markram, J. L. Ubke, M. Frotscher, and B. Sakmann, "Regulation of synaptic efficacy by coincidence of postsynaptic APs and EPSPs," *Science*, vol. 275, pp. 213–215, 1997.

68. G. Bi and M. Poo, "Synaptic modifications in cultured hippocampal neurons: Dependence on spike timing, synaptic strength, and postsynaptic cell type," *Journal of Neuroscience*, vol. 18, pp. 10464–10472, 1998.

69. J. A. Hertz, A. S. Krogh, and R. G. Palmer, *Introduction to the Theory of Neural Computation*. Addison-Wesley, 1991.

70. J. J. Hopfield and D. W. Tank, "Computing with neural circuits: A model," *Science*, vol. 233, pp. 625–633, 1986.

71. D. O. Hebb, *The Organization of Behavior*. Wiley, 1949.

72. J. Pearl, *Probabilistic Reasoning in Intelligent Systems: Networks of Plausible Inference*. Morgan Kaufmann, 1988.

73. G. E. Hinton and R. R. Salakhutdinov, "Reducing the dimensionality of data with neural networks," *Science*, vol. 313, pp. 504–507, 2006.

74. R. B. Tootell, M. S. Silverman, E. Switkes, and R. L. D. Valois, "Deoxyglucose analysis of retinotopic organization in primate striate cortex," *Science*, vol. 218, pp. 902–904, 1982.

75. K. Ohki, S. Chung, Y. H. Ch'ng, P. Kara, and R. C. Reid, "Functional imaging with cellular resolution reveals precise microarchitecture in visual cortex," *Nature*, vol. 433, pp. 597–603, 2005.

76. G. G. Blasdel, "Differential imaging of ocular dominance and orientation selectivity in monket striate cortex," *Journal of Neuroscience*, vol. 12, pp. 2115–3138, 1992

77. M. J. Swain and D. H. Ballard, "Color indexing," *International Journal of Computer Vision*, vol. 7, pp. 11–32, 1991.

78. M. Mishkin and L. G. Ungerleider, "Contribution of striate inputs to the visuospatial functions of parieto-preoccipital cortex in monkeys," *Behavioral Brain Research*, vol. 6, pp. 57–77, 1982.

79. M. A. Goodale and A. D. Milner, "Separate visual pathways for perception and action," *Trends in Neurosciences*, vol. 15, pp. 20–25, 1992.

80. S. Grossberg, "A solution of the figure-ground problem for biological vision," *Neural Networks*, vol. 6, pp. 463–484, 1993.

81. H. Zhou, H. S. Friedman, and R. von der Heydt, "Coding of border ownership in monkey visual cortex," *Journal of Neuroscience*, vol. 20, pp. 6594–6611, 2000.

82. K. Tsunoda, Y. Yamane, M. Nishizaki, and M. Tanifuji, "Complex objects are represented in macaque inferotemporal cortex by the combination of feature columns," *Nature Neuroscience*, vol. 4, pp. 832–838, 2001.

83. W. A. Freiwald and D. Y. Tsao, "Functional compartmentalization and viewpoint generalization within the macaque face-processing system," *Science*, vol. 330, pp. 845–851, 2010.

84. Freiwald, D. Y. Tsao and M.S. Livinstone, "A face feature space in the macaque temporal lobe.," *Nature Neuroscience*, 12 1187-1196, 2009.

85. A. G. Huth, S. Nishimoto, A. T. Vu, and J. L. Gallant, "A continuous semantic space describes the representation of thousands of object and action categories across the human brain," *Neuron*, vol. 76, pp. 210–224, 2012.

86. R. P. N. Rao and D. H. Ballard, "Predictive coding in the visual cortex: A functional interpretation of some extra-classical receptive-field effects," *Nature Neuroscience*, vol. 2, no. 1, pp. 79–87, 1999.

87. T. S. Lee and D. Mumford, "Hierarchical Bayesian inference in the visual cortex," *Journal of the Optical Society of America*, vol. 20, pp. 1434–1447, 2003.

88. Y. Weiss, E. P. Simoncelli, and E. H. Adelson, "Motion illusions as optimal percepts," *Nature Neuroscience*, vol. 5, pp. 598–603, 2002.

89. H. Lee, R. Grosse, R. Ranganath, and A. Y. Ng, "Convolutional deep belief networks for scalable unsupervised learning of hierarchical representations," in *Proceedings of the 26th International Conference on Machine Learning*, 2009.

90. F. Anselmi, J. Z. Leibo, L. Rosasco, J. Mutch, A. Tacchetti, and T. Poggio, "Unsupervised learning of invariant representations with low sample complexity: themagic of sensory cortex or a new framework for machine learning?" MIT Center for Brains, Minds and Machines, Memo 001, 2014.

91. L. Isik , E. M. Meyers , J. Z. Leibo, and T. Poggio, "The dynamics of invariant object recognition in the human visual system," *Journal of Neurophysiology*, vol. 111, pp. 91–102, 2013.

92. D. L. K. Yamins, H. Hong, C. F. Cadieu, E. A. Solomon, D. Seibert, J. J. DiCarlo, "Performance-optimized hierarchical models predict neural responses in higher visual cortex." *Proceedings of the National Academy of Sciences USA*, [abstract] 2014.

93. J. J. DiCarlo, D. Zoccolan, and N. C. Rust, "How does the brain solve visual object recognition?" *Neuron*, vol. 73, pp. 415–434, 2012.

93a. S. Wagner, E. Winner, D. Cicchetti, and H. Gardner, "'Metaphorical' mapping by human infants," *Child Development*, vol. 52, pp. 728–731, 1981.

93b. J. B. Tenenbaum, C. Kemp, T. L. Griffiths, and N. D. Goodman, "How to grow a mind: Statistics, structure and abstraction," *Science*, vol. 331, p. 1279, 2011.

93c. C. Kemp and J. B. Tenenbaum, "The discovery of structural form," *Proceedings of the National Academy of Sciences U.S.A.*, vol. 105, p. 10687, 2008.

94. J. C. Houk, J. T. Buckingham, and A. G. Barto,"Models of the cerebellum and motor learning," *Behavioral and Brain Sciences*, vol. 19, no. 3, 1996.

95. J. C. Houk and S. P. Wise, "Feature article: Distributed modular architectures linking basal ganglia, cerebellum, and cerebral cortex: Their role in planning and controlling action," *Cerebral Cortex*, vol. 5, no. 2, pp. 95–110, 1995.

96. S. N. Haber, "The primate basal ganglia: Parallel and integrative networks," *Journal of Chemical Neuroanatomy*, vol. 26, pp. 317–330, 2003.

97. A. M. Graybiel, "Habits, rituals, and the evaluative brain," *Annual Review of Neuroscience*, vol. 31, pp. 359–387, 2008.

98. A. Graybiel, "Neurotransmitters and neuromodulators in the basal ganglia," *Trends in Neurosciences*, vol. 13, pp. 244–254, 1990.

99. R. A. Rescorla and A. R. Wagner, *Classical Conditioning II: Current Research and Theory*. Appleton Century Crofts, 1972.

100. Y. Kubota, J. Liu, D. Hu, W. E. DeCoteau, U. T. Eden, A. C. Smith, and A. M. Graybiel, "Stable encoding of task structure coexists with flexible coding of task events in sensorimotor striatum," *Journal of Neurophysiology*, vol. 102, pp. 2142–2160, 2009.

101. A. M. Graybiel, "The basal ganglia and chunking of action repertoires," *Neurobiology of Learning and Memory*, vol. 70, pp. 119–136, 1998.

102. J. C. Houk, C. Bastianen, D. Fansler, A. Fishbach, D. Fraser,P. J. Reber, S. A. Roy, and L. S. Simo, "Action selection and refinement in subcortical loops through basal ganglia and cerebellum," *Philosophical Transactions of the Royal Society B* doi:10.1098/rstb.2007.2063, 2007.

103. K. Doya and M. Kimura, "The basal ganglia and the encoding of value," in *Neuroeconomics: Decision Making and the Brain*, ed. P. W. Glimcher and E. Fehr, pp. 407–414. Academic Press, 2009.

104. N. Schweighofer, M. Bertin, K. Shishida, Y. Okamoto, S. C. Tanaka, S. Yamawaki, and K. Doya, "Low-serotonin levels increase delayed reward discounting in humans," *Journal of Neuroscience*, vol. 28, pp. 4528–4532, 2008.

105. G. Tesauro, "Temporal difference learning and td-gammon," *Communications of the ACM*, vol. 8, pp. 58–68, 1995.

106. D. E. Rumelhart, G. E. Hinton, and R. J. Williams, "Learning representations by back-propagating errors," *Nature*, vol. 323, pp. 533–536, 1986.

107. C. F. Chabris, "Cognitive and Neuropsychological Mechanisms of Expertise: Studies with Chess Masters," PhD dissertation in Psychology, Harvard University, 1999.

108. J. K. O'Reagan, "Solving the 'real' mysteries of visual perception: The world as an outside memory," *Canadian Journal of Psychology/Revue Canadienne de Psychologie*, vol. 46, pp. 461–488, 1992.

109. S. Ullman, "Visual routines," *Cognition*, vol. 18, pp. 97–157, 1984.

110. R. Feifer and J. Bongard, *How the Body Shapes the Way We Think: A New View of Intelligence*. MIT Press, 2007.

111. M. A. Fischler and O. Firschein, *Intelligence: The Eye, the Brain, and the Computer*. Addison Wesley, 1987.

112. K. Rayner, "Eye movements in reading and information processing: 20 years of research," *Psychological Bulletin*, vol. 124, pp. 372–422, 1998.

113. C. Chabris and D. Simons, *The Invisible Gorilla*. Crown Press, 2010.

114. B. Hölldobler and E. Wilson, *The Ants*. Harvard University Press, 1990.

115. M. A. Just and P. A. Carpenter, "Eye fixations and cognitive processes," *Cognitive Psychology*, vol. 8, pp. 441–480, 1976.

116. S. M. Kosslyn, *Image and Brain*. MIT Press, Bradford Books, 1987.

117. T. Serre, L. Wolf, S. Bileschi, M. Riesenhuber, and T. Poggio, "Robust object recognition with cortex-like mechanisms," *IEEE Transactions on Pattern Analysis and Machine Intelligence*, vol. 29, pp. 411–426, 2007.

118. A. M. Treisman, "A feature-integration theory of attention," *Cognitive Psychology*, vol. 12, pp. 97–136, 1980.

119. J. Wolfe, K. Cave, and S. Franzel, "Guided search: An alternative to the feature integration model for visual search," *Journal of Experimental Psychology: Human Perception and Performance*, vol. 15, pp. 419–433, 1989.

120. G. Zelinsky, R. Rao, M. Hayhoe, and D. Ballard, "Eye movements reveal the spatio-temporal dynamics of visual search," *Psychological Science*, vol. 8, pp. 448–453, 1997.

121. M. Eckstein, "The lower visual search efficiency for conjunctions is due to noise not serial attentional processing," *Psychological Science*, vol. 9, pp. 111–118, 1998.

122. J. Palmer, P. Vergese, and M. Pavel, "The psychophysics of visual search," *Vision Research*, vol. 40, pp. 1227–1268, 2000.

123. J. Najemnik and W. S. Geisler, "Optimal eye movement strategies in visual search," *Nature*, vol. 434, pp. 387–391, 2005.

124. J. Schmidt and G. J. Zelinsky, "Search guidance is proportional to the categorical specificity of a target cue," *Quarterly Journal of Experimental Psychology*, vol. 6, no. 10, pp. 1904–1914, 2009.

125. J. Triesch, D. H. Ballard, M. M. Hayhoe and B. T. Sullivan, "What you see is what you need," *Journal of Vision*, 3(1),9, 2003.

126. J. A. Droll, M. M. Hayhoe, T. Triesch and B. T. Sullivan, "Task Demands Control Acquisition and Storage of Visual Information," *Journal of Experimental Psychology: Human Perception and Performance*, 31, 1416-1438, 2005.

127. C. Koch and S. Ullman, "Shifts in selective visual attention: Towards the underlying neural circuitry," *Human Neurobiology*, vol. 4, pp. 219–227, 1985.

128. N. D. B. Bruce and J. K. Tsotsos, "Saliency, attention, and visual search: An information theoretic approach," *Journal of Vision*, vol. 9, pp. 1–24, 2009.

129. C. A. Rothkopf, D. H. Ballard and M. M. Hayhoe, "Task and context determine where you look," *Journal of Vison*, vol. 7, p. 16. 2007.

130. A. Oliva and A. Torralba, "The role of context in object recognition," *Trends in Cognitive Science*, pp. 520–527, 2007.

131. A. Borji and L. Itti, "State-of-the-art in visual attention modeling," *IEEE Transactions on Pattern Analysis and Machine Intelligence*, vol. 35, pp. 185–207, 2013.

132. C. D. Gilbert and W. Li, "Top-down influences on visual processing," *Nature Reviews Neuroscience*, vol. 14, pp. 350–363, 2013.

133. M. Kinoshita, C. D. Gilbert, A. Das, "Optical Imaging of Contextual Interactions in V1 of the Behaving Monkey." *Journal of Neurophysiology*,102, 1930-1944 2009.

134. R. P. N. Rao, "An optimal estimation approach to visual perception and learning," *Vision Research*, vol. 39, pp. 1963–1989, 1999.

135. Y. Huang and R. P. N. Rao, "Reward optimization in the primate brain: a probabilistic model of decision making under uncertainty." *PLoS ONE*, vol. 8, p. e53344, 2013.

136. A. Feldman, "Once more on the equilibrium-point hypothesis (lambda model) for motor control," *Journal of Motor Behavior*, vol. 18, pp. 17–54, 1986.

137. A. Feldman and M. Levin, "The origin and use of positional frames of reference in motor control," *Behavioral and Brain Sciences*, vol. 18, no. 4, pp. 723–744, 1995.

138. F. A. Mussa-Ivaldi, "Modular features of motor control and learning," *Current Opinion in Neurobiology*, vol. 9, pp. 713–717, 1999.

139. F. A. Mussa-Ivaldi and E. Bizzi, "Motor learning through the combination of primitives," *Philosophical Transactions of the Royal Society of London, Series B*, vol. 355, pp. 1755–1769, 2000.

140. G. Torres-Oviedo and L. H. Ting, "Subject-specific muscle synergies in human balance control are consistent across different biomechanical contexts," *Journal of Neurophysiology*, vol. 103, pp. 3084–3098, 2010.

141. R. N. Jazar, *Theory of Applied Robotics: Kinematics, Dynamics, and Control*. Springer, 2005.

142. L. Squire, D. Berg, F. E. Bloom, S. duLac, A. Ghosh, and N. C. Spitzer, *Fundamental Neuroscience*. Academic Press, 2012.

143. F. Crevecoeur and S. H. Scott, "Priors engaged in long-latency responses to mechanical perturbations suggest a rapid update in state estimation," *PLoS Computational Biology*, vol. 9, no. 8, e1003177.

144. S. F. Giszter, F. A. Mussa-Ivaldi, and E. Bizzi, "Convergent force fields organized in the frog's spinal cord," *Journal of Neuroscience*, vol. 13, pp. 461–491, 1993.

145. A. Frigon, "Central pattern generators of the mammalian spinal cord," *Neuroscientist*, vol. 18, no. 1, pp. 56–69, 2011.

146. S. H. Collins, M. Wisse, and A. Ruina, "A three-dimensional passive-dynamic walking robot with two legs and knees," *International Journal of Robotics Research*, vol. 20, pp. 607–615, 2001.

147. T. Geijtenbeek, M. van de Panne, and A. F. van der Stappen, "Flexible muscle-based control for bipedal creatures," *ACM Transactions on Computer Graphics*, vol. 32, 2013.

148. C. T. John, F. C. Anderson, J. S. Higginson, and S. L. Delp, "Stabilisation of walking by intrinsic muscle properties revealed in a three-dimensional muscle-driven simulation." *Computer Methods in Biomechanics and Biomedical Engineering*, DOI: 10.1080/10255842.2011.627560, 2011.

149. P. Lennie, "Single units and visual cortical organization," *Perception*, vol. 27, pp. 889–935, 1998.

150. M. Iacoboni, I. Molnar-Szakacs, V. Gallese, G. Buccino, J. C. Mazziotta, and G. Rizzolatti, "Grasping the intentions of others with one's own mirror neuron system," *PLOS Biology*, 2005.

151. M. S. A. Graziano and T. N. Aflalo, "Mapping behavioral repertoire onto the cortex," *Neuron*, vol. 56, pp. 239–251, 2007.

152. A. Afshar, G. Santhanam, B. M. Yu, S. I. Ryu, M. Sahani, and K. V. Shenoy, "Single-trial neural correlates of arm movement preparation," *Neuron*, vol. 71, pp. 555–564, 2011.

153. J. Diedrichsen, R. Shadmehr, and R. B. Ivry, "The coordination of movement: Optimal feedback control and beyond," *Trends in Cognitive Sciences*, vol. 14, pp. 31–39, 2009.

154. R. Shadmehr and S. Mussa-Ivaldi, *Biological Learning and Control: How the Brain Builds Representations, Predicts Events, and Makes Decisions.* MIT Press, 2011.

155. S. H. Scott, "Inconvenient truths about neural processing in primary motor cortex," *Journal of Physiology*, vol. 586.5, pp. 1217–1224, 2008.

156. S. H. Scott, "The computational and neural basis of voluntary motor control and planning," *Trends in Cognitive Sciences*, vol. 11, pp. 541–549, 2012.

157. J. Adams, "A closed-loop theory of motor control learning," *Journal of Motor Behavior*, vol. 3, pp. 110–150, 1971.

158. M. Kawato, "Internal models for motor control and trajectory planning," *Current Opinion in Neurobiology*, vol. 9, pp. 718–727, 1999.

159. S. L. Hooper, *Central Pattern Generators.* Wiley Online Library, DOI: 10.1038/npg.els.0000032, 2001.

160. S. Chen, D. Donoho, and M. Saunders, "Atomic decomposition by basis pursuit," *SIAM Journal of Scientific Computing*, vol. 20, no. 1, pp. 33–61, 1999.

161. I. Daubechies, "Time-frequency localization operators: A geometric phase space approach," *IEEE Transactions on Information Theory*, vol. 34, pp. 605–612, 1988.

162. N. Bernstein, *The Coordination and Regulation of Movement.* Pergamon, 1967.

163. A. Feldman, "Functional tuning of the nervous system with control of movement or maintenance of a steady posture. II. Controllable parameters of the muscle," *Biophysics*, vol. 11, pp. 565–578, 1966.

164. A. Feldman, "Functional tuning of the nervous system with control of movement or maintenance of a steady posture. III. Mechanographic analysis of execution by man of the simplest motor tasks," *Biophysics*, vol. 11, pp. 766–775, 1966.

165. M. H. Raibert, "Symmetry in running," *Science*, vol. 231, pp. 1292–1294, 1986.

166. M. H. Raibert and J. Hodgins, "Animation of dynamic legged locomotion," in *Proceedings of SIGGRAPH*, ed. T. W. Sederberg, pp. 349–358, 1991.

167. M. Gunther and H. Ruder, "Synthesis of two-dimensional human walking," *Biological Cybernetics*, vol. 8, pp. 89–106, 2003.

168. X. Gu and D. H. Ballard, "An equilibrium point based model unifying movement control in humanoids," in *Robotics: Science and Systems*, 2006.

169. E. B. Torres and D. Zipser, "Reaching to grasp with a multi-jointed arm. I. Computational model," *Journal of Neurophysiology*, vol. 88, no. 5, pp. 2355–2367, 2002.

170. H. Gomi and M. Kawato, "Equilibrium-point control hypothesis examined by measured arm-stiffness during multi-joint movement," *Science*, vol. 272, pp. 117–120, 1996.

171. A. Polit and E. Bizzi, "Processes Controlling Arm Movements in Monkeys," *Science*, 201,1235-1237, 1978

172. A. J. Ijspeert, J. Nakanishi, H. Hoffman, P. Pastor, and S. Schaal, "Dynamical movement primitives: Learning attractor models for motor behaviors," *Neural Computation*, vol. 25, pp. 328–373, 2013.

173. S. Delp and P. Loan, "A computational framework for simulating and analyzing human and animal movement," *Computing in Science and Engineering*, vol. 2, pp. 46–55, 2000.

174. S. L. Delp, F. C. Anderson, A. S. Arnold, P. Loan, A. Habib, C. T. John, E. Guendelman, and D. G. Thelen, "OpenSim: Open-source software to create and analyze dynamic simulations of movement," *IEEE Transactions on Biomedical Engineering*, vol. 11, pp. 1940–1950, 2007.

175. A. Erdemir, S. McLean, W. Herzog, and A. J. van den Bogert, "Model-based estimation of muscle forces exerted during movements," *Clinical Biomechanics*, vol. 22, pp. 131–154, 2007.

176. V. B. Zordan, A. Majkowska, B. Chiu, and M. Fast, "Dynamic response for motion capture animation," in *ACM SIGGRAPH 2005 Papers*, pp. 697–701. ACM, 2005.

177. Y. Tassa, T. Erez, and E. Todorov, "Synthesis and stabilization of complex behaviors through online trajectory optimization," in *IROS*, pp. 4906–4913. 2012.

178. J. M. Cooper and D. H. Ballard, "Physically-based humanoid animation from human demonstration data," in *SIGGRAPH 2012*, submitted 2012.

179. F. Faure, G. Debunne, M.-P. Cani-Gascuel, and F. Multon, "Dynamic analysis of human walking," in *Proceedings of the Eurographics Workshop on Animation and Simulation*, 1997.

180. J. Cooper, *Analysis and Synthesis of Bipedal Humanoid Movement: A Physical Simulation Approach*. PhD thesis, University of Texas at Austin, 2013.

181. E. Tolman, "Cognitive maps in rats and men," *Psychological Review*, vol. 55, pp. 189–208, 1948.

182. U. Neisser, *Cognitive Psychology*. Appleton-Century-Crofts, 1967.

183. R. Sun, ed., *Cognition and Multi-Agent Interaction*, pp. 79–99. Cambridge University Press, 2006.

184. J. Anderson, *The Architecture of Cognition.* Harvard University Press, 1983.

185. P. Langley and D. Choi, "Learning recursive control programs from problem solving," *Journal of Machine Learning Research,* vol. 7, pp. 493–518, 2006.

186. J. E. Laird, A. Newell, and P. S. Rosenblum, "Soar: An architecture for general intelligence," *Artificial Intelligence,* vol. 33, pp. 1–64, 1987.

187. S. Ritter, J. R. Anderson, M. Cytrynowicz, and O. Medvedeva, "Authoring content in the pat algebra tutor," *Journal of Interactive Media in Education,* vol. 98, no. 9, 1998.

188. L. M. Trick and Z. W. Pylyshyn, "Why are small and large numbers enumerated differently? A limited-capacity preattentive stage in vision," *Psychological Review,* vol. 101, pp. 80–102, 1994.

189. M. A. Arbib, *The Handbook of Brain Theory and Neural Networks,* pp. 830–834. MIT Press, 1998.

190. E. Ruthruff, H. E. Pashler, and E. Hazeltine, "Dual-task interference with equal task emphasis: Graded capacity-sharing or central postponement?" *Attention Perception and Psychophysics,* vol. 65, pp. 801–816, 2003.

191. R. Brooks, "A robust layered control system for a mobile robot," *IEEE Journal of Robotics and Automation,* vol. 2, no. 1, 1986.

192. R. J. Firby, R. E. Kahn, P. N. Prokopowicz, and M. J. Swain, "An architecture for vision and action," in *International Joint Conference on Artificial Intelligence,* pp. 72–79, 1995.

193. J. J. Bryson and L. A. Stein, "Modularity and design in reactive intelligence," in *International Joint Conference on Artificial Intelligence,* 2001.

194. R. Arkin, *Behavior Based Robotics.* MIT Press, 1998.

195. R. P. Bonasso, R. J. Firby, E. Gat, D. Kortenkamp, D. P. Miller, and M. G. Slack, "Experiences with an architecture for intelligent reactive agents," *Journal of Experimental and Theoretical Artificial Intelligence,* vol. 9, pp. 237–256, 1997.

196. R. Pfeifer and C. Scheier, *Understanding Intelligence.* Bradford Books, 1999.

197. F. Adams, "Embodied cognition," *Phenomenology and Cognition,* vol. 9, pp. 619–628, 2010.

198. L. Shapiro, *Embodied Cognition.* Routledge, 2011.

199. J. K. O'Regan and A. Noe, "A sensorimotor approach to vision and visual consciousness," *Behavioral and Brain Sciences,* vol. 24, pp. 939–973, 2001.

200. A. Noe, *Action in Perception.* MIT Press, 2005.

201. A. Clark, "An embodied model of cognitive science?" *Trends in Cognitive Sciences,* vol. 3, pp. 345–351, 1999.

202. D. K. Roy and A. P. Pentland, "Learning words from sights and sounds: A computational model," *Cognitive Science*, vol. 26, pp. 113–146, 2002.

203. C. Yu and D. Ballard, "A multimodal learning interface for grounding spoken language in sensorimotor experience," *ACM Transactions on Applied Perception*, vol. 1, pp. 57–80, 2004.

204. D. H. Ballard, M. M. Hayhoe, P. Pook, and R. Rao, "Deictic codes for the embodiment of cognition," *Behavioral and Brain Sciences*, vol. 20, pp. 723–767, 1997.

205. F. J. Vareala, E. Thompson, and E. Rosch, *The Embodied Mind: Cognitive Science and Human Experience*. MIT Press, 1991.

206. J. Stewart, O. Gapenne, and E. D. Paolo, eds., *Enaction: Toward a New Paradigm for Cognitive Science*. MIT Press, 2010.

207. M. Nordfang, M. Dyrholm, and C. Bundesen, "Identifying bottom-up and top-down components of attentional weight by experimental analysis and computational modeling," *Journal of Experimental Psychology: General*, vol. 142, pp. 510–536, 2012.

208. S. Hurley, "The shared circuits model (SCM): How control, mirroring, and simulation can enable imitation, deliberation, and mindreading," *Behavioral and Brain Sciences*, vol. 31, pp. 1–22, 2008.

209. J. R. Anderson, *Cognitive Psychology and its Implications*, 7th ed. Worth, 2010.

210. S. E. Petersen and M. I. Posner, "The attention system of the human brain: 20 years after,"*Annual Reviews of Neuroscience*, vol. 35, pp. 73–89, 2012.

211. J. Fan, B. D. McCandliss, J. Fossella, J. I. Flombaum, and M. I. Posner, "The activation of attentional networks," *NeuroImage*, vol. 26, pp. 471–479, 2005.

212. M. I. Posner and M. K. Rothbart, "Research on attention networks as a model for the integration of psychological science," *Annual Review of Psychology*, vol. 58, pp. 1–23, 2007.

213. A. Baddeley, "Working memory," *Science*, vol. 255, pp. 556–559, 1992.

214. S. J. Luck and E. K. Vogel, "The capacity of visual working memory for features and conjunctions," *Nature*, vol. 390, pp. 279–281, 1997.

215. L. Itti and C. Koch, "A saliency based search mechanism for overt and covert shifts of attention," *Vision Research*, vol. 40, no. 10–12, pp. 1489–1506, 2000.

216. L. Itti, "Quantifying the contribution of low-level saliency to human eye movements in dynamic scenes," *Visual Cognition*, vol. 12, no. 6, pp. 1093–1123, 2005.

217. A. Torralba, A. Oliva, M. Castelhano, and J. M. Henderson, "Contextual guidance of attention in natural scenes: The role of global features on object search," *Psychological Review*, vol. 113, no. 4, pp. 766–786, 2006.

218. S. Ullman, "Visual routines," *Cognition*, vol. 18, pp. 97–157, 1984.

219. P. R. Roelfsema, P. S. Khayat, and H. Spekreijse, "Subtask sequencing in the primary visual cortex," *Proceedings of the National Academy of Sciences USA*, vol. 100, pp. 5467–5472, 2003.

220. M. Hayhoe, A. Shrivastava, R. Mruczek, and J. Pelz, "Visual memory and motor planning in a natural task," *Journal of Vision*, vol. 3, pp. 49–63, 2003.

221. W. Yi and D. H. Ballard, "Recognizing behavior in hand-eye coordination patterns," *International Journal of Humanoid Robotics*, vol. 6, pp. 337–359, 2009.

222. D. Terzopoulos and T. F. Rabie, "Animat vision: Active vision in artificial animals," *Videre: Journal of Computer Vision Research*, vol. 1, no. 1, pp. 2–19, 1997.

223. P. Stone. *Layered Learning in Multiagent Systems: A Winning Approach to Robotic Soccer.* MIT Press, 2000.

224. N. Sprague, D. Ballard, and A. Robinson, "Modeling embodied visual behaviors," *ACM Transactions on Applied Perception*, vol. 4, 2007.

225. R. E. Suri and W. Schultz, "Temporal difference model reproduces anticipatory neural activity," *Neural Computation*, vol. 13, pp. 841–862, 2001.

226. L. Maloney and M. Landy, "When uncertainty matters: The selection of rapid goal-directed movements [abstract]," *Journal of Vision* (to appear).

227. C. J. C. H. Watkins and P. Dayan, "Q-learning," *Machine Learning Journal*, vol. 8, pp. 279–292, 1992.

228. M. F. Land and P. McLeod, "From eye movements to actions: How batsmen hit the ball," *Nature Neuroscience*, vol. 3, pp. 1340–1345, 2000.

229. E. Ruthruff, and H. Pashler, "Mental timing and the central attentional bottleneck." In *Attention and Time*, ed. A. C. Nobre and J. T. Coull, pp. 123–135. Oxford University Press, 2010.

230. S. Subramanian, I. Biederman, and S. Madigan, "Accurate identification but no priming and chance recognition memory for pictures in RSVP sequences," *Visual Cognition*, vol. 7, pp. 511–535, 2000.

231. S. M. Crouzet, H. Kirchner, and S. J. Thorpe, "Fast saccades towards faces: Face detection in just 100 ms," *Journal of Vision*, vol. 4, pp. 1–17, 2010.

232. H. Shinoda, M. M. Hayhoe and A. Shrivastava, "What controls attention in natural environments?" *Vision Research*, vol. 41, pp. 3535–3545, 2001.

233. M. Humphrys, "Action selection methods using reinforcement learning," in *Proceedings of the Fourth International Conference on Simulation of Adaptive Behavior*, 1996.

234. J. Karlsson, *Learning to Solve Multiple Goals*. PhD thesis, University of Rochester, 1997.

235. R. E. Kalman, "A new approach to linear filtering and prediction problems," *Transactions of the ASME—Journal of Basic Engineering*, vol. 82, Series D, pp. 35–45, 1960.

236. L. Johnson, B. Sullivan, M. Hayhoe, and D. Ballard, "Predicting human visuomotor behaviour in a driving task," *Philosophical Transactions of the Royal Society B*, vol. 369, p. 20130044, 2014.

237. L. Itti and C. Koch, "A saliency-based search mechanism for overt and covert shifts of visual attention," *Vision Research*, vol. 40, pp. 1489–1506, 2000.

238. G. A. Miller, "The magical number seven, plus or minus two: Some limits on our capacity for processing information," *Psychological Review*, vol. 63, pp. 81–97, 1956.

239. H. Pashler, *The Psychology of Attention*. MIT Press, 1998.

240. S. D. Whitehead, D. H. Ballard, "Learning to perceive and act by trial and error," *Machine Learning*, vol. 7, pp.45–83, 1991.

241. C. A. Rothkopf and D. H. Ballard, "Credit assignment in multiple goal embodied visuomotor behavior," *Frontiers in Psychology*, doi: 10.3389/fpsyg.2010.00173, 2010.

242. C. A. Rothkopf, and D. H. Ballard, "Modular inverse reinforcement learning for visuomotor behavior," *Biological Cybernetics*, vol. 107, pp. 477–490, 2013.

243. P. Cisek and J. F. Kalaska, "Neural mechanisms for interacting with a full world of action choices," *Annual Review of Neuroscience*, vol. 33, pp. 269–298, 2010.

244. D. Ballard and N. Sprague, "Attentional resource allocation in extended natural tasks [abstract]," *Journal of Vision*, vol. 2, no. 7, p. 568a, 2002.

245. P. Stone and M. Veloso, "Multiagent systems: A survey from a machine learning perspective," *Autonomous Robots*, vol. 8, pp. 345–383, 2000.

246. R. Romo, A. Hernandez, and A. Zainos, "Neuronal correlates of a perceptual decision in ventral premotor cortex," *Neuron*, vol. 41, pp. 165–173, 2004.

247. R. O. Deaner, A. V. Khera, and M. L. Platt, "Monkeys pay per view: Adaptive valuation of social images by rhesus macaques," *Current Biology*, vol. 15, pp. 543–548, 2005.

248. K. H. Britten, M. N. Shadlen, W. T. Newsome, and J. A. Movshon, "The analysis of visual motion: A comparison of neuronal and psychophysical performance," *Journal of Neuroscience*, vol. 12, no. 12, pp. 4745–4765, 1992.

249. M. N. Shadlen and W. T. Newsome, "The variable discharge of cortical neurons: Implications for connectivity, computation and information coding," *Journal of Neuroscience*, vol. 18, p. 3870, 1998.

250. E. P. Cook and J. H. R. Maunsell, "Dynamics of neural responses in macaque MT and VIP during motion detection," *Nature Neuroscience*, vol. 5, pp. 985–994, 2002.

251. V. Navalpakkam, C. Koch, A. Rangel, and P. Perona, "Optimal reward harvesting in complex perceptual environments," *Proceedings of the National Academy of Sciences USA*, vol. 107, pp. 5232–5237, 2010.

252. A. Rangel and T. Hare, "Neural computations associated with goal-directed choice," *Current Opinion in Neurobiology*, vol. 20, no. 2, pp. 262–270, 2010.

253. J. Trommershäuser, L. T. Maloney, and M. S. Landy, "Decision making, movement planning and statistical decision theory," *Trends in Cognitive Science*, vol. 12, pp. 291–297, 2008.

254. R. Dawkins, *The Selfish Gene*. Oxford University Press, 1976.

255. J. von Neumann and O. Morgenstern, *Theory of Games and Economic Behavior*. Princeton University Press, 1944.

256. W. Poundstone, *Prisoner's Dilemma*. Doubleday, 1992.

257. R. Axelrod, *The Evolution of Cooperation*. Basic Books, 1985.

258. D. Lee, H. Seo, and M. W.Jung, "Neural basis of reinforcement learning and decision making," *Annual Rview of Neuroscience*, vol. 35, pp. 287–308, 2012.

259. P. Glimcher, *Decisions, Uncertainty, and the Brain*. MIT Press, 2003.

260. H. Seo and D. Lee, "Temporal filtering of reward signals in the dorsal anterior cingulate cortex during a mixed-strategy game," *Journal of Neuroscience*, vol. 27, pp. 8366–8377, 2007.

261. D. Fudenberg and D. K. Levine, *The Theory of Learning in Games*. MIT Press, 1999.

262. J. Robinson, "An iterative method for solving a game," *Annals of Mathematics*, vol. 54, pp. 296–301, 1951.

263. S. Singh, M. Kearns, and Y. Mansour, "Nash convergence of gradient dynamics in general-sum games," in *Proceedings of the Sixteenth Conference on Uncertainty in Artificial Intelligence*, pp. 541–548, Morgan Kaufmann, 2000.

264. M. Bowling and M. Veloso, "Convergence of gradient dynamics with a variable learning rate," in *Proceedings of the 18th International Conference on Machine Learning*, pp. 27–34, Morgan Kaufmann, 2001.

265. A. B. Markman, G. C. Baldwin, and W. T. Maddox, "The interaction of payoff structure and regulatory focus in classification," *Psychological Science*, vol. 16, pp. 852–855, 2005.

266. S. Zhu, *Learning to Cooperate*. PhD thesis, University of Rochester, 2003.

267. C. Hauert, S. D. Monte, J. Hofbauer, and K. Sigmund, "Volunteering as red queen mechanism for cooperation in public goods games," *Science*, vol. 296, pp. 1129–1132. 2002.

268. C. D. Batson, L. L. Shaw, and K. C. Oleson, "Differentiating affect, mood and emotion, toward functionally based conceptual distinctions," in *Emotion. Review of Personality and Social Psychology*, ed. M. S. Clark, vol. 13, pp. 294–326. Sage Publications, 1992.

269. B. Mayton, L. LeGrand, and J. R. Smith, "Robot, feed thyself: Plugging in to unmodified electrical outlets by sensing emitted ac electric fields," in *2010 IEEE International Conference on Robotics and Automation (ICRA)*, pp. 715–722, 2010.

270. W. James, "What is an emotion?" *Mind*, vol. 9, pp. 188–205, 1884.

271. C. Darwin, *The Expression of the Emotions in Man And Animals*. Fortana Press, 1872.

272. E. Dunlap, ed., *The Emotions*. Williams and Wilkins, 1885/1922.

273. W. B. Cannon, "The James-Lange theory of emotions: A critical examination and an alternative theory," *American Journal of Psychology*, vol. 39, pp. 106–124, 1929.

274. J. Panksepp, "Neurologizing the psychology of affects," *Perspectives on Psychological Science*, vol. 2, no. 281–296, 2013.

275. A. Damasio, *Descartes' Error: Emotion, Reason, and the Human Brain*. Gosset/Putnam, 1994.

276. J. E. Le Doux, *Synaptic Self: How Our Brains Become Who We Are*. Viking, 2002.

277. D. M. Wegner, *The Illusion of Conscious Will*. MIT Press, 2002.

278. E. T. Rolls, *Emotion Explained*. Oxford University Press, 2005.

279. P. Ekman, "Basic Emotions," in *Handbook of Cognition and Emotion*, ed. T. Dalgleish and M. Power, pp. 45–60. John Wiley & Sons, 1999.

280. P. D. MacLean, *The Triune Brain in Evolution*. Plenum, 1990.

281. G. F. Striedter, *Principles of Brain Evolution*. Sinauer, 2005.

282. J. Panksepp, "Empathy and the laws of affect," *Science*, vol. 334, pp. 1358–1359, 2011.

283. A. Damasio, "Feelings of emotion and the self," *Annals of the New York Academy of Sciences*, vol. 1001, pp. 253–261, 2003.

284. I. B.-A. Bartal, J. Decety, and P. Mason, "Empathy and pro-social behavior in rats," *Science*, vol. 334, pp. 1427–1230, 2011.

285. K. S. Kassam, A. R. Markey, V. L. Cherkassky, G. Loewenstein, and M. A. Just, "Identifying emotions on the basis of neural activation," *PLoS ONE*, vol. 8, p. e66032, 2013.

286. J. Panksepp, "Affective consciousness: Core emotional feelings in animals and humans," *Consciousness and Cognition*, vol. 14, pp. 30–80, 2005.

287. S. Panzeri, S. R. Schultz, A. Treves, and E. T. Rolls, "Correlations and the encoding of information in the nervous system," *Proceedings of the Royal Society, Series B*, vol. 266, p. 1001, 1999.

288. R. G. Phillips and J. E. LeDoux, "Differential contribution of amygdala and hippocampus to cued and contextual fear conditioning," *Behavioral Neuroscience*, vol. 106, pp. 274–285, 1992.

289. R. Adolphs, D. Tranel, H. Damasio, and A. Damasio, "Fear and the human amygdala," *Journal of Neuroscience*, vol. 15, pp. 5879–5891, 1995.

290. C. Breazeal, J. Gray, and M. Berlin, "An embodied cognition approach to mind-reading skills for socially intelligent robots," *The International Journal of Robotics Research*, vol. 28, pp. 656–680, 2009.

291. R. W. Picard, *Affective Computing*. MIT Press, 1997.

292. C. Breazeal, "Role of expressive behaviour for robots that learn from people," *Philosophical Transactions of the Royal Society, Series B*, vol. 364, pp. 3527–3528, 2009.

293. V. S. Ramachandran and S. Blakesee, *Phantoms in the Brain: Probing the Mysteries of the Human Mind*. Quill William Morrow, 1998.

294. J. LeDoux and A. Damasio, "Emotions and Feelings," in *Principles of Neural Science*, 5th ed., ed. Eric Kandel et al., McGraw Hill, 2013.

295. D. Kahneman, *Thinking, Fast and Slow*. Farrar, Strauss and Giroux, 2011.

296. G. Gigerenzer, *Gut Feelings: The Intelligence of the Unconscious*. Viking Press, 2007.

297. D. C. Dennett, *Consciousness Explained*. Little, Brown, 1991.

298. A. Damasio, *Self Comes to Mind: Constructing the Conscious Brain*. Pantheon, 2010.

299. B. Libet, *Mind Time: The Temporal Factor in Consciousness*. Harvard University Press, 2004.

300. T. Norretranders, *The User Illusion*. Penguin Books, 1999.

301. R. Sekuler, A. B. Sekuler, and R. Lau, "Sound alters visual motion perception," *Nature*, vol. 385, p. 308, 1997.

302. J. S. DeLoache, "Rapid change in the symbolic functioning of very young children," *Science*, vol. 238, pp. 1556–1557, 1987.

303. S. Wagner, E. Winner, D. Cicchetti, and H. Gardner, "'Metaphorical' mapping by human infants," *Child Development*, vol. 52, pp. 728–731, 1981.

304. F. Heider and M. Simmel, "An experimental study of apparent behavior," *American Journal of Psychology*, vol. 57, pp. 243–259, 1944.

305. J. N. Bailenson, J. Blasovich, A. C. Beal, and J. M. Loomis, "Interpersonal distance in immersive virtual environments," *Personality and Social Psychology Bulletin*, vol. 29, pp. 1–15, 2003.

306. S. Baron-Cohen, *Mindblindness: An Essay on Autism and Theory of Mind*. MIT Press, 1997.

307. C. Yu, D. Ballard, and R. Aslin, "The role of embodied intention in early lexical acquisition," *Cognitive Science*, vol. 29, pp. 961–1005, 2005.

308. A. Newell, *Unified Theories of Cognition*. Harvard University Press, 1990.

309. B. A. Olshausen and D. J. Field, "Sparse coding with an overcomplete basis set: A strategy employed by v1?" *Vision Research*, vol. 37, pp. 3311–3325, 1997.

310. R. P. N. Rao and D. H. Ballard, "Predictive coding in the visual cortex: A functional interpretation of some extra-classical receptive field effects," *Nature Neuroscience*, vol. 2, pp. 79–87, 1998.

311. J. Searle, "Who is computing with the brain?" *Behavioral and Brain Sciences*, vol. 13, pp. 632–634, 1990.

312. C. Koch, *The Quest for Consciousness*. Roberts, 2003.

313. V. Lamme, "Towards a true neural stance on consciousness," *Trends in Cognitive Sciences*, vol. 10, pp. 494–500, 2006.

314. M. A. Graziano, *Consciousness and the Social Brain*. Oxford University Press, 2013.

315. T. Norretranders, *The User Illusion: Cutting Consciousness Down to Size*. Penguin Press, 1998.

316. *New York Times Book Review*, November 26, 2010.

317. V. S. Ramachandran and S. Blakeslee, *Phantoms in the Brain: Probing the Mysteries of the Human Mind*. HarperCollins, 1999.

318. O. Sacks, *The Man Who Mistook His Wife for a Hat*. Simon and Schuster, 1985.

319. W. Schultz, "Getting formal with dopamine and reward," *Neuron*, vol. 36, pp. 241–263, 2002.

320. D. F. Wallace, *Consider the Lobster*. Abacus, 2007.

321. C. M. Bishop, *Pattern Recognition and Machine Learning*. Springer, 2008.

322. S. Thrun, W. Burgard, and D. Fox, *Probabilistic Robotics*. MIT Press, 2005.

323. G. Rizzolati and S. Sinigalia, *Mirrors in the Brain*. Oxford University Press, 2008.

324. M. Merleau-Ponty, *Phenomenology of Perception*. Routledge & Kegan Paul, 1962.

325. L. W. Barsalou, "Perceptions of perceptual symbols," *Behavioral and Brain Sciences*, vol. 22, p. 637, 1999.

326. N. Humphreys, *A History of the Mind*. Vintage, 1992.

327. N. Block, "Consciousness, accessibility, and the mesh between psychology and neuroscience," *Behavioral and Brain Sciences*, vol. 30, pp. 481–499, 2007.

328. L. Young, F. Cushman, M. Hauser, and R. Saxe, "The neural basis of the interaction between theory of mind and moral judgment," *Proceedings of the National Academy of Sciences USA*, vol. 104, pp. 8235–8240, 2007.

329. D. Dennett, *The Intentional Stance*. MIT Press, 1989.

330. A. Raine, *The Anatomy of Violence*. Pantheon Books, 2013.

331. D. Hofstadter, *I Am a Strange Loop*. Basic Books, 2007.

Index

Computational Neuroscience

Terence J. Sejnowski and Tomaso A. Poggio, editors

Computational Vision: Information Processing in Perception and Visual Behavior, Hanspeter A. Mallot, 2000

Graphical Models: Foundations of Neural Computation, edited by Michael I. Jordan and Terrence J. Sejnowski, 2001

Self-Organizing Map Formation: Foundations of Neural Computation, edited by Klaus Obermayer and Terrence J. Sejnowski, 2001

Neural Engineering: Computation, Representation, and Dynamics in Neurobiological Systems, Chris Eliasmith and Charles H. Anderson, 2003

The Computational Neurobiology of Reaching and Pointing, Reza Shadmehr and Steven P. Wise, 2005

Dynamical Systems in Neuroscience, Eugene M. Izhikevich, 2006

Bayesian Brain: Probabilistic Approaches to Neural Coding, edited by Kenji Doya, Shin Ishii, Alexandre Pouget, and Rajesh P. N. Rao, 2007

Computational Modeling Methods for Neuroscientists, edited by Erik De Schutter, 2009

Neural Control Engineering, Steven J. Schiff, 2011

Understanding Visual Population Codes: Toward a Common Multivariate Framework for Cell Recording and Functional Imaging, edited by Nikolaus Kriegeskorte and Gabriel Kreiman, 2011

Biological Learning and Control: How the Brain Builds Representations, Predicts Events, and Makes Decisions, Reza Shadmehr and Sandro Mussa-Ivaldi, 2012

Principles of Brain Dynamics: Global State Interactions, edited by Mikhail Rabinovich, Karl J. Friston, and Pablo Varona, 2012

Brain Computation as Hierarchical Abstraction, Dana H. Ballard, 2015

Printed in the United States
by Baker & Taylor Publisher Services